T0179115

Handbook of Pre-clinical Continuous Intravenous Infusion

Edited by
Guy Healing
David Smith

London and New York

First published 2000
by Taylor & Francis
11 New Fetter Lane, London EC4P 4EE

Simultaneously published in the USA and Canada
by Taylor & Francis Inc,
29 West 35th Street, New York, NY 10001

Taylor & Francis is an imprint of the Taylor & Francis Group

© 2000 Guy Healing and David Smith

Typeset in Sabon by
Prepress Projects, Perth, Scotland
Printed and bound in Great Britain by
TJ International Ltd, Padstow, Cornwall

British Library Cataloguing in Publication Data
A catalogue record for this book is available
from the British Library

Library of Congress Cataloging in Publication Data

Handbook of pre-clinical continuous intravenous infusion/
   edited by Guy Healing, David Smith
      p. cm.
   Includes bibliographical references and index.
   ISBN 0–7484–0867–3 (alk. paper)
   1. Toxicology – Animal models – Handbooks, manuals, etc. 2. Toxicity
   testing – Handbooks, manuals, etc. 3. Infusion therapy – Testing –
   Handbooks, manuals, etc. 4. Animal experimentation – Handbooks,
   manuals, etc. I. Title: Pre-clinical continuous intravenous infusion. II.
   Healing, Guy, 1961–. III. Smith, David, 1944–.
   [DNLM: 1. Drug Screening – Handbooks. 2. Infusions, Intravenous –
   methods – Handbooks. 3. Animals, Laboratory – Handbooks. QV 735
   H2365 2000]
   RA1199.4.A54 H357 2000
   615'.6–dc21 00–037718

Contents

Contributors

Editors

Guy Healing
AstraZeneca R&D Charnwood
Loughborough, UK

David Smith
AstraZeneca R&D Charnwood
Loughborough, UK

Authors

Paul C. Barrow
Phoenix International
L'Arbresle
France

John G. Evans
AstraZeneca R&D Charnwood
Loughborough
UK

Peter Brinck
Novo Nordisk AS
Maaloev
Denmark

Teresa R. Gleason
WIL Research Laboratories
Ashland, Ohio
USA

Roger Burnett
Phoenix International
L'Arbresle
France

Elaine Gordon
AstraZeneca R&D Charnwood
Loughborough
UK

Christopher P. Chengelis
WIL Research Laboratories
Ashland, Ohio
USA

Owen P. Green
LeVerts Ltd
Peterborough
UK

Christine Copeman
Clinical Trials Bio-Research Ltd
Senneville, Qubec
Canada

Kevin Hickling
AstraZeneca R&D Charnwood
Loughborough
UK

Stephen T. Cornell
Primedica Corporation
Worcester, Massachusetts
USA

Ikuo Horii
Nippon Roche KK
Kamakura, Kanagawa
Japan

Andrew Jacobson
Instech Solomon
Plymouth Meeting
USA

Philip J. Kerry
AstraZeneca R&D Charnwood
Loughborough
UK

Sarah Lynch
AstraZeneca R&D Charnwood
Loughborough
UK

Peter A. McAnulty
Scantox
Lille Skensved
Denmark

Alan M. McCarthy
AstraZeneca R&D Alderley
Macclesfield, Cheshire
UK

H. Vince Mendenhall
Primedica Corporation
Worcester, Massachusetts
USA

Ed S. Miedel
Hilltop Lab Animals Inc.
Scottdale, Pennsylvania
USA

Peg Murphy-Hackley
ALZA Corporation
Mountain View, California
USA

Duncan Patten
Huntingdon Life Sciences
Huntingdon
UK

Clarisa E. Peer
DURECT Corporation
Cupertino, California
USA

Lorri Perkins
DURECT Corporation
Cupertino, California
USA

Nigel Pickersgill
Phoenix International
L'Arbresle
France

David Robb
Inveresk Research
Tranent
Scotland

Keith J. Robinson
Clinical Trials Bio-Research Ltd
Senneville, Quebec
Canada

Mary Ann Scalaro
Primedica Corporation
Worcester, Massachusetts
USA

Christian R. Schnell
Novartis Pharma AG
Basle
Switzerland

Hans van Wijk
Inveresk Research
Tranent
Scotland

Mark D. Walker
Covance Laboratories Inc.
Vienna, Virginia
USA

Foreword

When I was asked to write a foreword to this book, I did so willingly, knowing that the editors are committed to the same maxim as myself – 'Good animal welfare can only contribute to good science'. It may be better paraphrased as 'compromising an animal's welfare as little as possible will lead to better reproducibility of results and therefore more reliable science, and the use of fewer animals with less suffering'. This is not only in keeping with the three Rs of replacement, reduction and refinement but is a more responsible way of using animals and can justifiably lead to the claim to practise humane science. Quite frankly, I was amazed and gratified about the level of detail and the practical emphasis on technical expertise and animal well-being. This book is unique in having that as an overt aim, and I much welcome it – better late then never!

I am sure that this book will make a positive contribution to both science and animal welfare, particularly outside the pharmaceutical industry. It contains an abundance of detail (sound practical advice based on a wealth of practical experience) and technical tips that will make all the difference between a successful and unsuccessful scientific procedure on animals. It also, quite rightly, emphasises the importance of manual skills. The scientific data are there for our sceptical colleagues as well, with paragraphs on physiological and pathological side-effects (but unfortunately with little on behavioural effects indicating how the animal itself is coping with the cannulation etc.). Read carefully: it also provides benchmarks for ethics committee audit, which are extremely useful for scientists who want their experiment to succeed as well as to do their best for the animals.

David B. Morton
University of Birmingham, UK

Introduction

D. Smith and G. Healing

Numerous pharmaceuticals on the market or undergoing clinical trials are administered by intravenous infusion, either for short or intermittent periods or continuously for longer periods. Intravenous infusion is used mainly for the administration of chemotherapeutic drugs in the treatment of different forms of cancer (Skubitz 1997; Vallejos *et al.* 1997; Ikeda *et al.* 1998; Patel *et al.* 1998; Stevenson *et al.* 1999; Valero *et al.* 1999), but it is also employed in the treatment of human immunodeficiency virus (HIV) (Levy *et al.* 1999) and hepatitis C infection (Schenker *et al.* 1997) and in cardiovascular disease (Phillips *et al.* 1997), as well as during and following problematical surgical procedures (Bacher *et al.* 1998; Llamas *et al.* 1998; Menart *et al.* 1998).

To enable these pharmaceuticals to be safely tested in humans and then subsequently to be marketed, it is a regulatory and ethical requirement that they must first be tested on both rodents and non-rodents. Although this book focuses on toxicity testing during pre-clinical development, many aspects, particularly the surgical techniques and husbandry/animal room procedures, will also be of interest to the academic researcher who needs to use this route of administration. We have also chosen to concentrate on the intravenous route rather than the other routes for continuous infusion that are less commonly used, such as subcutaneous (Weckbecker *et al.* 1992) or intrathecal (Burns *et al.* 1999).

The techniques of continuous infusion in both rodents and non-rodents have been developed principally over the past 30 years, and there are a number of resultant publications. However, no attempt has been made to collate these techniques into a single volume. The aim of this book is to fill this gap in the literature and provide a single, definitive reference source from which those new to the field can immediately compare the range of methodology and equipment available. It is also intended as a mechanism by which those already working in the field can refine or update their procedures. It is hoped that this book will also prevent unnecessary development work and hence potentially decrease animal usage. This is particularly pertinent with regard to the more politically sensitive issue of testing in non-rodent species.

Continuous intravenous administration presents a number of individual problems compared with other routes. For example, there are specific welfare issues that arise from the necessary restrictions imposed on the animals related to the equipment (e.g. backpacks, tethering, individual housing). The pathology associated with these studies is also unique in that there will be at least minor localised effects in the vessel adjacent to the catheter tip resulting from the simple presence of the catheter, and implantation of the catheter itself involves invasive surgery, thus necessitating good aseptic technique to prevent infection and associated pathology. These effects can influence data interpretation, as indeed will the effects of infusing relatively large volumes on indices such as water consumption and clinical pathology. However, as long as these factors are considered when interpreting the study findings, so that effects resulting from the infusion technique alone are discounted, there is no reason why the findings should cause any confusion. Of course, there is no substitute for good aseptic technique, and a well-trained and experienced surgery and study team will make all the difference to the success of a study.

In addition to describing the technical requirements for conducting continuous intravenous infusion studies, we have chosen to focus on animal welfare considerations. Not only can a manual such as this prevent unnecessary method development work and animal usage, through workers sharing their experiences, but it is also clear that, despite the physical constraints imposed by infusion apparatus, it is possible in many cases to provide the animals with a range of environmental enrichment. This can compensate for the burden of tethering or wearing a backpack or jacket, which is required in many cases, and should also positively improve the scientific model by providing a less stressed animal for testing.

The authors in this book have been selected for their established expertise in the field. It should be noted that many authors' laboratories conduct procedures in more species than those covered in their chapter(s), but to gain the widest possible selection of opinions, techniques and background data it was necessary to limit each to a specific aspect. There are also variations in the techniques used in different countries, and this has been reflected in the truly international selection of authors. There is, consequently, a certain amount of overlap between some of the chapters. This has been minimised as far as practicable, but some has been retained deliberately in order that each chapter covering a specific technique can stand alone without too much unnecessary cross-referencing.

Another current issue is the terminology used for the introduction of the catheter to the vessel. Two terms are used synonymously: cannulation and catheterisation. The latter is probably the most accurate description, since the flexible tubes used for most species are, strictly speaking, catheters. However, the use of both terms has been permitted within this book.

The book is organised by species, namely rat, mouse, rabbit, dog, primate and minipig, these being the six species most commonly used in pre-clinical studies. Within each section the techniques are essentially divided into surgical and in-life study conduct procedures. As well as techniques, awareness of vehicles suitable (and, equally pertinent, those that are unsuitable) for continuous infusion is covered in its own chapter. Common histopathology findings are discussed separately, as is the latest equipment and where to obtain it. A separate chapter is also dedicated to the use of mini-pumps in infusion studies, an important possible alternative method.

We have received much encouragement in advance of preparing this book from many sources, indicating the need for a reference volume of this nature. It is hoped that it will indeed be of assistance to workers and researchers alike.

References

Bacher, H., Mischinger, H.J., Supancic, A., Leitner, G. and Porta, S. (1998) Dopamine infusion following liver surgery prevents hypomagnesemia. *Magnesium Bulletin*, 20, 49–50.

Burns, L.H., Jin, Z. and Bowersox, S.S. (1999) The neuroprotective effects of intrathecal administration of the selective N-type calcium channel blocker ziconotide in a rat model of spinal ischaemia. *Journal of Vascular Surgery,* 30, 334–343.

Ikeda, K., Terashima, M., Kawamura, H., Takiyama, I., Koeda, K., Takagane, A., Sato, N., Ishida, K., Iwata, T., Maesawa, C., Yoshinari, H. and Saito, K. (1998) Pharmacokinetics of cisplatin in combined cisplatin and 5-fluorouracil therapy – a comparative study of three different schedules of cisplatin administration. *Japanese Journal of Clinical Oncology,* 28, 168–175.

Levy, Y., Capitant, C., Houhou, S., Carriere, I., Viard, J.P., Goujard, C., Gastaut, J.A., Oksenhendler, E., Boumsell, L., Gomard, E., Rabian, C., Weiss, L., Guillet, J.G., Delfraissy, J.F., Aboulker, J.P. and Seligmann, M. (1999) Comparison of subcutaneous and intravenous interleukin-2 in asymptomatic HIV-1 infection: a randomised controlled trial, *Lancet,* 353, 1923–1929.

Llamas, P., Cabrera, R., Gomezarnau, J. and Fernadez, M.N. (1998) Hemostasis and blood requirements in orthotopic liver transplantation with and without high-dose aprotonin. *Haematologica,* 83, 338–346.

Menart, C., Petit, P.Y., Attali, O., Massignon, D., Dechavanne, M. and Negrier, C. (1998) Efficacy and safety of continuous infusion of mononine® during five surgical procedures in three hemophilic patients. *American Journal of Hematology,* 58, 110–116.

Patel, S.R., Vadhanraj, S., Burgess, M.A., Plager, C., Papadopoulos, N., Jenkins, J. and Benjamin, R.S. (1998) Results of two consecutive trials of dose-intensive chemotherapy with doxorubicin and ifosfamide in patients with sarcomas. *American Journal of Clinical Oncology,* 21, 317–321.

Phillips, B.G., Gandhi, A.J., Sanoski, C.A., Just, V.L and Bauman, J.L. (1997) Comparison of intravenous diltiazem and verapamil for the acute treatment of atrial fibrillation and atrial flutter. *Pharmacotherapy,* 17, 1238–1245.

Schenker, S., Cutler, D., Finch, J., Tamburro, C.II., Affrime, M., Sabo, R. and Bay, M. (1997) Activity and tolerance of a continuous subcutaneous infusion of interferon-alpha-2b in patients with chronic hepatitis C. *Journal of Interferon and Cytokine Research,* 17, 665–670.

Skubitz, K.M. (1997) A phase I study of ambulatory continuous infusion of paclitaxel. *Anti-Cancer Drugs,* 8, 823–828.

Stevenson, J.P., DeMaria, D., Sludden, J., Kaye, S.B., Paz-Ares, L., Grochow, L.B., McDonald, A., Selinger, K., Wissel, P., O'Dwyer, P.J. and Twelves, C. (1999) Phase I pharmacokinetic study of the topoisomerase I inhibitor GG211 administered as a 21-day continuous infusion. *Annals of Oncology,* 10, 339–344.

Valero, V., Buzdar, A.U., Theriault, R.L., Azarnia, N., Fonseca, G.A., Willey, J., Ewer, M., Walters, R.S., Mackay, B., Podoloff, D., Booser, D., Lee, L.W. and Hortobagyi, G.N. (1999) Phase II trial of liposome-encapsulated doxorubicin, cyclophosphamide, and fluorouracil as first-line therapy in patients with metastatic breast cancer. *Journal of Clinical Oncology,* 17, 1425–1434.

Vallejos, C., Solidoro, A., Gomez, H., Castellano, C., Borriga, O., Galdos, R., Casanova, L., Utero, J. and Rodriguez, W. (1997) Ifosfamide plus cisplatin as primary chemotherapy of advanced ovarian cancer. *Gynecologic Oncology,* 67, 168–171.

Weckbecker, G., Tolcsvai, L., Liu, R. and Bruns, C. (1992) Preclinical studies on the anticancer activity of the somatostatin analogue octreotide (SMS 201–995). *Metabolism,* 41, 99–103.

Part 1

Continuous intravenous infusion in the rat

The rat is probably the most commonly used animal model for pre-clinical continuous intravenous infusion. As a result, a number of different techniques have been developed, all with their own advantages and potential disadvantages.

The favoured catheterisation site is certainly the femoral vein, and all the authors within this section use this route of administration. The jugular vein can be used, and has been used in the past at the editors' laboratory, but a number of disadvantages were identified, such as the accessibility of the site to the animal, the less secure anchorage and the smaller size of the vessel. The choice of vessel is important, as local effects such as irritation can result from this procedure.

The major difference between laboratories is in the exit site of the catheter. While some workers exit via the subscapular region on the back, secured by a jacket or harness, others exit via the tail, utilising a tail cuff attached to the tether. This method certainly seems to provide greater freedom of movement, but it can lead to unwanted lesions at the site of the tail cuff. In addition, rats can also be tethered to an implantable skin button, with or without a jacket or harness. All these methods are described in this section.

Other differences to be considered are the method of administration of anaesthesia (gaseous or injectable), design of the surgical suite, surgical technique, post-surgical analgesia and use of antibiotics. We would encourage all workers new to the field to consider all the options before making their selection. For example, gaseous methods of anaesthesia result in more rapid induction and recovery than administration of injectable agents, but this can result in the animals being too active post surgery for the fitting of a harness and can increase the chances of them interfering with their surgical stitches or staples.

Finally, the design and conduct of the two main regulatory infusion study types – general toxicology and reproductive toxicology – are described in detail in Chapters 4 and 5 respectively, each having their own specific parameters to record.

1 Femoral cannulation using the jacket/harness model in the rat

O. P. Green and D. Patten

1.1 Introduction

Effective rodent models of continuous intravenous infusion of haemocompatible materials have been developed and used for the study of the physiological processes of cardiovascular function and pharmacology of various agents for well over 20 years.

To adapt this route of administration for use in pre-clinical toxicological evaluation of agents, the model needed to be more durable to allow administration for periods of up to 90 days or even longer. Also, the model had to be relatively easy to set up surgically and also to maintain, since such pre-clinical studies often involve large numbers of rats.

There has also appeared to be a growing requirement for the use of this mode of delivery in the clinical field during the past two decades, during which time the technology and procedures involved have advanced exponentially. Consequently, by adapting the advancements in the clinical field of this technology for pre-clinical/animal use, similar progress has been made in the development of such animal models. For rats, and even mice, there are now several effective models of chronic continuous intravenous infusion, most of which differ in only a small way. This chapter describes, in detail, one of the more popular rat models for the mode of delivery together with all the support services necessary to conduct successful pre-clinical studies with the model.

1.2 Facilities and equipment

Gregory (1995) alludes to the finding that significant expense can go into the creation of an ideal facility in which to perform continuous intravenous infusion rat studies. The term 'facility' in this case is used to embrace the surgical suite and the study room incorporating the necessary specialist housing, equipment and staff resource.

Clearly with limitless resource, a utopian view could be presented of the ideal facility but, no doubt, there would be several different opinions on this matter, creating doubt as to whether such a facility exists. However, there are a number of general cases which are described in this chapter that need to be considered when setting up such a facility, and these can be taken as a minimum requirement to ensure the safe conduct of regulatory standard pre-clinical studies by this route of administration.

1.3 Surgical suite

It is important that this area, whether it is a purpose-built facility or animal room conversion, is set up as a dedicated area for the surgical preparation of rats. This is, after all, where the process of succeeding at the conduct of continuous intravenous infusion studies begins, and therefore, the standard of asepsis is paramount. This surgical suite should be located within the building that is to ultimately house the study. It is widely accepted that such a suite should have four components.

A changing room, in addition to the changing facilities upon entry into the building, should be located close to the surgical suite. This area should incorporate scrub-up facilities with foot- or elbow-operated taps and space where sterile gowns, gloves, hats and masks can be stored and put on prior to entering the theatre. The theatre should, therefore, be easily accessible from this area without compromising the sterile status of the individual.

The remaining three areas of the surgical suite all involve the throughput of animals and ideally should be located in relation to each other in such a way as to allow the process to take place. These three components are the preparation room, the surgical theatre and the recovery room. All three rooms should have smooth walls and ceiling and an impervious floor. The theatre should be positively ventilated with filtered air flowing out from the clean to the less clean areas. The heating system should maintain a comfortable ambient temperature which, ideally, should be a couple of degrees higher in the recovery room. If room controls are unable to provide this increased temperature then this may be achieved by strategic placement of infra-red lamps. In the theatre, a portable cold-illuminating system or special theatre lights should be provided. Artificial lighting and electrical fittings should be spark-proof if explosive volatile anaesthetic agents are used. Light fittings should be watertight to allow for cleaning with water and exposure to sterilising vapours such as formaldehyde. It may be necessary to provide special exhaust and scavenging facilities to remove surplus anaesthetic agents, which could be cumulatively hazardous. All three rooms should be kept clean with regular removal of dust and organic debris from all surfaces. A minimal amount of experimental apparatus should be kept in these rooms, but instruments and small pieces of equipment are best kept in drawers and cupboards in the preparation room, where anaesthetising and preparation of the animal (described in detail later) takes place. In times of activity, appropriate caging with bedding material should be in readiness in the recovery room.

1.4 Equipment

1.4.1 *Surgical instruments*

The specific details of the content of a surgical kit to perform a femoral vein catheter implant in rats will, of course, vary from surgeon to surgeon depending significantly on personal preference. A typical set of instrumentation is shown in Figure 1.1 and is considered to be the minimum requirement. Scissors, scalpels and suture needles should be maintained sharp, and all surgical instruments should be kept ready for use (Anderson and Romfh, 1980). This will usually involve having a number of sets of instruments already sterilised by steam autoclave in pre-packed protective wrappings. Maintenance of the sterility of these instruments during the surgical procedure and from animal to animal may be achieved by washing the instruments with Haemo-Sol (Hoechst, Germany) or a similar instrument-cleaning agent and then heat sterilising the instrument tips with a bead steriliser. This procedure avoids the need to have

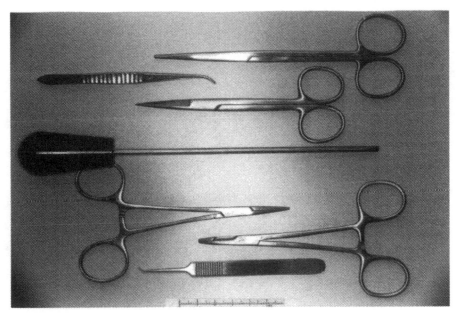

Figure 1.1 Surgical instruments. From top to bottom: blunt scissors, delicate forceps (curved), sharp scissors, trocar, mosquito haemostat forceps (×2), needle holder, iris forceps (curved).

excessive numbers of complete sets of expensive instruments and potential delays between cannulation of animals while instruments are cleaned, resterilised and cooled.

1.4.2 Specialist items

The remaining equipment necessary to complete this procedure involves the specialist items for rat intravenous infusion (Figure 1.2). These items include the implanted catheter itself, which can be of variable design and manufactured from a range of materials. The main requirements of a good catheter for this purpose are the correct dimensions (inner and outer diameters and length) and positioning of retention beads or suture stops for the job and that they are manufactured from materials that optimise tissue compatibility. A number of catheters for rat femoral vein cannulation are now commercially available, most of which are manufactured from various forms of polyurethane or silicone-based materials.

The exteriorisation site is protected by either a jacket or harness, which also provides the means for connecting a spring tether (12–16 inches) that protects the exteriorised catheter from the attentions of the animal while also providing the connection to the external swivel system. The length of the tether should be such as to allow the animal to reach and lie down in all parts of the cage. All of these items are also commercially available specifically for this animal model. There is a great deal of choice in the model of swivel for this procedure, ranging from disposable Teflon swivels through aluminium and brass reusable items. The relative merits and potential advantages and disadvantages of the variations in these items are discussed later.

A further specialist item essential to the success of the model is the infusion device itself. Many suitable models that comply to the relevant European standards are

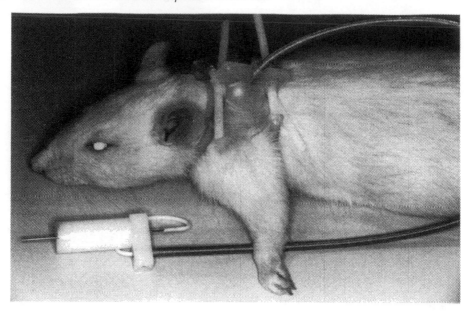

Figure 1.2 Anaesthetised rat with harness, steel spring tether, single-channel swivel and retention
       block and cannula.

available and the choice should be based on the required rate, volume, time period
and pattern of delivery.

1.5 Staffing resource and expertise

The technical resource necessary to perform continuous infusion studies and the
organisation of these technicians will vary from laboratory to laboratory. Factors
that need to be considered include size, duration and frequency of the studies to be
performed. To a lesser extent, the infusion model that has been adopted also needs to
be considered.

As a minimum, in order to adequately set up and maintain a small regulatory
toxicology study by continuous intravenous infusion delivery in rats, a small group of
three or four animal technicians could provide the necessary requirements. The essential
element is, of course, appropriate training and competence in the required skills. For
example, within this small group of people it would be important to include the
necessary surgical skill to implant the catheter; essential skills in anaesthetising rodents
and pre- and post-operative care of the animals; and also the skills to maintain the
systems for the duration of a study. As a consequence, it is possible to realise that such
a capability could be resourced with a small number of individuals who could be
drawn from other areas of your organisation, as long as they possess the necessary
skills and are documented as competent under good laboratory practice (GLP)
legislation. However, in a commercial toxicology environment, the demand is often
for several studies of this type to be conducted simultaneously. Therefore, in these
circumstances it is advantageous to have a dedicated team devoted to infusion work
in small animals, who can then undertake general rodent and reproductive toxicology
studies.

1.5.1 *The infusion team*

A team being established to undertake the cannulation of rodents will invariably include a veterinary surgeon supported by experienced animal technicians. The inclusion of a pathologist or toxicologist as part of the surgical team in the early stages of developing this technology will prove beneficial since an in-depth knowledge of the surgical procedures will help when interpreting results.

As the animal technicians gain experience with the techniques and procedures necessary to cannulate blood vessels, the practical involvement of vets and pathologists/toxicologists in the set-up of studies tends to recede. It is commonplace to find a senior technician leading the infusion team in facilities where the technique has become established.

The animal technicians that constitute such a team will usually be drawn from among the most experienced and technically capable technicians available. These technicians will have consistently displayed a high level of manual dexterity and a meticulous attention to detail to be considered for inclusion in the infusion team. Their skills will usually be developed further by first acting as assistants to those cannulating the animals and then gradually taking over the role for themselves under the careful supervision of the veterinary surgeon or technicians already experienced with the procedure. It helps greatly if the technicians caring for the animals and maintaining the infusion systems following cannulation have themselves gained some experience in the surgical procedure.

UK legislation (the Animals (Scientific Procedures) Act 1986) precludes the use of live animals for the practising of invasive techniques. Basic surgical procedures (making the initial incision, suturing etc.) are developed using cadavers. However, the manipulation and cannulation of 'full' blood vessels with associated control of blood flow can only be experienced with live animals. The development of phlebotomy techniques is therefore routinely gained on study animals while under close supervision. The frequency of such studies can lead to a protracted training period before full competence is finally achieved. Some facilities send their technicians to the USA or mainland Europe for 'fast-track' training.

1.5.2 *Technical resource*

The number of technicians required will vary according to study design. One equation that seldom changes is that you will require more technicians for the surgery and setting up of the study than to actually run the study itself.

Many establishments restrict the number of animals that each surgeon cannulates on a daily basis, the designation of a cut-off time during the afternoon after which no more animals are prepared or a limit of eight animals per surgeon per day being common. This is usually put in place to prevent tiredness compromising the quality of surgery and therefore the welfare of the animal. Sufficient time must also be allowed post-operatively to ensure detailed observation of each animal until fully recovered.

As previously mentioned, the preparation of smaller studies can be handled by a team of perhaps four technicians: one cannulating, one assisting, one preparing animals for surgery and one technician monitoring the animals post-operatively. Given the above limit of eight animals per surgeon per day, this team could produce 40 cannulated animals per week. However, for larger studies this level of output would lead to a large variance between the time the first and last animals were cannulated unless staggering the start of study for different batches of animals is considered.

When larger studies (80+ animals) are being performed on a regular basis, then a surgery team of three or four technicians cannulating, one or two assisting, two preparing the animals and two or three monitoring the animals post-operatively, giving a total of 8–11 technicians, would not be uncommon.

Post-operatively and during the course of the study, procedures become protracted with a tethered animal. Checking, weighing and observing clinical signs are all labour intensive when compared with non-tethered animals. This needs to be taken into account when allocating technicians to perform the work. A figure of one technician for every 40–60 cannulated animals seems to work quite well, although this will obviously vary according to the technicians available and the study design.

Assuming a 120-animal study, 8–11 technicians may be utilised to prepare the animals over the course of the week, but only two or three technicians will be required to run the study. This disparity in staffing requirements often leads to a nucleus of technicians forming an infusion team to run the studies but supported by technicians drawn from other areas to assist during the time of preparation.

1.6 Surgical procedures

There are many important procedures to go through before the animal is brought to the surgical suite:

- preparation by the surgeon (Markowitz *et al.* 1964; Lang 1982)
- acclimatisation
- conditioning
- clinical examination
- food/water/body weight, other data.

1.6.1 Pre-operative procedures

The use of pre-medication for rats depends upon the excitability of the animal and the anaesthetic to be used (Flecknell 1987). With the method of anaesthesia described below, pre-medication in rats is thought to be unnecessary as long as there is adequate control of body temperature.

In the preparation room, the animal receives an intraperitoneal injection of the combination anaesthetic agent known as Hypnorm®/Hypnovel®. Hypnorm® (supplied by Janssen Pharmaceuticals, Belgium) is a member of the neuroleptanalgesic class of drug and is a combination of the narcotic analgesic fentanyl citrate and the tranquilliser fluanisone. It is a safe and effective anaesthetic for the rat which induces a state of mental apathy and mental detachment, the animal being sedated and uncaring about its surroundings and unaware of pain (Green 1975). To improve the degree of muscle relaxation, the drug is often given in combination with a water-soluble benzodiazepine (Flecknell and Mitchell 1984), in this case Hypnovel® (midazolam, supplied by Roche Products, UK). A routine dosage in rats is 2.8 ml/kg body weight injected intraperitoneally of a stock solution of two parts water for injection, one part Hypnovel® and one part Hypnorm® (2.5 mg of fluanisone, 0.097 mg of fentanyl and 1.25 mg of midazolam per ml).

In addition to the anaesthetic agent, antibiotic cover is commenced on the day of surgery, prior to the operation, by the subcutaneous administration of Baytril® (Enrofloxacin, supplied by Bayer) at a dose of 4 mg/kg body weight.

Once anaesthetised, an ocular lubricant is applied to both of the animal's eyes to prevent drying of the cornea during the surgical procedure, which can often lead to permanent corneal scars and opacities if not accommodated for. The sites of cannulation, left or right inguinal region, and exteriorisation, the dorsal interscapular region, are clipped free of hair using electric clippers, taking care not to damage the skin and removing all loose hair. The operation sites are then washed with an antiseptic solution, in this case Hibiscrub® (Zeneca, UK)/isopropyl alcohol. It is important at this stage to keep the animal warm using heat pads, lamps or Vetbeds and to closely monitor its status while awaiting transfer to the surgery.

Once received in the surgery, the animal is placed ventral surface uppermost on a heated operating table and the rectal temperature probe associated with the table is inserted. This system allows automatic temperature control of the animal undergoing the surgery. At this stage, before the actual surgical procedure begins, further swabbing of the incision site with antiseptic solution may be undertaken along with the application of a sterile drape to isolate the skin incision site in order to reduce the risk of microbial contamination.

1.6.2 Surgical cannulation

Surgery is carried out using sterilised instruments, which are usually autoclaved prior to the procedures with the potential of further cold sterilisation between procedures. Tissue damage during surgery should be minimised (Waynforth 1980; 1987). The sterile cannula is also readily available for implantation already connected to a 2.0-ml syringe primed with heparinised saline solution for patency checking and flushing.

Damage can be caused by drying, cooling, overheating or manual handling, as well as by microbial infection. Tissues must be handled gently, and warm, moistened sterile swabs can be used to reduce cooling and drying. Bleeding must also be minimised to reduce surgical shock and prevent the accumulation of blood in the wound. Ligatures should be firmly tied to obtain haemostasis but should not include more tissue than the vessel concerned. The extent of tissue trauma affects the level of post-operative pain perceived by the animal and the need for relief with sedatives and analgesics. Trauma also delays wound healing and return to normal health.

The first incision is made in the skin proximal to the right or left inguinal region depending on whether the right or left femoral vein is to be cannulated. This incision is approximately 1.25 cm in length and is made with a scalpel to provide clean, straight edges to the incision, which will aid in the viability of the eventual wound closure. By a process of gentle dissection with blunt instruments, the femoral vein, femoral artery and saphenous nerve complex are located in a sheathed bundle within the muscle layers of the leg. Connective tissue is gently teased away from the bundle and, using blunt dissection with a pair of fine forceps, the sheath is penetrated and the femoral vein separated from the artery and nerve. At this stage all remaining connective tissue is removed from the vein. A single ligature of grade 3.0 braided Mersilk, folded double, is passed under the vein, whereupon it is cut in half to create two ligatures. Each ligature is clamped in both directions with artery forceps, which are used to maintain the tension on the ligatures to control the blood flow through the vein (Figure 1.3). A

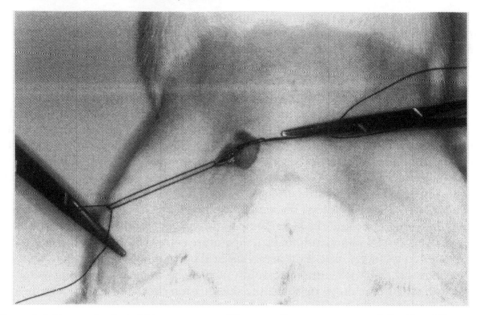

Figure 1.3 Femoral vein held between two silk suture ligatures to control the blood flow.

small incision is made in the blood vessel wall using a hypodermic needle (23G), the incision being made between the two ligatures. The cannula is then passed through the incision into the blood vessel (Figure 1.4). This can be achieved in a number of ways: the vein may be held in the forceps during the passing of the cannula or the vein may be released and blood flow controlled by pressure on the two ligatures, while a pair of vessel dilators is used to guide the cannula through the incision. The cannula is advanced to the first suture stop, which is 4 cm from the cannula tip. The positioning of this 'bead' of polyurethane or silicone on the catheter allows the tip of the cannula to reach the posterior vena cava and enables sutures to be securely tied to the cannula and vein. Should the surgeon encounter resistance when introducing the cannula, then gentle manipulation of the hindlimb or abdomen and flushing or rotating the cannula may be undertaken to advance the catheter. Under no circumstances should the advancing of the cannula be forced, and the positioning must be up to the first suture stop. At this stage a patency check is made on the cannula by drawing blood into the cannula. In the event that blood cannot be readily drawn back, repositioning of the cannula is performed until satisfactory patency is established.

Once the cannula is in the correct position, the ligatures are tied in place around the first suture stop, one anterior and one posterior to the stop (Figure 1.5). The anterior ligature is tied around the blood vessel, encompassing the cannula and so preventing leakage of blood from the phlebotomy site. The posterior ligature is used to tie off the now redundant distal femoral vein and to tie around the cannula to provide an anchoring point for the cannula. A small loop is made from the cannula, which is loosely sutured to the underlying musculature of the inner thigh in a small subcutaneous pocket. This additional loop of cannula is designed to allow for the natural growth and movement of the animal without putting unnecessary strain on the securing sutures at the site of venous access.

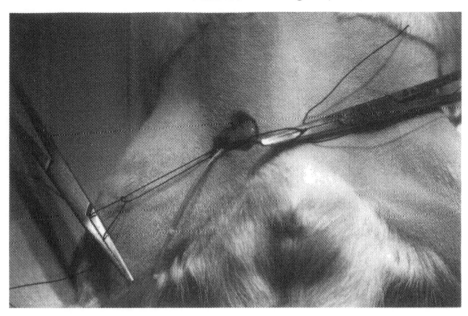

Figure 1.4 Cannula secured into the femoral vein through an incision made between the two ligatures.

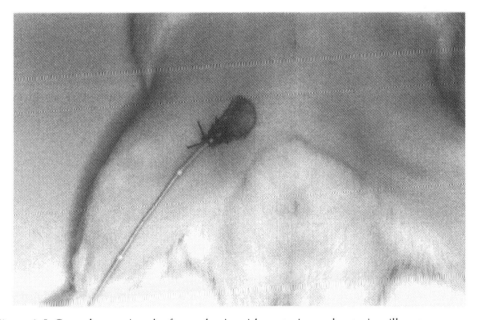

Figure 1.5 Cannula entering the femoral vein with posterior and anterior silk sutures.

The cannula is now clamped off just above the incision site to allow manipulation of the animal, and the syringe of heparinised saline is disconnected. The animal is turned over and a small, approximately 0.25-cm, incision is made in the subscapular region to allow the introduction of a trocar (Figure 1.6). The trocar is passed subcutaneously down the animal to exit at a point just above the cannulated vein and

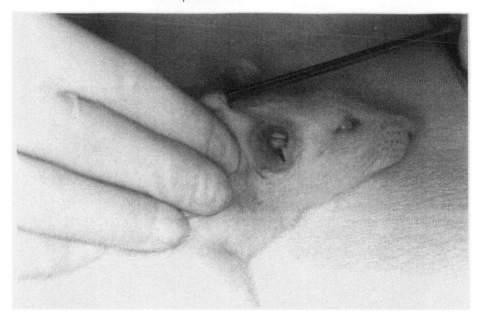

Figure 1.6 Introduction of a trocar into the subcutaneous tissue in the subscapular region.

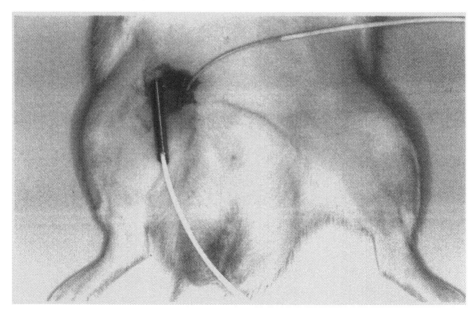

Figure 1.7 The trocar passes subcutaneously to exit at a point just above the cannulated vein and the free end of the cannula is passed along it.

within the original incision site (Figure 1.7). The stylet of the trocar is removed and the free end of the cannula passed up it, whereupon the trocar is removed to leave the exteriorised cannula in position (Figure 1.8). The cannula is then reconnected to the syringe. The position of the cannula in the 'pocket' is checked to ensure that it is not kinked and the incision site is closed. The objective here is to bring the cut edges of

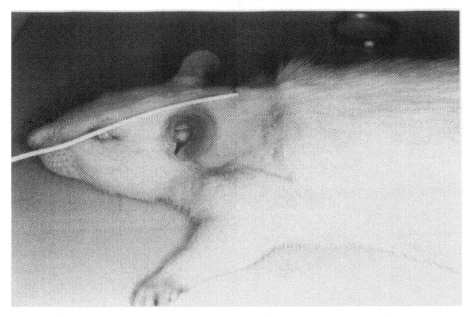

Figure 1.8 The trocar is removed to leave the exteriorised cannula in position.

tissues into opposition without restricting their viability (Douglas 1963; Peacock and van Winkle 1970). The wound must be healthy before closure; tissues should be moistened with warm saline if drying has occurred, and blood clots and damaged tissue should be removed. Restoration of the original disposition of the tissue layers as well as the skin can be achieved in some incisions by closure of muscles in one layer followed by a subcutaneous and skin layer. Animals will attempt to remove skin sutures, which should, therefore, be of the 'interrupted' variety. Interrupted or short continuous suturing of the other layers is also advisable. Metal clips, preferably of the automatically applied staple-type, may be used for skin closure (Figure 1.9).

At this stage, the animal is again turned over, the cannula is clamped just above the dorsal incision site, and the syringe is again disconnected. The end of the cannula is passed up the tether and reconnected to the swivel/infusion line. The clamp is removed and the dorsal incision site closed with a suture. The jacket or harness is then fitted to the animal (Figure 1.2).

1.6.3 Post-operative care

Following cannulation and fitting of the infusion system, the animal is passed to the recovery area, where it receives a second application of ocular lubricant and a subcutaneous injection of buprenorphine (Vetergesic® 0.3 mg/ml at 0.05 mg/kg). The animal's home cage in the recovery room is furnished with a Vetbed, on which the animal is laid on its side. The infusion, usually of saline during the pre-treatment period, is commenced with minimal delay at a maintenance rate of 1.0 ml/h, which has been determined as the lowest rate to ensure maintenance of cannula patency. The background temperature for the recovery room is approximately 27 °C. Additional local heat may be provided by heat lamps positioned a few feet from the cages while the anaesthetised condition persists and until consciousness is regained.

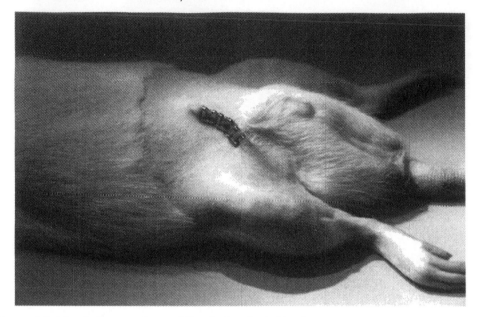

Figure 1.9 Metal clips may be used for the inguinal skin closure.

Post-operatively, all animals recovering from anaesthesia are closely monitored. They also receive post-operative broad-spectrum antibiotic treatment (Baytril®-Enrofloxacin at 0.025%) infused intravenously at 1.0 ml/h for 3 days following surgery.

The metal clips or sutures securing the inguinal skin incision should be removed, usually within 7 days of surgery, and a minimum of 7 days' recovery is recommended before the animals are allocated to the study.

1.7 Discussion

This model of continuous intravenous infusion in rats has now been in use for a number of years in regulatory toxicology and, as a consequence, is well characterised. Also, the surgical procedures and standards of asepsis are tried and tested and are set firm in the training documents supporting the standard operating procedures for the technique. Many of the problems/issues surrounding this model, of which investigators must have a knowledge in order that interpretation of generated data is not affected, mainly concern equipment issues and formulation of the product under test.

We are certain that advancements in catheter technology for use in rodents will continue since it would appear that the ideal catheter that can be commercially produced and sterilised has yet to be found. Polyurethane and silicone are the materials of choice for the manufacture of such catheters for tissue compatibility. There are several grades and types, ranging from the soft materials to quite rigid catheters. It is well known that in rats venous catheters, of all types, do routinely cause 'mechanical' damage to the blood vessel walls, resulting in a variety of pathological manifestations. However, the more rigid polyurethane can only exacerbate this reaction, which is primarily of an inflammatory nature. There is also the issue of whether to 'bevel' the tip of the implanted catheter or have a 'straight-cut' or even 'rounded' end, and it would appear that there is very little information published as to the merits and dangers of this practice. Perhaps the ideal catheter for use in this animal model would be a combination catheter. The majority of the catheter, particularly the length of tubing outside the animal, should be of a resilient material to withstand potential damage from the tether itself, and to reduce the potential of leakage often experienced with the softer materials. At least the tip, the last 5 cm or so, should be manufactured from a softer, more tissue-compatible material such as silicone or Silastic® in order to minimise vascular damage over the period of chronic implantation.

The rodent jacket has been invaluable as a reliable design feature for the protection of the exteriorised catheter in this model but has some disadvantages. The jacket itself does cause some irritation to the animal because of the need for a 'snug' fit and, since rats grow quite considerably during the course of a toxicology study, needs replacing or adjusting on several occasions during the study. The harness now available goes a considerable way to solving a number of these problems. It is fully adjustable so rarely needs replacing and is made from rounded soft plastic material, causing much reduced skin irritation.

When considering some of the other essential equipment, the swivel is one element over which there is considerable choice depending on the model that is to be employed. Single-, dual- and triple-channel swivels that allow both delivery and extraction of fluids in the animal model are available. Also, as with other elements of the equipment involved, swivels are available either as reusable items made from brass or aluminium or as cheaper, disposable versions made from Teflon or plastic.

In the current era of intravenous infusion technology in rats, once the technical skill of surgical cannulation has been mastered, there is now available, from commercial sources, the necessary equipment to suit most needs and budgets.

References

Anderson, R.M. and Romfh, R.F. (1980) *Technique in the Use of Surgical Tools*. Appleton Century-Crofts: New York.

Douglas, D.M. (1963) *Wound Healing and Management*. E & S Livingstone: Edinburgh.

Flecknell, P.A. (1987) *Laboratory Animal Anaesthesia*. Academic Press: London.

Flecknell P.A. and Mitchell, M. (1984) Midazolam and fentanyl-fluanisone: assessment of anaesthetic effects in laboratory rats and rabbits. *Laboratory Animals*, 18, 143–146.

Green C.J. (1975) Neuroleptanalgesic drug combinations in the anaesthetic management of small laboratory animals. *Laboratory Animals*, 9, 161–178.

Gregory, D.J. (1995) Practical aspects of continuous infusion in rodents. *Animal Technology*, 46, 115–130.

Lang C.M. (1982) *Animal Physiologic Surgery*. Springer-Verlag: New York.

Markowitz, J., Archibald, J. and Downie, H.G. (1964) *Experimental Surgery*, 5th edn. Williams & Wilkins: Baltimore.

Peakcock, E.E. and van Winkle, W. (1970) *Surgery and Biology of Wound Repair*. W.B. Saunders: Philadelphia.

Waynforth, H.B. (1980) *Experimental and Surgical Technique in the Rat*. Academic Press: London.

Waynforth, H.B. (1987) Standards of surgery for experimental animals. In Tuffery A.A. (ed.) *Laboratory Animals. An Introduction for New Experimenters*. John Wiley & Sons: Chichester, pp. 303–332.

2 Femoral cannulation using the tail cuff exteriorisation method in the rat

A. M. McCarthy

2.1 Introduction

The Small Animal Toxicology Unit at AstraZeneca R&D Alderley Park has a dedicated surgical team to carry out a wide range of surgical procedures. This chapter concentrates on the methods carried out to surgically cannulate the femoral vein in a rat and exteriorise the cannula using the tail cuff method. This method of cannula exteriorisation was adapted from a method observed at Notox of Holland and developed in-house over a number of years. We feel that the standard achieved offers excellent short- to mid-term results (up to 8 weeks) with the minimum amount of stress possible on the animals, while offering them considerably more freedom of movement than any of the subscapular methods of exteriorisation.

2.1 Animal acclimatisation

Depending on the length of the study involved, animals usually arrive at the unit weighing between 180 and 200 g. This allows us to offer the animals a sufficient level of acclimatisation to ensure that they are approximately 200–250 g in weight at surgery. The animals are housed individually in clear polycarbonate plastic cages ($48 \times 27 \times 20$ cm) with stainless-steel mesh lids and given free access to food and water. The bedding material used is Square-a-sorb (Datesand Ltd). After arrival in the unit, the animals are handled daily by trained technicians to accustom them to the extra level of handling required on an infusion-type study.

2.3 Preparation for surgery

Once the surgery suite has been prepared, equipment and instruments sterilised and all personnel are aware of their roles for the day, the surgery can commence. The animals are weighed in the infusion suite and an analgesic is administered subcutaneously at 0.005 mg/kg or 1 ml/kg (buprenorphine, Temgesic®, Reckitt and Coleman, UK). Each animal is then transported from the infusion suite to the animal preparation area in its allotted cage, which has been previously prepared with swivel, external catheter, syringe filter, syringe and fresh bedding. Gaseous anaesthetics are used and the animals are placed singly into the anaesthetic chamber (Figure 2.1). Anaesthesia is induced using halothane (Fluothane®, Zeneca Pharmaceuticals Ltd, UK) in conjunction with a Fluotec vaporisor and scavenger system. When a sufficient depth of anaesthesia has been induced, the animal is removed from the inhalation

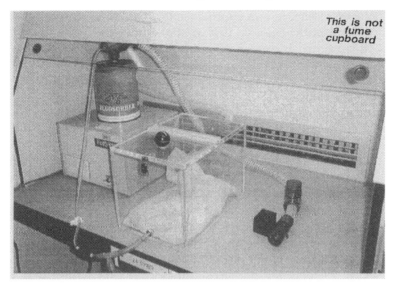

Figure 2.1 Anaesthetic chamber for induction of gaseous anaesthesia

chamber and placed with its muzzle in the scavenger mask to maintain anaesthesia while it is shaved and prepared for surgery. An area on the ventral side of the animal from the leg to the peritoneum and surrounding area is shaved of all fur (approximately 5 cm²). The skin and tail are then swabbed using a Hibitane®/water solution to remove any loose hair and thoroughly clean the incision area.

The tail is then marked using a non-toxic permanent marker at 3.5 cm and 4.5 cm from its base, which will provide the surgeon with an exteriorisation point for the cannula and a guide for the subsequent position of the wire to enable fixation of the tail cuff. The shaved area of the animal is then sprayed with Hydrex S/S chlorhexidine skin disinfectant (Adams Healthcare, UK) and the anaesthetised animal is then passed through a wall hatch to the surgery suite, where it is collected and placed in a scavenger mask in front of the surgeon. The surgery suite has been previously cleaned and prepared using appropriate cleansing techniques, and aseptic techniques are used and maintained throughout all surgical procedures. The animal is placed ventral side uppermost on a heated mat in front of the surgeon and its muzzle placed in an anaesthetic scavenger mask. The animal's hind legs are extended slightly and held in position using Blenderm® (3M Healthcare, UK) adhesive tape. The surgeon then checks the level of anaesthesia, places a prepared surgical drape over the animal and commences surgery when the correct depth of anaesthesia is attained.

2.4 Surgical methods

A skin incision of approximately 1.5 cm is made starting at the point where the femoral vein enters the peritoneum towards the hind leg. The underlying fat is carefully teased apart using small forceps to expose the femoral vein. Regular flushing with 0.9% (w/v) sterile saline is carried out throughout the procedure. At this point, the vein is manipulated using fine forceps and the vein is gently bluntly dissected to remove any residual fat and to enable it to be raised by inserting the tip of the forceps beneath it.

When this has been achieved and approximately 1 cm of 'clear vein' is visible, it is possible to put in place the ligatures that will tie off the vein and hold the cannula in the vein once it is inserted. It is good procedure to use three ligatures until competency is achieved.

One piece of 3/0 gauge Mersilk® silk braid (Ethicon) is cut to approximately 20 cm in length and then folded to obtain a double ligature approximately 10 cm in length. Using fine forceps, the vein is raised and the double ligature fed beneath it and pulled through to half its length. Care should be taken not to allow the vein to 'roll' or twist. The fold in the ligature is now cut and the two halves separated, creating two 10-cm lengths, both of which are beneath the vein. Each length is used to stop blood flow by clamping the two ends of each ligature together and using the loop formed to keep the vein raised and exposed by gently pulling them in opposite directions along the vein, one towards the peritoneum and the other towards the hind leg. A third single ligature is placed under the vein and used to tie off the vein below the venous incision and to aid cannula position within the vein. When all the ligatures are in position and the vein raised and exposed, a venous incision/puncture can be made using either micro-scissors or sterile 23 gauge × 1 inch needle fitted to a 1-ml syringe with the needle bent to an angle approximately 45° from the hub (the preferred method at this laboratory).

The puncture should be made in the lower half of the exposed vein by very carefully inserting the needle to a point just over its bevel. The needle is then removed from the vein (a little blood may appear but manipulation of the clamped ligatures should control any blood flow). The venous incision is then opened using micro-curved forceps and the cannula (Data Sciences Physiocath®), previously primed with heparinised saline, is gently inserted into the vein using the small curved forceps. When the tip of the cannula is inside the vein, the micro-forceps are removed and the cannula is slowly inserted 4.5 cm into the vein up to the stopper on the cannula. The ligature nearest the peritoneum may need to be loosened to enable the cannula to pass beyond it, and there may also be slight resistance encountered at the peritoneal wall. If this occurs, gentle manipulation of the peritoneum usually allows the cannula to pass through and up into its position within the vena cava. During this process the cannula should be checked for patency by drawing back on the syringe attached to the end of the cannula. No resistance should be apparent and blood flow into the cannula should be smooth. It is imperative to flush the blood back into the animal as any blood left in the cannula has the potential to clot. At this point in the procedure the vein must be tied off using the single ligature and a double knot below the incision and underneath the cannula. When this is completed, the same ligature may be tied around the cannula, which will help to keep it within the vein. The same procedure should be carried out using the other two ligatures, taking care to ensure that patency of the cannula has not been compromised during the tying of the knots. When the knots have been tied, the ligatures may be snipped off at the knot sites and the cannula checked once again for patency. Exteriorisation of the cannula may now begin.

2.5 Cannula exteriorisation

The next stage of the procedure is to exteriorise the cannula at the tail. Using a long stainless-steel trocar (International Market Supply) from the 4.5-cm mark on the tail, a subcutaneous tunnel is created up the tail between the left lateral vein and the ventral artery continuing under the skin. Care must be taken to ensure that the tip of the trocar remains subcutaneous and is directed away from the organs in the genital region and is fed up to the incision site, where the trocar should emerge underneath the fat previously teased away at the beginning of the procedure. When this position is reached, the patency of the cannula should be checked by drawing and flushing blood into and out of the cannula. With the cannula patent, the syringe and needle may be removed from the end of the cannula and the cannula passed down the trocar from the incision site. When the cannula becomes visible at the Luer lock end of the trocar, the trocar may be carefully removed. The cannula may then be gently pulled to remove any excess length at the incision site. To prevent the cannula twisting and thereby rolling the vein, it is held with a pair of forceps while the remaining 'loop'of cannula is pulled through. A little excess cannula length may be left at the incision site but the amount should be determined by the size of the animal. The cannula is now exteriorised, and the heparinised needle and syringe should be refitted to the cannula and a little administered to ensure patency.

2.6 Closure of the incision site

The incision site needs to be as clean as possible to inhibit the possibility of infection and therefore is flushed regularly with saline during the procedure. This also moistens the site and keeps it as similar as possible to its natural physiological state. The final flush is carried out and any excess saline removed using sterilised cotton buds. These are each used once only to clean the surrounding area of any blood that may have accumulated as a result of the incision or cannulation. The clean edges of the wound are then brought together using straight forceps and three or four sterile surgical skin clips (Proximate® Plus MD35 regular, Ethicon Endosurgery) are stapled along the incision. The skin at the incision site is lifted slightly before deploying the staples to ensure they do not penetrate the underlying cannula. Vet Bond is also administered between the staples as a secondary precaution in case the animal removes its clips. At this point, the animal's legs are untaped and it is transferred to a separate scavenging mask and heated pad for the fitting of the tail cuff. This removal to a separate surgical station keeps the cannulation area as sterile as possible. The tail cuff and external infusion equipment are fitted by a separate member of the infusion team and the animal is then returned in its cage to the animal room to be connected to the infusion pump.

2.7 Fitting the tail cuff

The animal is placed with the dorsal surface uppermost and the exteriorised cannula can be seen emerging from the side of the tail. The 3.5-cm mark on the tail is now used to achieve good cannula and tail cuff position. A sterile needle 22G × 1 inch (Sherwood Medical) is inserted at the tail mark above the right lateral vein and beneath the dorsal vein and excised above the left lateral vein. The needle is then used as a trocar and sterile stainless-steel monofilament wire (B/Braun) is inserted from the bevel end of the needle and carefully fed through the needle emerging from the Luer lock end. The emerging wire is held while the needle trocar is removed. The wire has now been positioned aseptically for the tail cuff to be fitted. The tail cuff (Figure 2.2) has recently been modified from an original length of 45 mm to the following

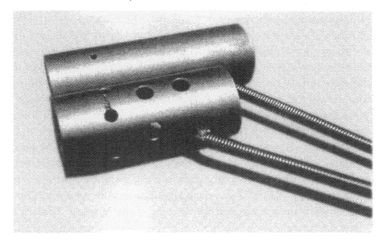

Figure 2.2 Tail cuff and spring attachment (showing revised, customised version in front of the original longer version).

dimensions: 30 mm long, 15 mm diameter. Small holes are also now drilled into the cuff to reduce the weight and improve air circulation. The spring attachment is 30 cm in length.

The animal is then turned over (i.e. ventral side uppermost) and the needle and syringe removed from the end of the cannula to enable the cannula to be fed inside the tail cuff and down the spring. This is made easier by using curved forceps to reach inside the cuff and feed the cannula down to the end of the spring until it emerges. This process should be completed promptly as the blood will begin to flow back down the cannula. This blood flow can be stopped by using a needle and syringe of heparinised saline which can be flushed through to deter any clot formation within the cannula. When this is complete, the tip of the animal's tail is fed into the tail cuff and the cuff positioned on the tail ensuring the cannula is not looped around the tail. Upon positioning of the tail cuff, each end of the surgical wire is fed through the appropriate holes on the cuff bent towards its underside and tied by twisting the wire together using a pair of needle forceps. This will achieve a strong tie around the cuff with minimal movement of the wire and cuff. Any excess wire is trimmed off using a pair of snips. The holes in the cuff are positioned so as to keep the wire straight within the tail and any pressure from the tying process is taken by the edge of the tail cuff and not the animal's tail. When the cuff is correctly fitted, the patency of the cannula is checked again as previously described. At this point, an antibiotic gel is administered (Bactroban®, Smith Kline Beecham) to the entry and exit points of the wire. This is repeated twice daily for 5 days to prevent any infection occurring at the site. The process of attaching the external infusion equipment may then commence.

2.8 Attachment of the external infusion equipment

The external infusion equipment (with the exception of the pump itself) is presented fully primed (saline 0.9%) with the animal on arrival at the surgery and its attachment is the final stage of the process within the surgery suite.

The external infusion equipment comprises a 50-ml syringe (B Braun) to which are attached a filter (Millex Millipore GS 0.22 ml) and an extensible coiled cannula (Vygon Lectro-cath). The Luer lock end of the cannula is screwed onto a three-way tap (Baxter Healthcare), which is pushed onto the hub of a modified mouse-dosing catheter (International Market Supply 4.5 FG, 60 mm), the gavage end of which is cut approximately 3 mm from the hub. The mouse catheter is then pushed onto the top of the infusion swivel (Lomir Biomedical 20G rodent swivel) and an 8-cm approximate length of silicone tubing (0.5 mm ID, 0.5 mm wall, 1.5 mm OD Esco tubing) is attached to the bottom of the swivel.

The needle and syringe are removed from the end of the internal cannula and the cannula is fed through the spring. The spring containing the internal cannula is then fed through the hole in the block of the swivel and secured by a grub screw. The free end of the internal cannula is looped around to meet the silicone tubing and both are cut down to form a loop approximately 7 cm in length when joined. The internal cannula is cut at a 45° angle, fed into the silicone and glued in place (Loctite Superglue). There must, of course, be no air within the system, which is now complete, and the animal may be replaced in its cage for transport back to the infusion suite. On arrival at the infusion suite, the syringe is attached to the infusion pump and a rate of 1.0 ml/h is set. The animal is closely monitored until fully recovered from the anaesthetic.

2.9 Adverse effects

Two adverse effects are observed in connection with the tail cuff. First is swelling around the cuff, which is rarely severe but has in extreme cases resulted in the animal being killed early for humane reasons. Second, open wounds are occasionally observed in the tail region, but these are now rare since a change of bedding was introduced.

2.10 Conclusion

The surgery technicians at this laboratory constantly seek refinements to the infusion system, its application and use. Any scientific advances in equipment specification or techniques that the surgery team considers beneficial to animal welfare and study integrity may be adopted. The tail cuff technique has been used successfully at the author's laboratory for continuous intravenous infusion for periods of up to 8 weeks in duration.

3 Tethering and the implanted button model in the rat

E. S. Miedel

Researchers often need to administer test materials to laboratory animals by intermittent or continuous infusion. Additionally, intermittent or continuous infusion may be required to study the physiological and pharmacological effects of synthetic or natural substances (Desjardins 1986), and is required in many areas of research, including pharmacokinetic studies (Bakar and Niazi 1983). Continuous infusion of heparin/saline has been shown to be useful in studies requiring continuous sampling, helping to ensure patency of the sampling cannulae. The continuous withdrawal method, when used along with the infusion of heparin/saline, has been demonstrated to be equivalent to the classic intermittent sampling method used in bioavailability studies (Humphreys *et al.* 1998). Infusion and sampling in the rat can be accomplished by surgically implanting cannulae. Intravenous, bile duct, urinary bladder, gastrointestinal and mesenteric lymph duct cannulae are often used.

One way to accomplish the infusion or sampling is to completely restrain the rat. This can be accomplished by using a Bollman cage. The Bollman cage is a system of rods that encase the rat, prohibiting free movement. Rats have been held in such cages for several days (Waynforth and Flecknell 1992). This method is extremely stressful to the rat. Rats under restraint produce altered levels of plasma concentrations of pituitary hormones (Tinsley *et al.* 1983). Studies in other animal models, such as non-human primates, indicate that this type of restraint may result in a decrease in haemoglobin, haematocrit and lymphocyte concentration (McNamee *et al.* 1984). Other stress conditions have also been reported. With this in mind, most laboratories have moved, or are moving to, less stressful, more humane and scientifically sound methods. One such method is the swivel and tether system (Figure 3.1).

Tethering of the rats permits the animals to move about the cage during infusion or sampling. Tethered animals must be housed individually. Depending on the needs of the researcher, the tethering system may include syringe or peristaltic pump, connection lines, swivel, tether, implanted button and rodent jacket.

Pumps are used to infuse test materials or to sample blood, urine or other body fluids. Connection lines are needed to connect between the infusion/sampling device and swivel, and to connect between the swivel and animal. Common connection line materials would include polyethylene, polyurethane, Silastic® (Dow Corning, Michigan, USA) or other medical tubing of the correct dimensions. The swivel allows the animal to move about the cage without twisting the cannula line or lines. In the late 1960s, Michael Loughnane developed a swivel for use with the laboratory rat. This swivel was adapted from an earlier model concept presented by Jim Weeks (Jacobson 1998). High-quality swivels have less friction and allow the animal to move

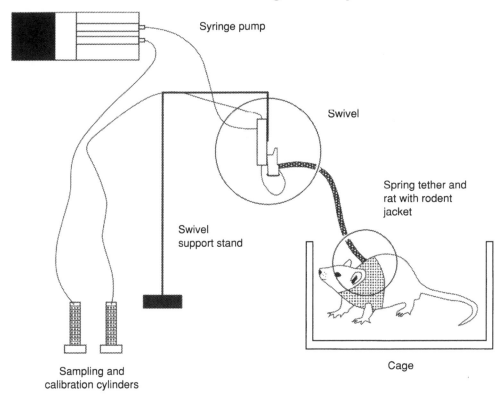

Syringe pump

Swivel

Spring tether and
rat with rodent
jacket

Swivel
support stand

Cage

Sampling and
calibration cylinders

Figure 3.1 Rat tether and swivel system. Reprinted with permission from Hilltop Lab Animals,
Inc., Scottdale, PA, USA.

about the cage more easily. The tether, typically a stainless-steel spring appropriately
sized, attaches the swivel to the implanted button. The tether provides protection to
the connection line or lines while continuing to permit the animal movement. The
implanted button permits the researcher to connect the animal to the tether.

There are a number of different types of implantable buttons available. Tethering
buttons are generally attached by suturing in place under the skin or on top of the
skin in the scapular area. The button allows the catheter to be externalised and attached
to the spring tether (Jacobson 1998). The button may also be held in place with the
rodent jacket without suturing the button to the rat.

The type of button used depends on the needs of the research. Stainless-steel buttons
have been used for many years. They can be autoclaved and reused virtually without
wear. Stainless-steel buttons appear to be best suited for short-term studies – those
studies lasting less than 7 days. Longer use of the stainless-steel buttons can cause
adverse tissue reaction.

Polysulphone buttons are rigid buttons that can be autoclaved and can be reused.
Polysulphone is an extremely durable plastic material capable of withstanding
extremely high temperatures. Both the stainless-steel and the polysulphone buttons
are rigid and are sutured into the tissue of the animal. Another type of button is the
Dacron mesh button. This button has a Dacron mesh screen base and a soft plastic
tip. The mesh base allows the tissue to infiltrate the material on longer studies. This

produces an excellent attachment between the button and the animal. The flexible, soft, plastic tip creates a simple, yet effective, method to attach the button to the spring tether. Dacron mesh buttons are not practically reusable, however they are relatively inexpensive (Figure 3.2).

Tethers can be attached directly to the rat via the implanted button. Attaching the rat directly to the tether is simple and quick. The undesirable side to attaching the rat directly to the tether is stress at the site of the button. As the animal moves about the cage, the spring tether tugs on the wound site where the cannula is exposed and the button is attached. One way to minimise the stress on the wound site is to use a rodent jacket or rodent harness.

There are rodent jackets that are designed to hold the button securely to the animal while allowing the researcher access to the button tip. Although this helps to minimise the stress on the wound site, it adds the stress of the rodent jacket. Rats will, at the very least, be stressed initially by the rodent jacket. Rodent jackets can also be used to store additional length of cannula to reduce the number of connections necessary to connect to the swivel. Fewer connections result in greater patency of the cannula. It is possible to implant longer cannulae either coiled and stored in a subcutaneous pocket or stored in the rodent jacket placed on the animal at the time of surgery.

Once the decision has been made regarding the type of button, the researcher needs to determine the amount of time needed prior to the start of the study. The cannula can be implanted and button attached. The rat can be connected to the tether at a later date. The cannula can be plugged and exposed through the button. Keeping the unplugged portion of the cannula in the subcutaneous pocket improves patency. Without the implanted button, it is often necessary to do some blunt dissection to open the subcutaneous pocket and remove the cannula. If it is required to have a long holding time prior to the start of the study, it is helpful to glue the plug into place within the cannula. This will help prevent the plug from accidental removal. Once the plug is glued into place, extra length of cannula under the skin will be necessary. The extra length will make it possible to cut off the section of cannula containing the glued plug.

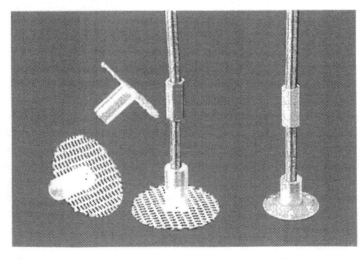

Figure 3.2 Tethering buttons and spring tethers. Reprinted with permission from Instech Laboratories, Inc., Plymouth Meeting, PA, USA.

Control animals should be subject to the same surgical procedures. This includes implanting the same cannula and implanted button. Simply implanting a lightweight plastic button on the back of rats has been determined to be stressful to the rat. In experiments comparing the weight gain of control rats with that of rats with an implanted button, the button was determined to be stressful. Fasting the animals for 24 h prior to the start of the experiment reduced the weight difference between the two groups (Birkhahn *et al.* 1979).

Animals that have been surgically fitted with cannulae should be housed individually. This will prevent the animals from damaging the surgical sites of other animals. Regularly handling the animals will reduce stress to the animals and make manipulations easier for the researcher. Housing rats in clear cages next to each other appears to allow more normal socialisation and result in calmer, easier-to-handle rats.

Those tether systems that allow the rat the most movement with the least effort, have the rat tethered for the shortest amount of time and permit the animal the easiest access to feed and water can be considered most suitable. Tethered animals should be given adequate time to acclimate to the tether, jacket and caging systems (Waynforth and Flecknell 1992). Even the best tethering systems will sometimes present problems. The animals should be checked regularly to ensure that they have not become entangled and that the lines have not become twisted.

Institutional animal care and use committees should closely review the need to tether rats and should consider the tethering process to be at least minimally stressful to the animal. Researchers should consider closely the differences between tethered and non-tethered animals. A tethered rat should not be considered to be a physiologically normal animal model. While tethering is vastly superior to total physical restraint, it should be considered as partial physical restraint. Any form of restraint will cause stress to the rat. As such, additional animal care observation time will be necessary (Institute of Laboratory Animal Resources Commission on Life Sciences National Research Council 1996).

References

Bakar, S.K. and Niazi, S. (1983) Simple reliable method for chronic cannulation of the jugular vein for pharmacokinetic studies in rats. *Journal of Pharmaceutical Sciences*, 72, 1027–1029.

Birkhahn, R.H., Long, C.L., Fitkin, D. and Blakemore, W.S. (1979) Stress induced by light weight back button used to prepare the rat for continuous intravenous infusion. *Journal of Parenteral and Enteral Nutrition*, 3, 421–423.

Desjardins, C. (1986) Indwelling vascular cannulas for remote blood sampling, infusion, and long-term instrumentation of small laboratory animals. In *Methods of Animal Experimentation*, Vol. VII, Part A, Academic Press: San Diego, pp. 177–187.

Humphreys, W.G., Obermeier, M.T. and Morrison, R.A. (1998) Continuous withdrawal as a rapid screening method for determining clearance and oral bioavailability in rats. *Pharmaceutical Research*, 15, 1257–1261.

Institute of Laboratory Animal Resources Commission on Life Sciences National Research Council (1996) *Guide for the Care and Use of Laboratory Animals*. National Academy Press: Washington, DC, p. 11.

Jacobson, A. (1998) Continuous infusion and chronic catheter access in laboratory animals. *Laboratory Animals*, 27, 37–46.

McNamee, G.A., Wannemacher, R.W., Dinterman, R.E., Rozmiarek, H. and Montrey, R.D. (1984) A surgical procedure and tethering system for chronic blood sampling, infusion, and temperature monitoring in caged non-human primates. *Laboratory Animal Science*, 303–307.

Tinsley, F.C., Short, W.G., Powell, J.G. and Shaar, C.J. (1983) Preparation of a jugular vein cannula: use with a semiautomatic blood-sampling system. *Journal of Applied Physiology: Respiratory, Environmental and Exercise Physiology*, 54, 1422–1426.

Waynforth, H.B. and Flecknell, P.A. (1992) *Experimental and Surgical Technique in the Rat*. Academic Press: London, pp. 193–202.

4 Multidose infusion toxicity studies in the rat

G. Healing

4.1 Introduction

The development of novel pharmaceuticals that require continuous or intermittent infusion in patients has increased dramatically in recent years. Consequently, in order to test these pre-clinically, pharmaceutical companies must, as far as possible, mimic the clinical situation. In practice, this requires the surgical catheterisation of animals followed by a period of uninterrupted intravenous infusion. This chapter describes the technology and methodology required to conduct a 1-month continuous infusion study for regulatory submission. However, the principles apply equally to longer term studies, as rats have been successfully infused with saline for periods of 3 (Cave *et al.* 1995) and even up to 6 months (Francis *et al.* 1992).

At the author's laboratory, techniques have been developed and refined to enable this procedure to be performed routinely in the rat. Described below is the so-called jacket or harness model, but for the practical purposes of undertaking a repeat dose–toxicity study, the methods described apply equally to rats restrained by the tail (the 'tail cuff' method), as described in Chapter 2 and elsewhere (Jones and Hynd 1981). It has also been demonstrated that a variety of enrichment aids can be successfully employed, even with tethered animals, and that the need for additional male fertility studies (and hence the use of further animals) can be removed by combining the latter with the general toxicity study. These techniques will also be described.

Since the surgical technique has already been documented in detail in the previous chapter, I will concentrate on the conduct of the study from when the animals recover from surgery through to the necropsy.

4.2 Animals

The rats used at the author's laboratory are Sprague–Dawley (Crl: CD®BR VAF/ PLUS), supplied by Charles River UK Ltd, UK. On arrival at the laboratory and before assignment to this study, the rats are subjected to a visual inspection. Any animal showing signs of disability or ill-health is excluded. The animals are acclimatised for at least 5 days prior to surgery and then for at least 5 days after surgery before commencing dosing. The animals should be of sufficient size and weight (*c.* 250 g) to aid the surgical procedure.

4.3 Post-surgical care

The surgical suite is maintained as an aseptic area (Figure 4.1) with separate preparation and recovery rooms. Surgical preparation and facilities are described in detail in the previous chapters and elsewhere (Gregory 1995), but the need for strict aseptic procedures cannot be emphasised enough.

At this laboratory, the rats are anaesthetised using the gaseous agent isoflurane (Schering-Plough, UK) and oxygen. In our experience, this has the advantage over injectable anaesthetics, including those with a reversal agent, that recovery is more rapid, and hence there are fewer post-surgical complications such as decreased body temperature, body weight losses and drying of the eyes (often leading to opacities). The rats are given a combined post-operative analgesic regimen of Vetergesic (buprenorphine; Reckitt & Coleman Products Ltd, UK; 0.05 mg/kg s.c.) and Carprofen (CVet, UK; 5 mg/kg s.c.), both given 30 min prior to anaesthesia (Flecknell *et al.* 1999). This has been shown to be more effective than either analgesic given alone. Post-surgical administration of antibiotics is not considered necessary; in our experience, these is no substitute for good aseptic technique and administration of antibiotics may even interfere with the interpretation of the study data.

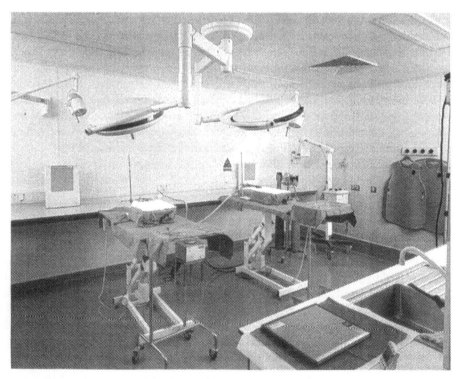

Figure 4.1 The rodent surgical facility at AstraZeneca R&D Charnwood.

4.4 Infusion equipment

4.4.1 Caging

The rats are housed singly in solid-bottomed cages (Techniplast UK Ltd, UK), dimensions 400 mm × 250 mm × 180 mm, with lids (raised an extra 60 mm in height) which have been adapted to allow the tether to move freely along the length of the cage (Figure 4.2). Originally grid-bottomed cages were used in this laboratory, as it was believed that the presence of bedding would adversely affect the catheterised animals as the bedding would become lodged in the jacket or harness. However, this has since been shown not to be the case and notable benefits in terms of animal welfare have arisen from the use of solid-bottomed cages: there is less likelihood of physical damage occurring to the feet/claws and enrichment items and bedding can be included in the cage. This is an advantage over the tail cuff method, in which it is possible for bedding to become trapped in the cuff and adversely affect the entry site of the cannula.

4.4.2 Giving set

The apparatus that delivers the test compound to the animal (the giving set) comprises a Monoject syringe (Sherwood, UK) attached to a pink Luer cannula (Sims Portex, UK), which is in turn connected to a disposable swivel and 'shoebox' (Lomir Biomedical, Canada). The latter is secured by a clamp above the cage, and is then connected to the catheter (see section 4.4.4) which enters the animal (see Figure 4.2). The catheter is covered by a protective tether (Instech Laboratories Inc., USA). The giving set is secured to the animal by means of a 'Covance' harness (Instech Laboratories Inc., USA). This is adjustable so that it can be loosened as the animal grows.

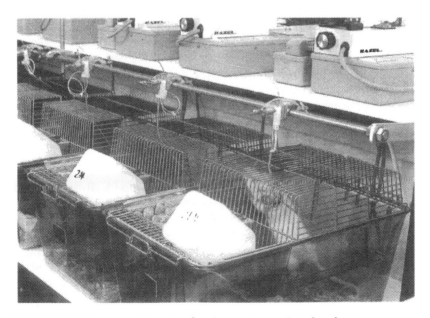

Figure 4.2 Typical giving set and caging for the continuously infused rat.

4.4.3 Syringe pumps

Syringe pump technology has advanced rapidly during the past few years, with numerous acceptable models now on the market. A number of different infusion pumps have been used in our laboratory. Until recently the syringe pumps of choice were the KDS 100 (KD Scientific Inc., USA) with a range of flow rates between 0.02 and 426 ml/h, dependent on syringe size, and the Razel A-99 (Razel Scientific Instruments Inc., USA) (range 0.25–143 ml/h). However, these have now been superseded by more advanced models. Among these are the Perfusor compact (B Braun, Germany), which has a digital display and numerous functions, plus the benefit of a battery back-up should an interruption in power occur. The range of flow rates with this model is 0.1–99.9 ml/h. It is recommended that any worker starting to undertake infusion studies for the first time researches the current status of new syringe pumps and purchases those best suited to the set-up of the room. Cost is also a consideration, as to perform a full regulatory infusion study requires approximately 100 pumps; relatively inexpensive pumps that are less technical but perform all the required functions are available. There are also multiheaded pumps on the market, which are capable of simultaneously dosing up to six animals, although the risk with these is that any malfunction will result in a number of animals being affected. In addition, individual flow rates cannot be set, only a mean if dosing relative to body weight. This could result in possible overdose if, owing to toxicity, the body weights of some of the animals were reduced compared with others on the same pump.

4.4.4 Catheters

The favoured catheter material in this laboratory and many others is silicone, but extensive experience has also been acquired with polyurethane. However, there are occasions when a drug formulation is incompatible with softer plastics such as these, and as a result harder plastics such as polythene or polyethylene have been utilised. There are a number of disadvantages when using a more rigid catheter, such as damage to the vessel during and after catheterisation and greater difficulty in securing the catheter in place. There are also differences in the resulting pathological findings, and a brief summary of our findings comparing the different catheters is shown in Table 4.1. Suppliers of the catheters used at this laboratory include Access Technologies, USA, Strategic Applications Inc., USA, and Data Sciences, USA, which are available in Europe through Uno, The Netherlands.

4.5 Husbandry and enrichment

The animals are housed throughout the study in a room that is ideally in close proximity to the surgical suite. In many laboratories, the study room is adjacent to the surgical suite, but when a number of studies are to be run simultaneously this is not always possible. The room should be organised to enable clear uninterrupted access to the cages and the infusion pumps for ease of checking the animals and the apparatus. In the author's laboratory, it has been found that this is best achieved by placing the cages on modified tables (Figure 4.3). Standard racking is also considered acceptable and is used in other laboratories (Gregory 1995). This is particularly appropriate

Table 4.1 A comparison of the pathology findings associated with different catheters

Catheter material	In-life findings	Histopathological findings
Polyethylene	Difficulty cannulating (e.g. torn vessels); problems securing catheter	Some vascular wall thickening; slight thrombus formation
Polythene	Difficulty cannulating (e.g. torn vessels); occasional leakage at connections; higher incidence of technical failure	Slight inflammation at infusion site; some vascular wall thickening; slight thrombus formation
Polyurethane	Occasional leakage at connections	Frequent tissue thickening along catheter; some vascular wall thickening and thrombus formation; slight fibrosis and oedema
Silicone	Occasional kinks in catheter	Slight vascular wall thickening; slight thrombus formation/ vasculitis
Vinyl	Occasional tearing of vessel; some leakage from swivel	Vascular/perivascular inflammation; occasional abscess formation

Figure 4.3 Typical animal room layout for a rat continuous infusion study.

when there is restricted space within the animal facility but can lead to reduced ease of observation of the animals and equipment.

The rats are housed under artificial light between 06.00 and 18.00 h and in darkness between 18.00 and 06.00 h, with temperature maintained between 19 and 23 °C and humidity of between 40% and 70%, although the temperature can be raised slightly during recovery from anaesthesia. Water is supplied *ad libitum* by a bottle and food (Rat & Mouse No. 1 Maintenance SQC expanded diet, Special Diet Services, UK) in a removable hopper. Bedding is in the form of aspen wood chips (B&K Universal Ltd, UK). Cages and bedding are changed twice a week.

A number of enrichment aids are included in the cages for the animals. These include Enviro-dri nesting material (B&K Universal Ltd, UK) and rabbit-type chew-sticks made from aspen wood (70 mm × 35 mm mm × 35 mm) which are supplied with certificates of analysis, as is the bedding (Figure 4.4). It has been determined in this laboratory that inclusion of bedding, nesting material and other enrichment aids do not adversely affect the surgically-prepared rat in any way. This is an obvious benefit of the harness method of tethering, since the entry point of the catheter is on the dorsal surface of the rat, away from the base of the cage (compared to the problems associated with the tail cuff). The practical benefit of including these materials in the cage, apart from the obvious enriched environment, is that the exploratory activity of the rats is increased, thus making it easier to observe abnormalities.

There is perhaps a welfare issue over the need to house continuously infused animals individually, but the potential problems of housing more than one tethered animal in a cage (entangled tethers, animals interfering with one another's giving sets, etc.) are considered to make group housing impracticable at the present time.

4.6 Study design

The design of the continuous infusion study for regulatory submission is by necessity similar to standard protocols by other routes of administration. However, there are a number of practical difficulties that need to be overcome with the use of tethered animals. Described below are the common procedures undertaken on a standard 1-month repeat dose–toxicity study. The practicalities are covered in the next section.

Figure 4.4 Cage showing enrichment aids (Enviro-dri nesting material and chew block).

Before commencing an infusion study it is essential to demonstrate that the compound is compatible with the giving set, in particular the catheter. Some compounds interact with or are adsorbed onto soft plastics, such as silicone or polyurethane in particular, and even interact with the rubber in syringes. If this is the case, alternative catheter materials must be identified and syringes without rubber used. In addition, any stability measurements must be made within the giving set under laboratory conditions, as the compound could potentially remain in the syringe for 3 days or more.

Another possible variable is that a test compound might only be required to be infused for part of every 24 h (e.g. 2- to 4-h periods are common). With this study design, a decision will need to be made as to whether it is beneficial to surgically prepare the animals, and to infuse with saline during the off-dose period at a low rate to maintain patency, or use a non-surgical method by restraining them and infusing daily via the tail vein. Here the relative stresses of repeated restraint and administration against undergoing a surgical procedure must be considered.

Daily clinical signs are relatively easy to monitor, as the rat can be removed from its cage (with extra care) and examined. There are a number of signs that may be observed as a result of the surgical procedure or method of restraint, such as dermal effects at the point of entry of the catheter, and abrasions due to overtightening of the jacket or harness. Food and water consumption are easily monitored on a daily or weekly basis. Blood sampling for clinical pathology is usually conducted on a single occasion prior to necropsy, and urinalysis is conducted prior to surgery and then again during the final week of the study. Toxicokinetic analysis is commonly conducted during the first and last weeks of the study and one or two samples per rat are usually sufficient to estimate the steady-state exposure. The time after syringe change needs to be taken into account though, as plasma levels of a drug with a short half-life could be affected when the infusion is temporarily halted. Ophthalmoscopy is also conducted prior to surgery and during the last week of the study. The former observation is easiest from a practical viewpoint prior to the rats being tethered, as by necessity the ophthalmologist must otherwise move between cages to examine the animals.

Finally, it is often a requirement that a period of off-dose observation for a subgroup of animals is included in the study design to assess reversibility of any compound-related effects. For infusion studies, the catheters are simply cut and sealed, so that the animals can move freely for the duration of the recovery period.

4.7 Animal room procedures and study conduct

As stated in the previous section, there are a number of the study procedures that are either unique to infusion studies or present practical difficulties resulting from the need to maintain continuous infusion of the test compound. These include changing syringes, collection and processing of the dosing data, body weighing, blood sampling and urinalysis. There is a clear need for specifically trained staff to undertake infusion procedures, and this must be formally addressed.

During the period of recovery following surgery, saline is infused at a rate of 2 ml/kg/h (in other laboratories, a standard rate of 0.1–1 ml/h is employed). This is also the standard preferred infusion rate when on study. There are occasions when practical considerations dictate that a higher or lower rate is employed, for example restricted solubility of the compound under investigation. It is not recommended at this laboratory to exceed 4 ml/kg/h for continuous infusion because of the stresses on the equipment (especially at the interfaces between sections of the giving set) and the resultant infused volume, which could also have an effect on the renal excretion of the compound.

At the start of the in-life phase of the study, the syringe pumps are temporarily halted and the syringes containing test solution or vehicle, as appropriate, are connected aseptically, the start volume recorded and the pump restarted. When the syringe needs changing (depending on the stability of the formulation and the syringe size), this procedure is repeated and the end volume noted. Alternatively, syringes can be weighed before and after infusion. Syringes must be filled aseptically (after filtration of the solution), ideally in a flow cabinet.

Body weighing tethered animals must take into account the weight of the apparatus attached. Previously at this laboratory the animals were disconnected from the infusion apparatus to be weighed, and then the weight of the tether and harness (measured before the study commenced) deducted. An alternative method has now been validated whereby the rats are not disconnected from the infusion apparatus. A portable balance (model PB3001; Mettler-Toledo Ltd, UK) is located adjacent to the cage and the tethered rat lifted carefully onto it (Figure 4.5). The readings have been found to be consistent and circumvent the need to halt the infusion.

Blood sampling from the retro-orbital sinus is relatively straightforward for tethered animals, but when the method of choice is via an alternative route such as the lateral tail vein, as at the author's laboratory, this presents practical difficulties. Prior to sampling the tail must be warmed, classically by placing the rat in a heated box, necessitating disconnection from the infusion pump (particularly undesirable for toxicokinetic sampling of compounds with a short half-life). An alternative is to warm the tail adjacent to the cage. This has been achieved through an infra-red lamp. The rat is then manually restrained on a mobile table next to the cage, and the phlebotomist can then take the blood sample without disconnecting the animal.

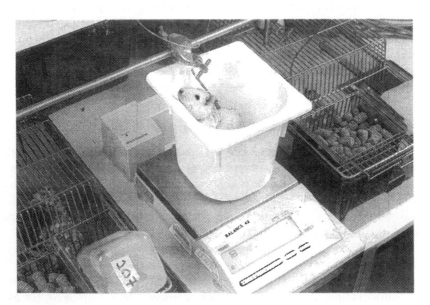

Figure 4.5 Body weighing procedure using a portable balance.

It is also a requirement to collect urine samples for biochemical analysis, usually for periods of between 6 and 18 h. For non-infused animals this is commonly performed by moving the animals to specially designed urinalysis batteries. For tethered rats, the tables on which the cages are located can be modified to accommodate a metal collection funnel leading to a glass graduated tube under the table. The rat is lifted from the solid-bottomed home cage, the cage replaced with a grid-bottomed cage for the collection period, and the rat replaced still attached to the infusion pump. At the end of the collection period the procedure is reversed, again removing any need to interrupt the infusion of the test compound.

Ophthalmoscopic investigation of tethered animals necessitates continual movement of the investigator from cage to cage, rather than the animals being brought to a central point. For this reason is it advisable to conduct pre-dose observations prior to surgery.

There are often minor problems with maintenance of the infusion line. In-line filters are not used at this laboratory, but solutions are filtered prior to use. Occasionally catheters can become blocked, and such occlusions should be cleared as rapidly as possible. This can be achieved by gentle administration of less than 1 ml of saline from a 1-ml syringe and needle into the detached catheter. There is a theoretical risk of dislodging a thrombus into the systemic circulation, but experience suggests that this does not occur even after repeated clearances.

Another possible problem is the use of light-sensitive compounds. This can be overcome simply by covering the syringes and catheters. To enable observation of the volume remaining in the syringe, a small flap can be cut in the foil that can be raised temporarily and then replaced.

4.8 Background data

There are a number of findings that can be expected as background resulting from the cannulation and continuous infusion of rats. Those parameters commonly affected during studies of 2 weeks to 1 month duration that have been conducted at this laboratory are clinical signs, water consumption, ophthalmoscopy, clinical pathology and histopathology. A brief overview of expected background findings is given below, with saline being the most common vehicle, but with mannitol, mannitol and glucose and polyethylene glycol also being used. The choice of vehicle can dictate the signs and effects observed, as some vehicles may not only result in minor toxicity when infused but also cause direct irritation at the site of infusion.

4.8.1 Clinical signs

A certain number of minor clinical signs in a small proportion of control animals is sometimes observed, including staining of the fur, pilo-erection, lachrymation and cold extremities, due to the surgical procedure. The degree and incidence depend on the quality of the surgical and aseptic procedures, and also on the vehicle used. For example, continuously infused polyethylene glycol at 20% (w/v) and above resulted in observations of noisy respiration, body and facial soiling, depressed activity, hunched posture and eyes half-closed in a number of rats, whereas infusion of mannitol (5% w/v) or a mannitol–glucose mix (2.5% w/v: 2.5% w/v) resulted in effects more consistent with saline infusion.

4.8.2 Body weight

Body weights typically fall or remain static immediately post surgery. However, the animals usually start to gain weight again soon afterwards and often catch up with the weights of non-tethered animals by the end of a 1-month study (see Chapter 22 for data).

4.8.3 Food and water consumption

Food consumption is similar in continuously infused rats to that in rats that are dosed by other routes. Water consumption is commonly observed to decrease slightly for rats being continuously infused (Table 4.2), compared with non-infused rats. This appears to compensate for the fluids received, while urine output has been observed to remain relatively constant. Use of a diuretic such as mannitol as a vehicle will almost certainly increase urine volume, as will increasing the infusion rate (see Chapter 22).

4.8.4 Ophthalmoscopy

There are a number of recognised ophthalmoscopic effects in the rat associated with continuous infusion (Loget *et al.* 1997), particularly in the cornea. Typically, anaesthetised animals were found to develop corneal opacities, principally localised in the medial region of the palpebral fissure. This finding could be either a direct (palpebral fissure staying open resulting in the absence of wetting by tears) or indirect (lack of tear secretion) effect of the injectable anaesthetic used.

These observations have been reflected in the author's laboratory, but have been overcome by minimising the time under anaesthesia, for example by using a rapidly reversible gaseous anaesthetic. This practice has eradicated the incidence of corneal opacities completely.

4.8.5 Clinical pathology

Continuous intravenous infusion of saline commonly results in only minor effects on haematological parameters, principally an acute-phase increase in fibrinogen. Other workers (Francis *et al.* 1992; Gregory 1995) have reported all clinical pathology values remaining within reference ranges. If there are problems with infection following surgery, increased white blood cell (WBC) counts and plasma protein levels may also be observed. Infusion of polyethylene glycol has been shown in the author's laboratory to markedly increase WBC and neutrophil counts, particularly at 40% (w/v), and also increase urine volume. Increased urine volume has been reported by Cave *et al.*

Table 4.2 A comparison of water consumption and urine volume between infused and non-infused rats (mean and range)

	Urine output*(ml/kg/day)	Water intake*(ml/kg/day)
Cage controls ($n = 5$)	38.3 (25.8–45.7)	126.9 (106.2–144.1)
Saline-infused rats ($n = 26$)	41.0 (21.4–79.4)	106.5 (69.6–140.4)

*Infusion rate: 2 ml/kg/h.

(1995) following saline infusion, but this was not always the case in the author's laboratory.

4.8.6 Histopathology

A detailed description of the expected pathology following continuous infusion is provided in Chapter 20. However, even a 'clean' study will result in some background histopathology, particularly at the infusion site, resulting from the presence of the cannula. Most common is thrombus formation, which takes the form of a small ring of thrombus around the tip of the cannula. Some thickening and perhaps slight irritation or even abscessation may also occur in a small percentage of the rats. Occasional pathology is also observed in the liver (e.g. hepatic inflammatory cell foci) and lungs (pulmonary pneumonitis), which is related indirectly to the effects at the infusion site. Continuously infused polyethylene glycol resulted in abscessation and necrosis of the infusion site, whereas this was not seen with either mannitol alone or in combination with glucose. There is also likely to be some inter-laboratory variation in the pathologists' definition of the threshold of significant change, which makes comparisons between laboratories difficult.

4.9 Combined general toxicology/male fertility studies

The assessment of male and female fertility is also a regulatory requirement (Barrow *et al.* 1996; see also Chapter 5). When planning the pre-clinical development programme for a pharmaceutical, any possibilities for reducing animal usage should be examined, particularly when the animals need to be surgically prepared. One such way is to coincide the assessment of male fertility with the general toxicology study. At the end of the study (for example after 28 days' dosing), all males on test can be mated with virgin, untethered females that are in pro-oestrus, which are introduced into the cage overnight. The males continue to be infused during mating, as this does not appear to affect their performance. Any animals which do not mate are given a second opportunity to do so, with another naïve female in pro-oestrus, during the subsequent night.

This strategy does not affect the integrity of the general toxicology study and has the twin benefits of reducing animal usage and accelerating the development programme.

4.10 Conclusion

The rat is probably the most common model for testing pharmaceuticals by continuous intravenous infusion during pre-clinical development. There are therefore numerous different approaches besides that described in this chapter, many of which will be perfectly acceptable. Workers setting up the technique should therefore follow the principles laid out above, while having the freedom to adapt the techniques to suit their own laboratory and budgets.

With the technological advances in this field, methodology and the equipment used should continually be appraised, and in the light of increasing awareness of animal welfare so should the immediate environment of the animal. Pathology findings should also be regularly reviewed and always communicated to the surgical team, so that any increase in the background histopathology can be acted upon rapidly. An animal in a good state of health and as little stressed as possible will result in a far more robust model for toxicity testing in general but particularly for those that are surgically prepared.

Acknowledgements

I would like to acknowledge the effort in developing the techniques described in this chapter by the technical staff in Safety Assessment at AstraZeneca R&D Charnwood, and in particular the assistance of Chris Larner and David Smith in the preparation of this chapter.

References

Barrow, P.C., Heritier, B. and Marsden, E. (1996) Continuous deep intravenous infusion in rat fertility studies. *Toxicology Methods*, 6, 139–147.

Cave, D.A., Schoenmakers, A.C.M., van Wijk, H.J., Enninga, I.C. and van der Hoeven, J.C.M. (1995) Continuous intravenous infusion in the unrestrained rat – procedures and results. *Human and Experimental Toxocology*, 14, 192–200.

Flecknell, P.A., Orr, H.E., Roughan, J.V and Stewart, R. (1999) Comparison of the effects of oral or subcutaneous carprofen or ketoprofen in rats undergoing laparotomy. *Veterinary Record*, 144, 65–67.

Francis, P.C., Hawkins, B.L., Houchins, J.O., Cross, P.A., Cochran, J.A., Russell, E.L., Johnson, W.D. and Vodicnik, M.J. (1992) Continuous intravenous infusion in Fischer 344 rats for six months: a feasibility study. *Toxicology Methods*, 2, 1–13.

Gregory, D.J. (1995) Practical aspects of continuous infusion in rodents. *Animal Technology*, 46, 115–130.

Jones, P.A. and Hynd, J.W. (1981) Continuous long-term intravenous infusion in the unrestrained rat – a novel technique. *Laboratory Animals*, 15, 29–33.

Loget, O., Nanuel, C., Le Bigot, J.F. and Forster, R. (1997) Corneal damage following continuous infusion in rats. In Green *et al.* (eds.) *Advances in Ocular Toxicology*. Plenum Press: New York, pp. 55–62.

5 Reproductive toxicology studies by infusion in the rat

P. C. Barrow

5.1 The use of intravenous infusion in reproductive toxicology

Non-conventional routes of administration are of interest in the regulatory reproductive toxicology testing of medicinal products for many of the same reasons that they are used in general toxicology (see Chapter 4). Continuous and intermittent intravenous (i.v.) infusion are of particular value in reproductive toxicology because of the crucial role that toxicokinetic considerations have been shown to play in mechanisms of developmental toxicity (see below). Until recently, i.v. infusion was often considered impracticable for reproductive toxicology, in view of the technical complexity of the study protocols. It was not unusual to use conventional methods of administration for the reproductive studies with a given test substance, even when continuous i.v. infusion had been used for the general toxicology studies. The argument is no longer valid, however, since methods of continuous i.v. infusion have now been developed and validated for all types of regulatory reproductive toxicology assessment in the rat and rabbit (Barrow and Heritier 1995; Barrow and Guyot 1996; Barrow *et al.* 1996).

5.1.1 Regulatory guidelines

All regulatory assessments of reproductive and developmental toxicity of medicinal agents are now performed according to the ICH harmonised guidelines (ICH 1994). These recommendations are intended for the primary detection of toxic influences on reproduction. Once an effect has been detected, further experiments are usually necessary to fully characterise the nature of the response. This chapter covers the use of continuous i.v. infusion in regulatory studies only.

The ICH guidelines allow some flexibility in the strategy used for the evaluation of potential reproductive and developmental effects, but all of the following phases of development need to be assessed: copulation and fertilisation, early embryonic formation and implantation, organogenesis, fetal development, birth, pre-weaning development and post-weaning development.

The most common ICH testing strategy, which is well adapted for studies by i.v. infusion, involves a three-segment study design comprising a fertility study, embryotoxicity studies (one rodent and one non-rodent), and a pre- and post-natal study. These studies are sometimes referred to as segments I, II and III respectively. Alternatively, the embryotoxicity and fertility studies can be blended into a single protocol (see section 5.5.3 below). The rat is the rodent of choice for reproductive toxicology because of the vast accumulated experience with this species since the first regulatory guidelines were issued more than 30 years ago (FDA 1966).

The results of the pre- and post-natal study are not normally required before initiating the clinical trials, so this study is most often performed some time after the embryotoxicity and fertility studies.

5.1.2 Toxicokinetic considerations

The pharmacokinetic properties of the test substance (and of any active metabolites) have a particular relevance in reproductive toxicology. Many of the physiological changes that occur during pregnancy and lactation in the rat may potentially influence the toxicokinetic and pharmacodynamic properties of the test article. These changes include: reduced gastric secretion, increased intestinal transit time, increased plasma volume, increased extracellular fluid volume, increased body fat, decreased xenobiotic hepatic transformation, increased kidney function and differences in protein binding (Clarke 1993).

Because of their stage-dependent nature, developmental effects may be differentially dependent on peak plasma levels and the total exposure (area under the curve, AUC) of a given toxicant. For example, high maternal plasma levels (C_{max}) of sodium valproate at a specific stage of gestation in rodents have been shown to cause exencephaly in the developing embryo, while sustained exposure to lower levels results in embryonic resorption (Nau 1985). Infusion experiments suggest that the total exposure, represented by the area under the curve of plasma or target tissue concentration with time (AUC), is the predominant pharmacokinetic factor in the teratogenesis of cyclophosphamide and retinoids (O'Flaherty and Clarke 1994).

The rat, being smaller than man, has a faster rate of metabolism and eliminates most potential toxicants much more rapidly, resulting in a less prolonged exposure. An opposing argument relevant to teratogenicity is that the gestation length is much shorter in the rat so a relatively short duration of exposure will cover more developmental stages in the rat than in man. Programmable intravenous infusion methods may be employed to simulate the human pharmacokinetic profile in the rat. In some cases, it may be feasible to maximise both the C_{max} and the AUC in a single study by using variable infusion rates, such as a high daily loading dose followed by prolonged infusion of a lower maintenance dose.

Because of the very stage-specific responses in reproductive toxicology, the time of day when the rats are dosed may also influence the results of the study, particularly for compounds that are rapidly metabolised or eliminated. A frequent oversight of this type involves the duration of exposure to the test substance with respect to the time of copulation. In studies using conventional methods of administration, rats are normally dosed once per day in the morning. The rat, however, being nocturnal, tends to copulate at night. Because of the delay between dosing and copulation, the plasma or target tissue levels of the test article may have declined or it may even have been completely eliminated at the time of mating, so possible influences on mating behaviour could remain undetected. Continuous infusion can be used to ensure that the animals are correctly exposed at the time of copulation.

5.1.3 Methods of infusion

For routine toxicity testing, a tether system with a remote pump is the only method of i.v. dose infusion that allows the administration of relatively high volumes of test article at a constant rate over long periods with minimum stress to the animal. The infusion catheter is most easily implanted in a peripheral vein, but the relatively slow blood flow often results in local pharmacological effects or irritation because of insufficient mixing and dilution of the test article. For this reason, deep venous catheterisation with subcutaneous routing of the catheter connected to a remote infusion pump via a tether system is preferred.

For long-term infusion, the type of catheter used is very important to the success of the administration method. The chosen material must have low thrombogenicity, otherwise the catheter will become blocked by clot formation. Other important criteria include chemical stability, toxicity, the sizes of tubing available and compatibility with the test formulation. The following materials have been used successfully in our laboratories: polyethylene, polyurethane, silicone rubber, polyvinyl chloride and Teflon®. Silicone rubber (e.g. Silastic®) is soft and flexible and causes the least tissue damage (Hodge and Shalev 1992); unfortunately, these same properties make the catheter difficult to insert into the vein (Wyman *et al.* 1994).

5.1.4 *Selection of dose levels*

Toxicokinetic principles must be taken into account in the selection of dose levels (see above). The ICH guidelines suggest that dose levels should be assessed in terms of total body burden (i.e. AUC) rather than as administered dosages. Peak plasma levels also influence developmental effects and need to be taken into account.

The emphasis in reproductive toxicology studies is on the detection of effects on reproduction which are more prominent than other toxicological hazards. Therefore, if toxic influences that will limit the use of the test article have already been identified at a given exposure level in general toxicity studies, there is usually no need to exceed these dose levels in the reproduction studies. When the test article causes little or no toxicity at very high doses (e.g. penicillin in rats) the inclusion of just one treated group and one control group may be justified. The low dose level should result in a total exposure that is comparable with or, preferably, slightly higher than the anticipated exposure in man.

The results from subchronic rat studies are often sufficient for the selection of dose levels in reproduction studies, provided that the same infusion methods have been used. Otherwise, preliminary reproduction studies may be performed. For a preliminary infusion study prior to an embryotoxicity study, it is usual to allocate about six pregnant rats per group (plus satellite animals for toxicokinetics, if required), which are then submitted to the same in-life and Caesarean examinations as will be employed in the main study. Detailed fixed fetal examinations are not usually necessary. The purpose of preliminary studies is essentially to evaluate dose levels based on parental toxicity – they should not be used to select non-developmentally toxic levels, which would undermine the objectives of the main studies. In preliminary studies, the emphasis is placed on generating data that facilitate the selection of the high dose levels for the main studies; the lower dose levels are generally imposed by other criteria, including the anticipated human exposure and comparative toxicokinetics. Therefore, it is advisable to choose preliminary dose levels that are relatively high with respect to the anticipated human exposure level.

5.2 Test system, routine examinations and environmental conditions

Most of our experience has been acquired using the Sprague–Dawley rat (strain reference OFA.SD. (IOPS Caw), supplied by Iffa Credo, L'Arbresle, France); all of the results presented in this chapter were generated with this strain.

The in-life phase of the studies is performed in a barrier-protected unit. The animal rooms are maintained at 22 °C (± 2 °C) and 55% relative humidity (± 15%), with a 12-h light cycle and at least 15 air changes per hour. The rats are allowed free access to a complete pelleted rodent diet (reference AO4, Usine d'Alimentation Rationnelle, Villemoisson, France) and filtered (0.2 μm) mains drinking water from an automatic watering system. The infused animals are individually housed in stainless-steel cages measuring 260 × 175 × 200 mm (supplied by Iffa Credo, L'Arbresle, France), which are also used during co-habitation. Larger stainless-steel cages measuring 260 × 350 × 200 mm or polycarbonate cages measuring 225 × 365 × 180 mm with sawdust and paper bedding are used for littering and post-natal phases.

The rats are monitored for changes in clinical condition throughout the study and body weights and the amount of food consumed are recorded at intervals of 3 or 4 days. The daily infusion rate for each rat is adjusted according to the most recent body weight (or according to the mean weight per group, depending on the study design).

5.3 Catheter implantation

The following technique is used routinely in our laboratories for all types of reproductive toxicology study by i.v. infusion. For embryotoxicity and pre- and post-natal studies, the female rats are implanted the day after mating. This procedure, surprisingly perhaps, has been shown over the years to have no adverse influence on the course and outcome of pregnancy (Table 5.1). For fertility studies, the male or female rats are implanted about 1 week before the start of infusion with the test solution.

The rats are deeply anaesthetised (induction and maintenance) by inhalation of isoflurane (Abbott, Rungis, France). In our initial validation studies, the rats were anaesthetised with 50 mg/kg ketamine (Imalgene®, Rhône Mérieux, Lyon, France) and 5 mg/kg diazepam (Valium®, Roche, Neuilly-sur-Seine, France) administered by the intramuscular and intraperitoneal routes respectively. Gaseous methods, however, have proven to allow a more controlled state of anaesthesia and a more rapid recovery of the animal.

The femoral cannulation procedure is covered in detail in Chapter 1 and so will be described only briefly here to identify the differences in methodology at this laboratory. An indwelling catheter is implanted into the posterior vena cava by introduction into the femoral vein. The catheter consists of two lengths of Silastic® tubing (Sigma Medical, Nanterre, France), 200 mm of 0.6-mm bore and 400 mm of 0.3-mm bore, joined end to end with the aid of a commercial adhesive (Raumedic Adhesive S1 1511, Rehau, Germany). Following disinfection with Vetedine (Vetoquinol, Lure, France), a 5-mm incision is made in the skin of one inner thigh of the anaesthetised rat and a 1-mm incision is made in the skin on the dorsal neck. The finer part of the catheter is then threaded subcutaneously between the two incisions. The adjacent

Table 5.1 Maternal performance and Caesarean data (Day 20) from embryotoxicity studies

	Infused		Not infused	
Number of studies	16		10	
Number of females mated	277		196	
Females pregnant	255	(92%)	183	(93%)
Females with viable fetuses	252	(91%)	182	(93%)
Corpora lutea – mean (SD)	15.6	(2.3)	15.2	(3.1)
Implantation sites – mean (SD)	13.5	(2.8)	13.3	(3.2)
Per cent pre-implantation loss – mean (SD)	13.1	(14.2)	13.8	(15.2)
Early resorptions – mean (SD)	1.3	(1.5)	1.1	(1.3)
Late resorptions – mean (SD)	0	(0.2)	0	(0.2)
Per cent post-implantation loss – mean (SD)	10.3	(11.6)	9.1	(13.4)
Live fetuses – mean (SD)	12.2	(3.0)	12.2	(3.4)
Fetal weight (g) – mean (SD)	4.0	(0.4)	4.0	(0.4)

tissue is delicately pared away to expose the femoral artery and vein, which are carefully lifted with the aid of a blunt metal hook (Figure 5.1). The vein is punctured with the tip of a curved 8/10 gauge hypodermic needle, which is inserted to about half the length of the bevel. The catheter is filled with sterile saline and the distal end is pushed into the vein through the opening formed by the cavity of the bevel of the hypodermic needle and is inserted to a depth of 40 mm, so that the open tip protrudes inside the posterior vena cava. The position of the tip of the infusion catheter inside the vena cava is verified at necropsy after the end of the study and is generally found to lie 1 to 2 cm distal to the branch of the left renal vein. The proximal part of the catheter is then sutured to the underlying muscle using surgical silk thread (Ethicon 2/0, Ethinor SA, Neuilly, France). The skin is closed with 7.5-mm surgical staples (Lepine, Lyon, France).

The exit site of the catheter in the nape of the neck is covered by a custom-made cotton gauze jacket placed around the thorax of the animal (Figure 5.2). The exposed catheter is protected with a flexible metal sleeve and swivel joint (type Ealing, Harvard Apparatus, USA). The swivel joint is then mounted in the top of a rodent cage. The rat is weighed and placed inside the cage to recover from the anaesthetic. The swivel joint is connected by another length of 0.6-mm-bore Silastic® tube to a syringe mounted on a programmable infusion pump (e.g. Medfusion 2010i, supplied by Medex Inc., Hillard, USA). Infusion is started immediately to prevent coagulation inside the indwelling catheter.

The flexible tether system and swivel joint allow the rat almost unrestricted movement inside the cage (Figure 5.3).

5.4 Infusion limitations

In our experience, the rat model of infusion has proven to be remarkably resistant to potential adverse influences on homeostatic mechanisms. Infusion volumes as high as 200 ml/kg/day have been successfully employed in pregnant rats without adverse effects on the mother or offspring. Providing that the test solution has a low buffering capacity, solutions of up to three times physiological osmolarity can be infused at rates of up to

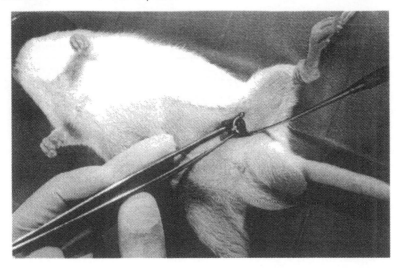

Figure 5.1 Exposure of the femoral vein in the anaesthetised rat.

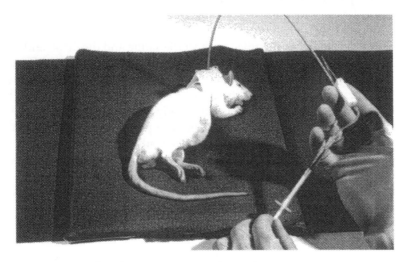

Figure 5.2 Rat equipped with jacket, tether and swivel joint.

48 ml/kg/day without any observable adverse effects on the course and outcome of pregnancy. Other physical properties of the test solution that need to be controlled are temperature (to avoid the risk of induced hypothermia, for instance) and surface tension (to avoid the risk of blood cell lysis).

One important limitation of the technique is the required physical and chemical stability of the test solution. The solution must be stable for at least the duration of each infusion. The solutions in the infusion syringes are changed at least daily (or more frequently if necessary), but in practice they must be stable for at least 8 h. Also, the test solutions must be chemically compatible with the materials used in the infusion equipment (i.e. syringes, tubing, swivel joint and catheter). Fortunately, several alternatives are available (see section 5.1.3 above).

Figure 5.3 Infusion of a lactating rat with litter.

The major disadvantage of tethered methods of infusion in fertility studies is that male and female rats have to be treated in separate studies, since the tethers would soon be entangled during mating. Therefore, the infused males are paired with untreated females and the infused females are paired with untreated males.

5.5 Study designs

5.5.1 *Embryotoxicity*

The ICH guidelines require treatment from implantation of the embryo until the end of the embryonic period. Palatal closure, which marks the end of the embryonic period, is normally complete in the Sprague–Dawley rat on Day 17 of gestation (Marsden and Roche 1997). Twenty-five mated female rats are allocated to each treatment group in order to provide at least 20 pregnant rats at term.

Time-mated female rats, 10–12 weeks old, are obtained from the supplier on the day of identification of a vaginal plug or sperm in a vaginal smear (Day 0 of gestation). The transport of the rats is very short since our laboratories and the animal supplier share the same site. The catheters are implanted the following day (Day 1 of gestation) and infusion of the test substance normally commences on Day 6 (i.e. soon after implantation of the embryos on the uterus). On the morning of Day 17 or 18 post coitum, infusion is stopped and the jacket removed. The catheter is cut off at the neck, leaving the implanted portion in place. A Caesarean examination is then performed on Day 20 of gestation (i.e. 1 or 2 days before parturition normally occurs) and litter parameters are recorded. The dams are given a macroscopic necropsy examination. Approximately half of the fetuses from each dam are given an external examination at necropsy and then preserved whole in Harrison's fixative prior to soft tissue examination using a modification of the Barrow–Taylor microdissection method (Barrow and Taylor 1967). The remaining fetuses are given an external and internal necropsy examination under low power magnification prior to staining for skeletal examination using a modification of the Wilson technique (Barrow 1990).

5.5.2 Male fertility studies

At least 20 male rats per group are treated for 4 weeks before pairing. They are usually about 9 weeks old at the start of treatment. Previous regulatory requirements specified an 8- to 10-week pre-pairing treatment period for males, based on the length of the spermatogenic cycle in the rat. It is a condition of the shortened ICH pre-pairing treatment period that histopathological data of the reproductive organs must be available from a general toxicology study of at least 13 weeks' duration (ICH 1996).

Infusion is continued throughout a 3-week co-habitation period with a single untreated female. The males are then euthanised and submitted to post-mortem examinations, including sperm analysis and histopathological examination of the reproductive organs. The vast majority of rats, infused or not, copulate within the first 4 days of pairing. If there is any concern over the tolerance of the test solution, the mating period can be reduced to 2 weeks in the interests of shortening the infusion period. On the other hand, however, it is preferable to treat the males for at least 8 weeks prior to necropsy, during which time additional influences on spermatogenesis may become apparent and thus be detected in the histological evaluation of the testes (Creasy 1997).

The untreated female is introduced into the home cage of the male. Vigilant monitoring is necessary during the excesses of the mating period, since it is not unusual for the infusion jackets to be torn and have to be replaced before the catheter can be dislodged or bitten through. Vaginal smears are taken from the paired females every morning. Once sperm are detected in a vaginal smear, the rats are separated and smearing is stopped. Any female that fails to copulate within 1 week may be replaced. The inseminated females are submitted to a Caesarean examination 2 weeks after mating and litter data are recorded.

5.5.3 Female fertility studies

At least 20 female rats per group are treated for 2 weeks before pairing, throughout mating and during the first week of gestation. They are usually about 11 weeks old at the start of treatment. The oestrus cycle of the implanted females is evaluated by daily vaginal smearing. Smearing is commenced 1 week before the start of infusion of the test solution in order to assess any adverse influences of the surgery or indwelling catheter. No such changes have been detected in our studies to date, however.

Mating takes place under similar conditions as in the male fertility study (see above), except that the rats are paired in the home cage of the female. Our experience has shown that, contrary to expectations, male cage dominance does not have a significant influence on mating performance (Table 5.2). Any male that fails to copulate within 1 week may be replaced. Infusion is stopped on the morning of Day 8 post coitum, the jacket is removed and the catheter cut and tied off. The females are then kept untreated until Day 13 post coitum, when they are euthanised and given a Caesarean examination.

5.5.4 Combined embryotoxicity and female fertility

Under some circumstances it is possible to blend the female fertility and embryotoxicity studies into a single experiment. The females are implanted, infused and paired as described for the female fertility study (above). The infusion period is then extended

Table 5.2 Mating performance and Caesarean data (Day 13) from fertility studies

	Infused males		Infused females		Not infused	
Number of studies	3		3		11	
Females paired	65		70		211	
Females inseminated	62	(95%)	68	(97%)	208	(99%)
Females pregnant	59	(95%)	63	(93%)	194	(93%)
Females with live embryos	59	(95%)	62	(91%)	194	(93%)
Corpora lutea – mean (SD)	16.7	(2.2)	16.2	(4.3)	16.6	(2.6)
Implantation sites – mean (SD)	15.5	(2.3)	14.7	(2.6)	15.2	(2.9)
Per cent pre-implantation loss – mean (SD)	7.0	(9.2)	9.6	(13.8)	8.5	(12.8)
Viable implantations – mean (SD)	14.7	(2.4)	13.8	(3.0)	14.4	(3.0)
Resorptions – mean (SD)	0.7	(0.7)	0.9	(1.1)	0.8	(1.1)
Per cent post-implantation loss – mean (SD)	4.8	(4.9)	6.9	(13.6)	5.2	(7.7)

until Day 17 post coitum, after which the pregnant females are submitted to a Caesarean examination on Day 20 with detailed examinations of the fetuses as required for embryotoxicity studies (see above).

5.5.5 ICH pre- and post-natal development study

In this study, mated females are infused from the start of the embryonic period, during littering and throughout lactation until weaning of the offspring. The group sizes, animal specifications and surgery are identical to those of the embryotoxicity study (see section 5.5.1 above).

Where possible, parturition is observed to detect any abnormalities in nesting or nursing behaviour. Gestation length and the numbers of live- and stillborn pups are noted for each female. Litters are examined daily for mortality and changes in clinical condition. Each pup is weighed on Days 1, 4, 7, 14 and 21 post partum.

The size of the litters is reduced to 10 pups on Day 4 by random elimination to yield, where possible, five male and five female pups. There has been much discussion recently concerning the validity of this practice (i.e. 'culling'), with arguments for and against its continued use (Agnish and Keller 1997; Chapin and Heck 1997; Palmer and Ulbrich 1997). The decision to standardise litter sizes in infusion studies was taken with the aim of reducing the influence of reduced milk availability on pup development caused by the partial occlusion of the cranial pair of nipples of the dam under the infusion jacket. The quantity of milk available to the pups has been shown to be the major influence on the rate of post-natal weight gain in the rat (Reddy and Donker 1964).

Pup development is assessed by daily monitoring for the occurrence of pinna unfolding, incisor eruption and eye opening. The surface righting reflex of each pup is assessed on Day 8, the gripping reflex on Day 17 and auditory and pupil reflexes on Day 21 post partum. One male and one female pup are randomly selected from each litter and retained untreated for developmental, behavioural and mating investigations.

The infused dams and unselected pups are euthanised after weaning and given a necropsy examination for structural or pathological changes. Dams that fail to produce a viable litter by Day 26 post coitum are necropsied.

5.6 Results

All data presented in this section were compiled from the results obtained for control groups of rats infused with physiological saline (with or without buffering agents) in experiments performed in the facilities of Phoenix International Preclinical Services Europe (Chrysalis), formerly Pharmakon Europe.

5.6.1 Lesions at infusion site

The indwelling catheter is usually very well tolerated in rat reproduction studies. Less than 5% of rats have lesions associated with the catheter. The types of lesion most often found are slight intimal hyperplasia or thrombus formation in the vena cava.

5.6.2 Mating performance

Catheterised male and female rats in infusion studies copulate almost as effectively as non-catheterised rats in conventional studies, with at least 95% of females inseminated. The proportion of inseminated females that become pregnant is similar whether or not the females or males are catheterised. The litter data from male and female fertility studies using infusion are similar to those from conventional studies (Table 5.2).

5.6.3 Influences during gestation

In the initial validation studies, infusion of physiological saline at a high rate of about 100 ml/kg/day appeared to cause an increase in embryonic resorption and retarded fetal development; no such effects were seen with a lower infusion rate of 24 ml/kg/day (Barrow and Heritier 1995). Experience over the last 5 years has demonstrated that this is no longer the case, probably owing to improved surgical techniques. Volumes of up to 200 ml/kg/day, albeit with innocuous solutions, have now been demonstrated to cause no adverse effects. Nonetheless, administration volumes of 48 ml/kg/day or less are strongly preferred.

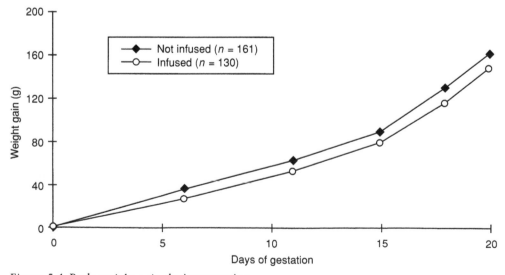

Figure 5.4 Body weight gain during gestation.

Maternal weight gain (Figure 5.4) is slightly reduced in the infused rats at the start of gestation, as could be expected following anaesthesia on Day 1. Thereafter, the rate of maternal growth is normal throughout the remainder of gestation. Pregnant females in our infusion studies tend, on average, to consume more food (Figure 5.5), but the difference is too small to be of biological significance.

More than 90% of catheterised time-mated rats are pregnant at term. The infusion procedures do not adversely influence litter parameters, including pre-implantation loss, resorption incidences, litter size and fetal weight (Table 5.1).

5.6.4 Fetal observations

The incidence of malformed fetuses is not noticeably increased in infusion studies. Among the 50 most common skeletal anomalies in this strain of rat, four particular

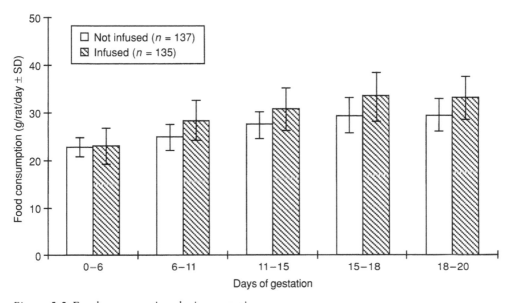

Figure 5.5 Food consumption during gestation.

Table 5.3 Fetal skeletal abnormalities observed more frequently in infusion studies

	Infusion studies				Not infused
Rate of infusion (ml/kg/day)	24	48	78	*Overall*	
No. of studies	2	4	2	8	15
No. of fetuses	325	653	290	1268	1411
Incidence (%) of fetuses with:					
Incomplete ossification of occipital(s)	15	16	15	16	11
Incomplete ossification of interparietal	26	16	27	21	13
Incomplete ossification of first four sternebrae	59	36	56	46	33
Unossified arches of caudal vertebrae	41	23	33	30	17

minor abnormalities have so far been detected more frequently in infusion studies than in conventional studies (Table 5.3). All of these anomalies are indications of slightly retarded ossification and are not normally considered to have any permanent physiological consequences. It is unlikely that these spurious findings are associated with infusion of the dams, but more data will be needed to assess their relevance.

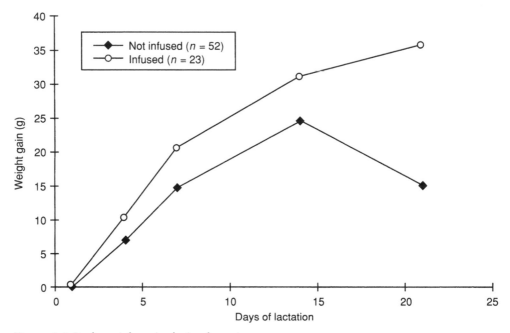

Figure 5.6 Body weight gain during lactation.

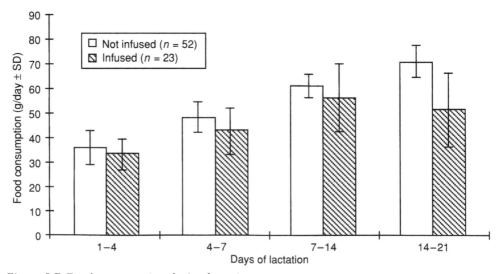

Figure 5.7 Food consumption during lactation.

5.6.5 Pre- and post-natal influences

The initial reduction in mean weight of the infused females during gestation is recovered at the end of lactation (Figure 5.6), despite the infused dams consuming slightly less food (Figure 5.7). Pre-birth loss (i.e. resorption) tends to be increased in the infused dams, resulting in a minor reduction in litter size (Table 5.4). This difference with respect to the results seen in embryotoxicity studies could be explained by the continued infusion of the dams during late gestation.

Viability of the pups is slightly reduced during the first 4 days post partum. Pup growth is slightly reduced until weaning (Figure 5.8). Both of these findings may be due to reduced access to the nipples of the dam since the infusion jacket tends to

Table 5.4 Parturition and post-partum data

	Dams not infused		Dams infused	
Number of females				
Mated	117		25	
Pregnant	105	(90%)	23	(92%)
Delivering live pups	104	(89%)	23	(92%)
Raising pups to weaning	103	(88%)	22	(88%)
Mean values per female				
Gestation length (days)	21.9		21.7	
Number of uterine implants	13.9		13.8	
Total pups born (live and dead)	12.7		12.2	
Number of liveborn	12.6		11.4	
Pre-birth loss	1%		13%	
Pup mortality at birth	1%		4%	
Pup mortality birth to Day 4	2%		16%	
Pup mortality Day 4 to Day 21	0%		7%	

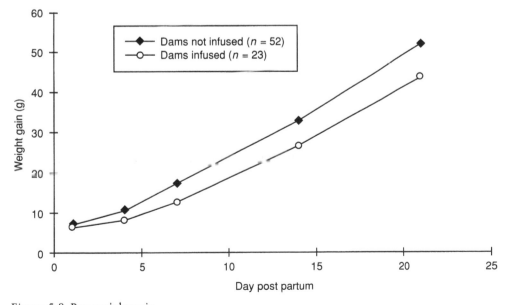

Figure 5.8 Pup weight gain.

occlude the cranial pair of nipples. A modified jacket design is presently under evaluation and is expected to eliminate these influences. Physical development of the pups is slightly delayed in accordance with the observed reduction in growth. There is no evidence of any increased incidence of abnormalities or defects in the offspring of the infused dams.

None of the observed influences on the dam and litter are sufficiently severe to interfere with the objectives of a pre- and post-natal study, provided that an adequate number of appropriately treated control animals is included in the study design.

References

Agnish, N.D. and Keller, K.A. (1997) The rationale for culling of rodent litters. *Applied Toxicology*, 38, 2–6.

Barrow, M.V. and Taylor, W.J. (1967) A rapid method for detecting malformations of rat foetuses. *Journal of Morphology*, 127, 291–306.

Barrow, P. (1990) *Technical Procedures in Reproduction Toxicology*. *Laboratory Animals Handbooks 11*, Royal Society of Medicine: London.

Barrow, P.C. and Guyot, J.Y. (1996) Continuous deep intravenous infusion in rabbit embryotoxicity studies. *Human and Experimental Toxicology*, 15, 214–218.

Barrow, P.C. and Heritier, B. (1995) Continuous deep intravenous infusion in rat embryotoxicity studies: the effects of infusion volume and two different infusion fluids on pregnancy. *Toxicology Methods*, 5, 61–67.

Barrow, P.C. and Heritier, B. (1996) Continuous deep intravenous infusion in rat pre- and post-natal studies. *Contemporary Topics in Laboratory Animal Science*, 35, 66–69.

Barrow, P.C., Heritier, B. and Marsden, E.K.S. (1996) Continuous deep intravenous infusion in rat fertility studies. *Toxicology Methods*, 6, 139–147.

Chapin, R.E. and Heck, H.d'A. (1997) Editorial: An exchange of views on culling. *Applied Toxicology*, 38, 1.

Clarke, D.O. (1993) Pharmacokinetic studies in developmental toxicology: practical considerations and approaches. *Toxicology Methods*, 3, 223–251.

Creasy, D.M. (1997) Evaluation of testicular toxicity in safety evaluation studies: the appropriate use of spermatogenic staging. *Toxicological Pathology*, 25, 119–131.

Food and Drug Administration of USA (1966) *Guidelines for Reproduction Studies for Safety Evaluation of Drugs for Human Use*.

Hodge, D.D. and Shalev, M. (1992) Dual cannulation: a method for continuous intravenous infusion and repeated blood sampling in unrestrained mice. *Laboratory Animal Science*, 42, 320–322.

ICH (1994) Step 4 tripartite harmonised guidelines. Detection of toxicity to reproduction for medicinal products. In D'Arcy, P.F. and Harron, D.W.G. (eds.) *Proceedings of The Second International Conference on Harmonisation Orlando 1993*. Queen's University: Belfast, pp. 557–578.

ICH (1996) Tripartite guideline on detection of toxicity to reproduction for medicinal products: addendum. *Federal Register* 61, 15359–15361.

Marsden, E.K.S. and Roche, F. (1997) Day of closure of the foetal palate in the SD rat. *Teratology*, 6, 393.

Nau, H. (1985) Teratogenic valproic acid concentrations: infusion by implanted minipumps vs conventional injection regimen in the mouse. *Toxicology and Applied Pharmacology*, 80, 243–250.

O'Flaherty, E.J. and Clarke, D.O. (1994) Pharmacokinetic/pharmacodynamic approaches for developmental toxicity. In Kimmel, C.A. and Buelke-Sam, J. (eds.) *Developmental Toxicology*, 2nd edn. Raven Press: New York, pp. 215–244.

Palmer, A.K. and Ulbrich, B.C. (1997) The cult of culling. *Journal of Applied Toxicology*, 38, 7–22.

Reddy, R.R., and Donker, J.D. (1964) Lactation studies. V. Effect of litter size and number of lactation upon milk yields in Sprague–Dawley rats with observations on rates of gain of young in litters of various sizes. *Journal of Dairy Science*, 47, 1096–1098.

Wyman, J.F., Moore, T.J. and Buring, M.S. (1994) Simple procedure for jugular vein cannulation of rats. *Toxicology Methods*, 4, 12–18.

Part 2

Continuous intravenous infusion in the mouse

The mouse is not as commonly used for continuous intravenous infusion studies as the rat but has emerged in recent years as a viable alternative owing to the successful miniaturisation of the equipment used to infuse the larger rodent. The main advantage of the mouse is its size, which enables compounds in early development and thus in short supply to be tested due to the smaller quantities required.

Like the rat, the usual site of catheterisation is the femoral vein, but the exit site of the catheter is usually via the tail, using a tail cuff. The chapters in this section reflect this established technique, but there is also a smaller version of the rat harness now on the market that would allow the exit site to be via the subscapular region.

In this section, the use of the mouse model for both general toxicology and efficacy studies is described, and the less favoured catheterisation models (jugular and tail vein) are also discussed in Chapter 8.

6 Femoral cannulation using the tail cuff model in the mouse

H. van Wijk and D. Robb

6.1 Introduction

A number of techniques for continuous intravenous infusion and/or venous blood sampling have been described for mice. A number of early techniques described percutaneous cannulation of the tail vein, and attachment of the tail to a device such as a splint or sheath to prevent the mouse from removing the cannula (Plager 1972; Paul and Cave 1975; Grindey *et al.* 1978; Connor *et al.* 1980; Braakhuis *et al.* 1995). Such techniques inevitably involve a considerable degree of restraint of the animals. Techniques for blood sampling from the jugular vein such as those described by Popovic *et al.* (1968) and Hodge and Shalev (1992) involve exteriorisation of a cannula from the dorsal neck. The patency of the cannula is maintained with heparin. Such techniques could be adapted for continuous infusion by connection of the cannula to a pump. Similarly, a skin button or cap, or a jacket, can be used to protect and anchor the dorsal cervical exit point of a jugular cannula, with connection to a spring tether (Lemmel and Good 1971; Desjardins 1986).

An alternative method widely described for continuous infusion of mice and other rodents is the use of implantable osmotic pumps (Pimm *et al.* 1987; Davol *et al.* 1995), for intravascular, subcutaneous and intraperitoneal infusions. These pumps have the advantage in that they require no tethering of the animals, but also have their disadvantages. The reservoir of pumps small enough to be injected into mice is necessarily very small, which either restricts the amount of administration required or requires regular injections into the pumps. There is also a restriction on potential vehicles that can be used with osmotic pumps, as they may not work well with viscous materials. The accuracy of delivery of pumps is also a crucial factor in terms of administration of set doses of test material in toxicity studies. Generally, the accuracy of such pumps is less than that of other types of pumps available for connection to exteriorised systems. In addition, administration of the test material starts as soon as the pump is inserted, allowing no recovery period from surgery prior to the start of the toxicity study. The stress of surgery and the interference of analgesics administered at surgery could be a complicating factor in determining the response of the animals to the test material.

Experience in the authors' laboratory in techniques for continuous intravenous infusion was gained initially in the rat, with cannulation of the femoral vein and exteriorisation of the cannula through the tail. The cannula is protected by a tail cuff and tether in preference to the more commonly used method of exteriorisation of cannulae from between the scapulae and attachment of a tether from the thorax, with or without the use of a jacket. It was considered that the development of a tail cuff method, as described in Jones and Hynd (1981) and Cave *et al.* (1995), would provide benefits both in terms of the running of studies and the welfare of the animals.

It is now clear that the procedure as developed in rats is also applicable to mice. Difficulties presented by the use of the technique in mice are generally in terms of the scale of the animals and equipment required, and there is scope for further refinement of procedures. A validation of the surgical procedures and maintenance of surgically prepared mice has demonstrated the ability to maintain the animals for up to 28 days' dosing with test materials, by continuous intravenous infusion via the femoral vein. At the end of this dosing period, 40 out of 40 cannulae were still patent, although other undesirable findings such as leaking cannulae were noted in a number of cases.

This chapter will describe procedures followed and developed from the original validation study conducted in mice at the authors' laboratory and subsequent studies, giving indication of improvements and refinements where appropriate. The animals used to date were CD-1 mice from Charles River (UK) Ltd, Kent, UK, and NMRI and NMRI *nu/nu* mice from Mollegaard Breeding and Research Centre A/S, Skensved, Denmark. There are no known reasons at present as to why such techniques could not equally be applied to other mouse strains. The major limiting factors are the size of the animals and the size of the cannulae. Mice of 25–30 g can be successfully cannulated and maintained for at least 5 weeks.

6.2 Surgical facilities

As for all surgical procedures, provision of a dedicated surgical facility is highly desirable. This allows a dedicated surgical team to maintain the area in a clean and orderly fashion and to ensure availability of all equipment required. As with surgical suites for other species, separate areas for storage, animal preparation, surgery and recovery should be provided. The post-surgical recovery room should be maintained at approximately 24–26 °C.

There is no logical or ethical justification for compromising standards of asepsis for rodent surgery in comparison with standards for non-rodent surgery. This may, in fact, be crucial for surgery in immunodeficient animals. Therefore, constant attention must be paid to maintenance of the sterile field. The surgeon should wear sterile gloves and use sterile instruments. If a group of animals is to be surgically prepared in one session, the instruments can be sterilised between animals in an autoclave or they may be resterilised in a glass bead steriliser (e.g. Sterilquartz, Keller, Switzerland) when not in use or between using the instruments for successive animals. At least one assistant should be available to pass equipment and alter gaseous anaesthesia as required. The first surgeon may be responsible for all procedures including the attachment of the tail cuff but, if a group of animals is to be prepared, attachment of the tail cuff and tethering equipment may be undertaken by an assistant. This procedure should also be carried out using sterile surgical techniques, whether by the first surgeon or by an assistant.

It is essential that, as soon as the animal has been anaesthetised, it is transferred to a heated pad, thermostatically controlled if possible (e.g. Harvard UK, Lameris, The Netherlands), in order to minimise the possibility of anaesthetic death from hypothermia. During surgery the use of a binocular microscope is recommended to assist in the visualisation of the blood vessels and nerves. The authors use a microscope with ×10 and ×16 magnification (Carl Zeiss, Jena, Germany).

6.3 Cannulae

Only a limited number of commercially available cannulae are suitable for vascular cannulation in mice. Cannulae can be manufactured in house, but this is likely to be a major time investment in order to produce high-quality cannulae. It is recommended that a commercial supplier be used, although cannulae may still need to be altered prior to implantation, depending on the skill of the surgeon and the size of the animals. Where possible, a cannula with a smooth antithrombogenic tip should be used, which will reduce subsequent thrombosis in the cannula.

Small polyurethane cannulae, with an antithrombogenic bevelled tip, are available from Data Sciences International (DSI), St Pauls, MN, USA. A number of different sizes are available, and the authors currently use the 1.2 Fr size. These have an inner diameter (ID) of 0.18 mm and an outer diameter (OD) of 0.41 mm. This narrow-gauge cannula can be supplied sealed prior to sterilisation to a larger cannula (0.6 mm ID × 0.9 mm OD, 3 Fr). The narrow-gauge tubing is implanted into the vein, and the larger gauge tubing is external to the vein. The size of the cannula for implantation is limited both by the size of the vessel into which it can be passed and by the inguinal canal, where the cannula tends to stick as it is being fed up towards the vena cava. Care must be taken if this happens to avoid perforating the vessel by using excessive force.

The wider bore, 3-Fr cannula is used external to the vein as it is less fragile and is therefore less likely to be damaged when passing through the tail cuff and tether. In addition, this gauge of cannula is wide enough to allow connection to a 22-gauge swivel above the animal's cage.

Both silicone and polyurethane cannulae for mice are also commercially available from Access Technologies, IL, USA. This make of polyurethane cannula has the same dimensions as the DSI cannula, but it is softer and therefore less abrasive to tissues. It can also be obtained attached to a larger bore cannula (0.6 mm ID × 0.9 mm OD). The silicone cannula is also 0.18 mm ID × 0.41 mm OD, and can also be obtained attached to a 3-Fr silicone cannula. Silicone cannulae are soft but are felt by some to be too flexible.

Silicone cannulae of different sizes can be attached in house using silicone glue, which can also be used to create suture beads on silicone cannulae. These aid considerably in effective anchoring of the cannula in the vein, preventing movement of the cannula in and out of the vein if sutures are placed either side of the suture beads.

As mentioned, these small mouse cannulae have to be small enough to get into the femoral vein and through the inguinal canal. Passage through the inguinal canal is likely to be less of a problem with mice of 35 g or heavier. Even using the smallest size of cannulae commercially available, it may not be possible to cannulate the femoral veins of small mice of about 25 g or less.

If difficulty is experienced in successfully inserting cannulae, further alteration can be attempted. Previous experience has shown that the tips of the cannulae can be stretched and made more flexible with hot water to enable insertion into the vein. Care should be taken to maintain sterility if this step is found to be necessary.

The advantage of softer cannulae in terms of reduced potential tissue damage has to be countered by an increased tendency for them to be damaged and leak when exteriorised. Great care must also be taken during implantation to ensure that soft cannulae are not occluded by the sutures used to tie them in place.

6.4 Tail cuffs

Tail cuffs for mice are commercially available, e.g. from Uno Roestvaststaal bv, The Netherlands, but the weight of the cuff should be checked. Cuffs may weigh approximately 2 g or even greater, which can be a significant weight for a mouse. Alternatively, cuffs can be custom-built in house, with the same consequences for time and resources as for the manufacture of cannulae.

The tail cuff consists of a stainless-steel cylinder with an exit channel for the cylinder leaving the upper surface of the tube at an approximate 45° angle. By working with a local manufacturer, the authors have minimised the weight of tail cuffs in order to reduce physical effects on the animals of having the cuff attached. The cuffs are made from stainless steel (grade 316), of approximately 0.5 mm diameter and 15 mm long. Two diameters have been produced, approximately 5 mm for small mice and approximately 6 mm for mice of greater than 30 g in weight. The cuffs have had numerous holes drilled in the body, to result in a weight of approximately 1.1 g. This represents 3.7% of the weight of a 30-g mouse, which can be compared with the equivalent rat tail cuff of approximately 8.3 g (3.3% of a 250-g rat). The tail cuff is attached to the mouse's tail by a sterile wire suture passed through small fixation holes towards the proximal end of the tail cuff. The degree of movement of this tail cuff provides a great deal of the potential for adverse effects in the animals.

6.5 Tether system

A flexible stainless-steel spring tether is attached to the exit channel of the tail cuff by a thread. This means that tethers can be unscrewed and replaced if they become damaged during a dosing period. The spring is then attached at this other end to a swivel holder (Figure 6.1). The authors use a 22-gauge swivel, onto which the wider bore polyurethane tubing noted above will fit comfortably. Swivels are supplied by a number of companies, such as Lomir, Canada, and Instech, USA, and the 22-gauge swivel is that also appropriate for rat studies. A smaller, 27-gauge swivel will attach directly to the 1.2-Fr mouse cannula, but attachment of this cannula and swivel can be difficult because of their small diameter. It is, however, easier with the softer silicone cannulae than with polyurethane cannulae. It is possible to obtain other sizes of swivel on request (e.g. from Instech) that may fit the chosen size of cannula better.

The swivel is held in place above the cage by a swivel holder on a metal stand. Above the swivel is a further piece of polyethylene tubing (0.58 mm ID × 0.96 mm OD, Portex Ltd, UK, which will fit a 22-gauge swivel), and this is attached to a 0.2-μm pyrogen-free filter (e.g. Schleider & Scheull UK Ltd, UK; Braun AG, Germany; Medex Ltd, UK) by a 22-gauge needle. The filter is then connected directly to a syringe located in the infusion pump. The authors have been using the Medfusion 2001 syringe driver (Medex Ltd, UK; see Chapter 7).

6.6 Anaesthesia and surgery

Numerous standard texts exist on techniques for general anaesthesia and analgesia of mice (Flecknell 1987; Wixson and Smiler 1997; Wolfensohn and Lloyd 1998). Hypnorm®/Hypnovel® has been used by the authors, but isoflurane is preferred as it provides sufficient muscle relaxation and analgesia, rapid control of anaesthetic depth and rapid post-surgical recovery.

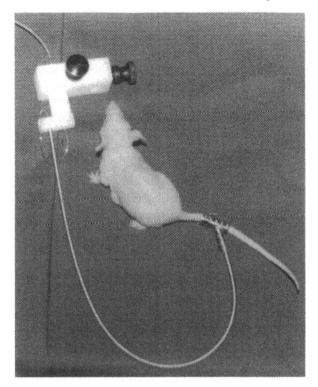

Figure 6.1 Nude (*nu/nu*) mouse with a tail cuff and tether, attached to a swivel and swivel holder.

Induction of anaesthesia is initiated by placing the animal in a clear-walled induction chamber. When voluntary movement ceases, the animal can be removed and placed on top of a heated pad. The animal's nose is placed within a nose cone supplying isoflurane in oxygen, with or without nitrous oxide. The use of a Fluovac and Fluosorber (IMS, UK) minimises any pollution. At this stage injectable analgesia is provided. The authors have used carprofen (Zenecarp, C-Vet Veterinary Products, UK), at a dose rate of 5 mg/kg s.c., diluted 1:5 with sterile water for injection. This injection is repeated 24 h post surgery.

The inguinal region overlying the femoral vein is shaved after induction of anaesthesia, using a scalpel blade (size 22), and the loose hair removed. The skin is then disinfected (e.g. 0.5% chlorhexidine (Hibitane®), ICI, UK, diluted in 70% ethyl alcohol). The tail is also cleaned with the Hibitane® solution at this time. The animal is then placed on its back on a sterile drape by an assistant, and the limbs secured to the surgical pad with tape. The animal is then covered with a sterile surgical Dermafilm (Adhesive Transparent Drape, Vygon, UK), leaving only the area overlying the femoral vein exposed.

An initial skin incision is made in the skin of the inguinal region, overlying the vein, and the underlying fascia gently prised apart to expose the femoral vein, which as in other species lies just caudal to the femur and next to the femoral artery and sciatic nerve. The vein is carefully separated from the other tissues over an area of approximately 5–8 mm. Micro-forceps should be used for the handling of these tissues during this delicate manipulation, and the vein handled as little as possible to reduce the occurrence of vasospasm. Micro-forceps, needle holders and scissors are available from a number of companies (e.g. Gimmi, Germany, Lawton, Germany, and John Weiss & Sons, UK).

Once the tissues have been gently separated and the vein identified and isolated, two lengths of sterile non-absorbable suture material are placed around the vein, about 3 mm apart. A non-braided suture material such as polyamide (Ethilon, Ethicon, UK) should be chosen for these pre-placed sutures, to reduce abrasive action on the vein during and after implantation. Polydioxanone (PDS, Ethicon, UK) may be used if cannulation is required for a relatively short period of time, but this will lose approximately half of its tensile strength in 28 days (Boothe 1993). The chosen suture material should be approximately 0.5 metric (7/0).

Although cannulae can be supplied with a bevelled tip, further modification may be necessary prior to insertion in the vein, particularly of mice smaller than 30–35 g in weight. This modification is to provide sufficient additional tapering (approximately 15 mm long) and flexibility for ease of cannulation of the vessel. Although this step may be found necessary, it does destroy the carefully prepared antithrombogenic tip and so should be avoided if at all possible.

The cannula is pre-filled with heparinised saline (25 iu/ml) and a syringe and needle left attached to the distal end in order to prevent leakage of fluid from the cannula. The pre-placed suture lengths are held tense and the vein incised between them with a sterile 30-gauge needle. The tip of the cannula is then advanced approximately 25 mm into the vein and through the inguinal canal, so that the tip will be lying in the caudal vena cava. The proximally placed suture is then tied around the vein and cannula. The distal length of suture is used to tie off the vein below the cannulation site and also to anchor the cannula to the vein. The excess ends of the suture material are then removed. Patency of the cannula is checked at this point in order to ensure that it has not been occluded by the sutures.

The distal end of the cannula is tunnelled towards the tail. This step can be carried out immediately before cannulation of the vessel rather than immediately after, which reduces by 1 the number of times the syringe has to be removed from the cannula after cannulation of the vessel. A trocar needle (e.g. Quincke-Babcock, 9 cm × 0.9 mm diameter, Van Straten Instrumenten, The Netherlands) is inserted bevel upwards in the ventrolateral area of the tail approximately 10 mm from the base, and advanced carefully towards the inguinal surgical site. The needle used should be just large enough to allow passage of the cannula through, or attached to, the needle. A metal pin can be passed through the trocar to clear any debris prior to passing the free end of the cannula through the trocar needle. If using a cannula incorporating a larger bore of cannula attached to the portion inserted in the vein, the larger bore may not fit through the trocar needle. If so, it should be possible to stretch the cannula over the tip of the needle, and it should then remain attached as the trocar needle is gently removed. As the cannula is drawn through to the exit point, care should be taken to prevent the cannula from kinking.

As mentioned above, provision of a suture bead at the cannulation site is a distinct advantage in ensuring that the cannula does not slip slowly out of the vein. If suture beads are not used, it may be advisable to anchor the cannula again, to the subcutaneous tissues. Sufficient slack should be left in the subcutaneous tissue to allow for the animal's movement, and for growth, although the latter is less of an issue for mice than for rats.

Immediately after the cannula has been drawn through to the tail, the syringe and needle should be reconnected and the cannula flushed with approximately 0.02 ml of heparinised saline to remove any blood which may have flowed back into the cannula. The syringe and needle are left attached.

The inguinal incision is then closed with a subcutaneous layer (e.g. PDS, 0.5 metric, Ethicon, UK), and a layer of skin sutures using a non-absorbable material such as Mersilk (0.5 metric, with TG 140G micropoint spatula needle, Ethicon, UK). Following wound closure, the area is swabbed again with Hibitane® solution. The Dermafilm and tape over the limbs are then removed.

The animal is then turned onto its front and a length of sterile surgical wire (USP 00 monofilament, Ethicon, USA) is inserted through the skin laterally from one side of the tail to the other, on the dorsal surface of the tail approximately 3 mm proximal to the exit point of the cannula. The tail cuff is then passed over the tail to overlie the cannula exit point. The animal is then turned onto its back again. It is important to have assistance in moving the animal or some other means of protecting the sterility of the surgeon. The syringe and needle are removed again from the free end of the cannula and the cannula passed through the exit channel of the tail cuff and the spring tether, which can be handled by the surgeon if they have previously been autoclaved or otherwise sterilised. The syringe and needle are again reattached and the cannula flushed with a further 0.02 ml of heparinised saline. The ends of the surgical wire are passed through the fixation holes in the tail cuff and twisted tight. The excess ends of wire are removed and the knot lain flush against the tail cuff.

Following completion of attachment of the tail cuff, it is essential to ensure that the wire suture is suitably tight. If this suture were too loose, this would encourage excess movement and irritation. If, on the other hand, the wire is too tight, it may be predisposed to break.

The animal is then returned to its individual home cage for connection of the free end of the cannula to a pump, by passing the tether and cannula through the top of the cage and connecting the cannula to the underside of a swivel. The authors currently use a 22-gauge swivel. The top of the swivel is connected to a pump via polyethylene tubing and an infusion filter. The swivel, tubing and filter are pre-filled with saline prior to attachment to the implanted cannula. A saline infusion is started immediately after connection, initially at a rate of 0.1 ml/h.

A post-surgery recovery period of at least 5 days, and preferably 7, should be allowed before the animal is used in toxicity studies. During this period the rate of administration of saline can be gradually increased to that required for the dosing period. Body weight and food consumption should be assessed daily to monitor post-surgical recovery, in addition to routine clinical observations of the animals and their surgical wounds.

6.7 Adverse effects at surgery

A number of adverse sequelae are possible during and following femoral vein cannulation, related to various aspects of the technique. A number of common causes of failure will be discussed below.

Principal among reasons for surgical failure is damage to the femoral vein or sciatic nerve during identification and separation of the structures, or failure to successfully cannulate the vein. In order to minimise these potential problems, handling of the tissues must be undertaken with extreme care, and any such handling minimised. Use of magnification has also been recommended.

Once the tissues have been separated, damage can also be minimised during cannulation and tying procedures by correct selection of non-braided suture material. Strategies for assisting entry to the vein of the cannula have been discussed.

Damage to the femoral vessels can be seen at surgery, but nerve damage may not be appreciated until the animal recovers from surgery and starts to move about. Adverse effects noted at this stage may include complete paralysis of the affected leg or variable degrees of reduced control over the leg muscles, reduced muscle tone and proprioceptive defects. Self-mutilation of the affected leg may be seen. Such clinical signs should result in the immediate sacrifice of the affected animal, but relatively mild defects such as knuckling of the foot have been noted to improve with time after surgery. A clinical decision on the fate of animals must be taken in such cases, depending on the animal's general condition and mobility, severity of the defect, secondary effects on the animal and the expected possibility of recovery.

Other incidences of self-mutilation such as removal of the sutures by animals have been noted in CD-1 mice and NMRI *nu/nu* mice after surgery. If only the skin sutures have been removed, this may be addressed by resuturing under general anaesthetic or the use of tissue glue. Where further suturing is required, the analgesia regime can be extended by a further 24 h.

Body weight loss post surgery is inevitable but can be minimised by making the anaesthetic period as short as possible and by using appropriate analgesia. It should take about 1 week for CD-1 mice to recover pre-surgical body weight and up to 14 days for immunodeficient mice (see Chapter 7).

Abscess formation at the surgical site has been noted occasionally in immunodeficient mice. Abscess formation can be minimised or prevented by strict attention to sterile procedures during surgery. If considered appropriate, additional safety equipment such as extraction hoods can be employed during surgery, but this must not be at the expense of good sterile technique.

The majority of the problems seen with maintenance of mice for continuous intravenous infusion relate to the cannulae and the tail cuff. Clinical effects arising from the use of tail cuffs will be discussed further in Chapter 7. With regard to cannulae, it has been noted above that there are a small number of commercially available cannulae for use in mice. Polyethylene and Silastic® tubing have commonly been used for vascular cannulation of mice and other species, but polyurethane tubing is also available in a very small gauge (1.2 Fr), with an anti-thrombogenic tip. This narrow-gauge cannula is fragile and is also smaller than that successfully used by a number of authors to cannulate the tail vein and the jugular vein of mice. However, when cannulating the femoral vein, the limiting factor is the width of the inguinal canal in small mice of approximately 25 g, which has resulted in the authors using the 1.2-Fr cannula.

As mentioned above, the disadvantage of the narrow-gauge cannula is its fragility, resulting in damage and leakage, normally at one of two sites. Firstly, cannula leakage can occur at the site of passage through the tail cuff. This is a function of time and friction between the cannula and the metal cuff. Every effort should be made to create a smooth internal surface and rounded edges on the tail cuff. As a result of this problem, and in addition to further attention to tail cuff design, it may be preferable to use a thicker-bore, more robust cannula external to the femoral vein and, as noted, such cannulae are also commercially available. The weak link then becomes the seal between the two gauges of cannula, and care should be taken at implantation to minimise any pressure on this seal by the way in which it is anchored to the animal's tissues.

Another alternative may be to use a wider-bore cannula as a protective sleeve overlying the narrow-bore cannula, but this has not been tried by the authors to date, and could conceivably cause problems with the proper functioning of the swivel. Also, if this method is used, a narrow-gauge swivel would have to be used, or the narrow-gauge cannula would need to be connected to a wider-bore cannula close to the swivel. This would have the advantage of having the connection in a visible area, so that any leakage would be seen immediately. Leakage from a seal buried within the animal would only show by fluid accumulation at the site, allowing no opportunity for repair.

If problems associated with leakage from narrow, fragile cannulae cannot be overcome, a more fundamental change in the technique can be undertaken, by cannulating the jugular vein, which can accommodate a wider-bore cannula without the restriction imposed by the inguinal canal. A 2-Fr cannula may fit the jugular vein of a large (approximately 35 g upwards) mouse. The cannula is then tunnelled subcutaneously to the tail as for femoral cannulation. This approach has its own restrictions, such as the relatively short length of vein into which a cannula can be inserted before the heart.

6.8 Conclusion

There are clearly technical issues that could be improved with time regarding femoral vein cannulation and infusion in mice. However it is considered by the authors that this technique has the advantage over other published tethering techniques for continuous intravenous infusion in mice, such as tail vein cannulation or jugular vein cannulation with or without the use of a jacket, in terms of the reduced amount of restriction imposed on the animals post surgery. It is certainly possible that continuous intravenous infusion can be conducted following warming mice and subsequent percutaneous cannulation of the tail vein. This technique decreases the preparation time per animal and removes the requirement for invasive surgery, but the tethering previously reported for such a procedure is certainly more restrictive than the use of a tail cuff. Validation of this cannulation method in addition to a tail cuff and tether may be an appropriate refinement to the procedure. However, the method of fixation of the cannula within the vein for long-term access could be problematical as it is likely to rely on direct fixation of the cannula to the skin of the tail.

In comparison with the use of osmotic pumps, use of this surgical technique allows a post-surgical recovery period before administration of test materials. In addition, it is possible to infuse relatively high volumes of test material with no additional direct interference with the mouse, for extended periods of time.

References

Boothe, H.W. (1993) *Textbook of Small Animal Surgery*, 2nd edn. W.B. Saunders: Philadelphia, p. 205.

Braakhuis, B.J.M., Ruiz van Haperen, V.W.T., Boven, E., Veerman, G. and Peters, G.J. (1995) Schedule-dependent antitumour effects of gemcitabine in *in vivo* model systems. *Seminars in Oncology* 22(4), 42–46.

Cave, D.A., Schoenmakers, A.C.M., Van Wijk, H.J., Enninga, I.C. and Van der Hoeven, J.C.M. (1995) Continuous intravenous infusion in the unrestrained rat – procedures and results. *Human and Experimental Toxicology* 14, pp. 192–200.

Connor, M.K., Dombroske, R. and Cheng, M. (1980) A simple device for continuous intravenous infusion of mice. *Laboratory Animal Science* 30, 212–214.

Davol, P., Beitz, J.G., Mohler, M., Ying, W., Cook, J., Lappi, D.A. and Frackelton, Jr., A.R. (1995) Saporin toxins directed to basic fibroblast growth factor receptors effectively target human ovarian teratocarcinoma in an animal model. *Cancer* 76, 79–85.

Desjardins, C. (1986) *Methods of Animal Experimentation,* Vol. VII, Part A, *Patient Care, Vascular Access and Telemetry.* Academic Press: Orlando, pp. 166–177.

Flecknell, P.A. (1987) *Laboratory Animal Anaesthesia.* Academic Press: London, pp. 93–94.

Grindey, G.B., Hoglind Semon, J. and Pavelic, Z.P. (1978) Modulation versus rescue of antimetabolite toxicity by salvage metabolites administered by continuous infusion. *Fundamentals in Cancer Chemotherapy* 23, 295–304.

Hodge, D.E. and Shalev, M. (1992) Dual cannulation: a method for continuous intravenous infusion and repeated blood sampling in unrestrained mice. *Laboratory Animal Science* 42, 320–322.

Jones, P.A. and Hynd, J.W. (1981) Continuous long-term intravenous infusion in the unrestrained rat – a novel technique. *Laboratory Animals* 15, 29–33.

Lemmel, E. and Good, R.A. (1971) Continuous long-term intravenous infusion in unrestrained mice – method. *Journal of Laboratory Clinical Medicine* 77, 1011–1014.

Paul, M.A. and Dave, C. (1975) A simple method for long-term drug infusion in mice: evaluation of guanazole as a model. *Proceedings of the Society for Experimental Biology and Medicine* 148, 118–122.

Pimm, M.V., Clegg, J.A. and Baldwin, R.W. (1987) Biodistribution and tumour localisation of radiolabelled monoclonal antibody during continuous infusion in nude mice with human tumour xenografts. *European Journal of Cancer and Clinical Oncology* 23, 521–527.

Plager, J.E. (1972) Intravenous, long-term infusion in the unrestrained mouse – method. *Journal of Laboratory Clinical Medicine* 79, 669–672.

Popovic, P., Sybers, H. and Popovic, V.P. (1968) Permanent cannulation of blood vessels in mice. *Journal of Applied Physiology* 25, 626–627.

Wixson, S.K. and Smiler, K.L. (1997) *Anesthesia and Analgesia in Laboratory Animals.*: Academic Press: New York, pp. 165–203.

Wolfensohn, S. and Lloyd M. (1998) *Handbook of Laboratory Animal Management and Welfare,* 2nd edn. Blackwell Science: Oxford, pp. 169–179.

7 Multidose infusion toxicity studies in the mouse

H. van Wijk and D. Robb

7.1 Introduction

Continuous intravenous administration of pharmaceuticals is practised in many fields of medicine. Preclinical safety testing of materials intended for continuous intravenous therapy must as far as possible emulate the clinical situation. The primary technical objective in such studies is to ensure that the time of administration of the test material is in line with the protocol requirements for dosing, in the same way as for any other type of toxicity study. Therefore, the equipment and conduct of procedures must be designed to be compatible with continuous administration, with no need for discontinuation. The only necessary exception to this is disconnection during changeover of syringes of an infusion pump.

This chapter will discuss the design of continuous infusion studies in mice based on recent developments in techniques at the authors' laboratory.

7.2 Caging and other equipment

The cages initially used for mouse infusion studies were clear MT1 polycarbonate cages (North Kent Plastic Cages, UK), of the same design as cages used for infusion studies in rats, in that they incorporated a linear grid bottom. This grid overlies a tray paper containing AlphaDri (Sheppard Speciality Papers Inc., USA), which soaks up urine, reducing ammonia concentration and the amount of allergen present. The use of a grid-bottomed cage was accepted as a welfare cost for the animals, but was justified on the basis that it has resulted in reduced adverse effects. Previous experience with solid-bottomed cages and bedding in rats demonstrated that bedding accumulates in the tail cuff of surgically cannulated animals, resulting in much increased incidence and severity of ulcerative lesions on the tail within the tail cuff, at the site of attachment of the cuff with a wire suture. However, continuous intravenous infusion studies in immunodeficient nude mice (*nu/nu*) have been conducted, and the design of these studies incorporated the use of a solid-bottomed cage, with shavings as bedding, to reduce the possibility of hypothermia in this strain. The use of bedding was not associated with any increase in adverse effects such as ulceration at the tail cuff. It is therefore considered both possible and desirable to use solid-bottomed cages for mouse infusion studies, although consideration should be given to the type of bedding used. Coarse bedding is less likely to become trapped within the tail cuff and so is preferable to fine bedding such as sawdust.

Cages are cleaned and bedding changed twice weekly as routine for continuous infusion studies, although the frequency of bedding change may need to be increased in studies with a high infusion rate, which will lead to an increase in urinary volume.

No additional enrichment beyond inclusion of bedding has been attempted to date in mouse continuous infusion studies, although there is no apparent reason why additional cage furniture could not be used in the cages, provided that such furniture did not interfere with the movement of the tether in the cage. It may be possible to pair or group-house animals, but this has not been attempted by the authors as yet given the difficulties of entanglement that could occur between two tethers in a cage and the increased risk of damage to the cannulae and tethers by the animals. One alternative previously suggested to improve the welfare of single-housed experimental animals is the addition of a companion, non-surgically prepared animal in the cage. Although it is true that this may increase the contentment of animals of sociable strains, it is highly unlikely that additional use would be found for the non-dosed animals on completion of the study, resulting in a doubling of the animals used in the studies for no increased scientific information. Nonetheless, improvements in the welfare of the animals in infusion studies should be continuously sought.

The lids used for the cages are of a standard design, with space for a food hopper and water bottle. The cage lids also have a linear grid. Mice are noted to climb up and along this grid in the same manner as non-tethered mice, demonstrating how little the tail cuff tether affects their movement.

The metal tether attached to the tail cuff must be passed through the lid of the cage and be attached to the underside of a swivel. This swivel is mounted in a Perspex block made locally for the authors, which is in turn attached to a metal post or swivel stand. This stand is attached to the lid of the cage. Creating a gap in the grid of the cage lid at the level of the swivel stand allows the tether to be slipped through the side of the lid. This modification allows the animals to be lifted completely clear of the cage for examination and other procedures. In addition, increased flexibility is provided by the use of a long section of polyethylene cannula between the top of the swivel and the infusion pump (Figure 7.1). Although the use of such a long section of cannula provides increased dead space, it allows the cages to be moved an appreciable distance from the pump, which is an advantage during the conduct of routine procedures. The cage can be moved onto a table or trolley and the animal removed from its cage for examination or other on-study procedures. The cages can also be placed directly into a heated cabinet prior to tail vein blood sampling. All of these procedures can be conducted without interruption of infusion.

7.3 Infusion pumps

The authors currently use the Medfusion 2001 pump (Medex Ltd, UK). This pump can dose as little as 0.01 ml/h, with an accuracy of ± 3%. It also contains a back-up battery, which will allow disconnection from the mains for up to 10 h, for the conduct of procedures. This pump has a pressure alarm (triggered at pressure greater than 17.3 psi).

The dose rate required for studies is a function of the dose of the test material to be administered and the concentration of the solution. Immediately after surgery, a low-volume infusion should be started with saline to avoid the cannula becoming clotted. The authors have used 0.08–0.1 ml/h to ensure patency, in line with previous work in rats. This infusion rate has been found to be sufficient to maintain the patency of the cannula without putting undue stress on the vein but is a high volume in relation to body weight for a mouse. It may be possible to reduce this infusion rate to closer to

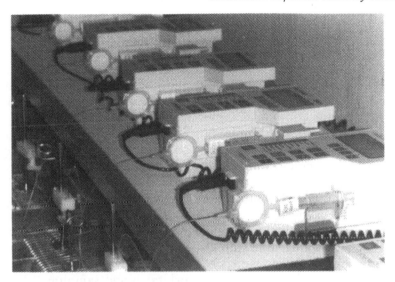

Figure 7.1 Room set-up for continuous intravenous infusion, showing connection of syringe pumps to cage lid equipment.

0.01 ml/h and still maintain patency although this has not been attempted by the authors to date. If it proved to be possible, this would further reduce the stress on the animals of the large daily volume of saline being administered.

During the test period, the dose rate should be reduced to the absolute minimum needed to achieve the daily dose of test material, but if the rate has to be increased this should be done gradually during the post-surgical recovery period, in steps of approximately 0.01 ml. During validation of this method, the authors achieved a dose rate of up to 8 ml/kg/h (0.24 ml/h for a 30-g mouse) in CD-1 mice for up to 28 days. There were no observable effects as a result of this treatment regimen other than an increased volume of dilute urine and decreased water consumption. Haematology, urinalysis and pathology investigations revealed no abnormalities.

Previous published work has demonstrated that it is possible to infuse volumes even higher than those quoted above. Up to 0.37 ml/h saline has been infused for up to 6 days via the tail vein of mice weighing approximately 20 g (Grindey *et al.* 1978) and rates of 0.03–0.3 ml/h for 7 days have been achieved in mice weighing 25–35 g (Plager 1972), with no observable effects on the mice, although it is unclear what investigations were carried out to demonstrate adverse effects. Similarly, others have infused test material via the tail vein of mice weighing approximately 18 g for up to 47 h at 0.3 ml/h, although there was no discussion of any effects as a result of this infusion rate (Paul and Dave 1975). The effect of longer-term dosing at such dose rates is uncertain. Lemmel and Good (1971) reported infusion via the jugular vein for up to 35 days at 0.2 ml/h and considered this rate to be well tolerated.

7.4 Animal room procedures

One procedure routinely required in toxicity studies is body weight measurement. It is anticipated that mice, like other small species, will lose weight post surgery, but ideally they should have regained their pre-surgery weight prior to use on a study. The authors have found that this is likely to take up to about 1 week in the case of CD-1 mice (Figure 7.2) but up to 12–14 days with NMRI *nu/nu* mice (Figure 7.3). However, in the case of any animal that has not regained its pre-surgical weight, a clinical judgement must be made as to whether its condition allows it to be used in a toxicity study. Other aspects to be taken into consideration in addition to body weight include the activity of the animal, its food consumption, the condition of the surgical wounds, the patency of the cannula and the degree and pattern of weight reduction.

The accuracy of the weighing procedure of tail cuff-tethered animals should also be assessed. The animals remain surgically attached to various pieces of equipment such as the swivel and holder during the weighing procedure. Therefore, every effort must be made to ensure that this equipment is accounted for in some way during the weighing procedure, by exclusion of the known weight from the gross weight, or by ensuring that the equipment is not included in the measured weight. This operation can be difficult to accomplish when dealing with a lively mouse on a weigh balance.

Tail vein blood sampling, as mentioned above, is not restricted by the presence of a tail cuff or by the attachment of the associated equipment. If preferred, animals can be anaesthetised in a chamber for retro-orbital blood sampling, also without interruption of infusion. Other standard procedures, such as detailed clinical examinations and ophthalmoscopy, do not present a problem with regard to continuous infusion. If urinalysis is required, the animals can be housed in a purpose-built cage such as a Jencons glass metabowl (Figure 7.4). By removing the glass lid and using a grid instead, the infusion line can be passed through the lid with no difficulty.

Other routine study procedures include maintenance of the infusion line. In-line filters are routinely changed at the authors' facility at 2-week intervals. In some cases, cannulae may become blocked. Immediate attention is required in such cases if a cannula is to be successfully unblocked. The cannula should be disconnected from the base of the swivel and a syringe and needle attached to the end of the cannula. Very

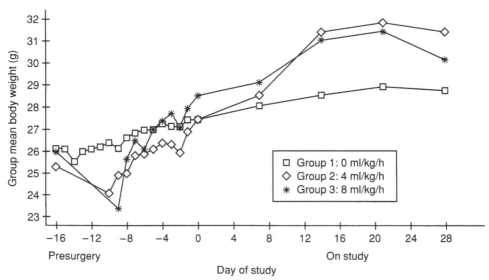

Figure 7.2 Body weight graph of CD-1 mice before and after femoral vein cannulation (reprinted with the kind permission of Animal Technology).

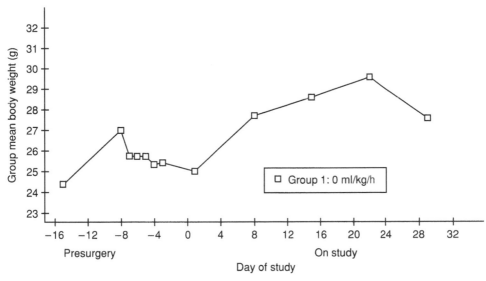

Figure 7.3 graph axes and labels:
- y-axis: Group mean body weight (g)
- x-axis: Day of study
- Presurgery / On study
- Legend: □ Group 1: 0 ml/kg/h

Figure 7.3 Body weight graph of *nu/nu* mice before and after femoral vein cannulation.

Figure 7.4 Urine collection from a CD-1 mouse during continuous intravenous infusion (reprinted with the kind permission of Animal Technology).

gentle pressure and suction, using a solution of heparinised saline (25 international units/ml), should remove any blockage. This has been undertaken on many occasions with no adverse effects for the animals.

7.5 Method validation

A validation study has previously been conducted at the authors' laboratory in which CD-1 mice were dosed by continuous intravenous infusion (Van Wijk 1997). The study comprised three groups of 10 animals per sex, of which 10 per sex were designated as cage controls. The remaining animals had a cannula surgically implanted into the femoral vein and exteriorised via the tail, and protected by a tail cuff and tether.

The animals were approximately 6–7 weeks old on arrival. They were acclimatised to the laboratory for a minimum of 1 week after arrival, and body weights were taken for at least 3 days before surgery. The animals weighed 20–30 g at the time of surgery. All procedures were carried out in strict accordance with the requirements of licences authorised under the provisions of the Animals (Scientific Procedures) Act 1986. Following recovery from surgery, for a minimum of 5 days, two groups of animals were dosed with saline, at 4 and 8 ml/kg/h, half for 7 days and half for 28 days.

The major considerations regarding the welfare of animals during the conduct of continuous infusion studies using tail cuff tethers relate to possible leakage of the cannula, as previously discussed, and the physical effect of the tail cuff. It is inevitable that some tissue damage is associated with this method of tethering, as the cuff is attached by a wire suture passing through the skin, therefore allowing a potential entry site for bacteria (Figure 7.5). In addition, a degree of movement of the cuff as the animal moves provides a source of irritation, and there is the possibility that the tail may become too large for the cuff, either by natural growth or by swelling. As mentioned previously, entry of bedding material into the space between the tail and

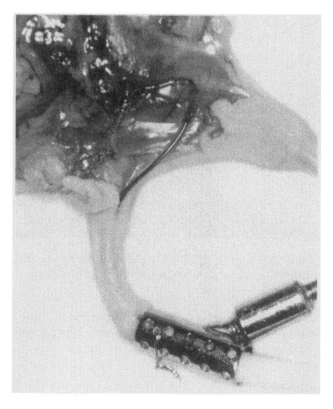

Figure 7.5 Photograph of a *nu/nu* mouse taken at necropsy showing swelling proximal to the tail cuff.

Table 7.1 Necropsy findings related to femoral vein cannulation and tail cuff tethering of CD-1 mice

Necropsy finding	Incidence	
	7 days	28 days
Tail ulceration/erosion	1/20	1/20
Subcutaneous thickening at cannulation site	1/20	1/20
Abnormal tail shape	0/20	13/20

Table 7.2 Histological findings related to femoral vein cannulation and tail cuff tethering of CD-1 mice

Histological finding	Incidence	
	7 days	28 days
Tail		
Fasciitis	18/20	17/20
Cellulitis	17/20	16/20
Epidermal hyperplasia	15/20	17/20
Osteomyelitis	2/20	0/20
Necrosis/ulceration	0/20	3/20
Phlebitis/periphlebitis	2/20	0/20
Dermal fibrosis	0/20	1/20
Femoral vein/		
Phlebitis/periphlebitis	4/20	10/20
Thickening/fibrosis	2/20	3/20
Vena cava		
Phlebitis/periphlebitis	16/20	6/20

the cuff is an additional potential source of irritation, leading to skin erosion and ulceration.

Tables 7.1 and 7.2 show typical necropsy and histological findings, taken from the validation study, which may be associated with the use of a tail cuff.

Despite the high incidence of local inflammation seen microscopically, other findings, such as body weight, are generally similar to those found in cage controls (see Figures 7.2 and 7.3). Tables 7.3 and 7.4 show the haematology results from mice on the 7- and 28-day validation study. A small increase in white blood cell count and percentage of neutrophils is apparent, presumably associated with the inflammation noted in the majority of animals, but this response was not marked.

Unsurprisingly for animals being given saline at 4 and 8 ml/kg/h, urinalysis generally demonstrated an increase in urine volume and a decrease in its specific gravity. A concomitant decrease in water consumption is also seen at high infusion rates, but no renal lesions have been associated with infusion at the above rates for up to 28 days.

Intravenous infusion of rats using a tail cuff tether has been validated for up to a year but has not yet been carried out in mice by the authors beyond 28 days. The likely primary limiting factors relate to tail pathology and the integrity of the cannulae.

Table 7.3 Haematology results (Day 8, individual and mean values) following 7 days' continuous intravenous infusion of various concentrations of 0.9% NaCl in CD-1 mice (reprinted with the kind permission of Animal Technology)

Animal	Hb (g/dl)	RBC (×10¹²/l)	Hct (l/l)	MCH (pg)	MCV (fl)	MCHC (g/dl)	WBC (×10⁹/l)	Neut (×10⁹/l)	Lymph (×10⁹/l)	Mono (×10⁹/l)	Eos (×10⁹/l)	Baso (×10⁹/l)	LUC (×10⁹/l)	Plat (×10⁹/l)
0.9% NaCl 0 ml/kg/h														
6	15.4	9.25	0.448	16.7	48.4	34.5	8.99	0.59	7.81	0.25	0.18	0.03	0.13	1245
7	15.0	9.49	0.459	15.8	48.4	32.7	3.26	0.40	2.59	0.15	0.09	0.01	0.01	1179
8	14.3	8.78	0.436	16.3	49.7	32.7	9.15	2.61	5.73	0.45	0.24	0.02	0.10	829
9	15.2	9.08	0.457	16.7	50.3	33.3	4.58	0.77	3.58	0.10	0.11	0.01	0.01	1081
10	13.5	8.34	0.404	16.2	48.4	33.4	8.08	0.96	6.60	0.24	0.21	0.03	0.04	1211
Mean	14.7	8.99	0.441	16.3	49.0	33.3	6.81	1.07	5.26	0.24	0.17	0.02	0.06	1109
0.9% NaCl 4 ml/kg/h														
16	14.0	9.38	0.429	14.9	45.8	32.6	9.73	1.82	7.21	0.29	0.32	0.03	0.05	1463
17	14.4	9.23	0.433	15.6	46.8	33.3	8.57	2.06	6.03	0.22	0.13	0.02	0.11	1392
18	12.7	8.24	0.390	15.4	47.3	32.6	9.51	5.05	3.83	0.28	0.20	0.01	0.15	1337
19	13.2	8.77	0.415	15.1	47.3	31.9	6.96	1.19	5.44	0.17	0.11	0.01	0.04	1473
20	14.8	9.23	0.455	16.0	49.3	32.5	8.23	1.49	6.18	0.28	0.23	0.02	0.04	1255
Mean	13.8	8.97	0.424	15.4	47.3	32.6	8.60	2.32	5.74	0.25	0.20	0.02	0.08	1384
0.9% NaCl 8 ml/kg/h														
26	14.6	8.99	0.456	16.3	50.7	32.1	7.20	3.37	3.26	0.31	0.18	0.01	0.08	1599
27	14.5	9.22	0.454	15.7	49.3	31.9	6.20	1.33	4.39	0.11	0.26	0.02	0.09	1443
28	14.1	9.06	0.435	15.6	48.1	32.5	8.87	2.14	5.97	0.34	0.30	0.02	0.09	1254
29	14.5	8.89	0.441	16.3	49.6	32.9	6.26	1.14	4.44	0.27	0.18	0.02	0.21	1024
30	14.2	9.13	0.445	15.5	48.7	31.8	12.46	8.28	3.44	0.31	0.19	0.02	0.20	1224
Mean	14.4	9.06	0.446	15.9	49.3	32.2	8.20	3.25	4.30	0.27	0.22	0.02	0.13	1309

LUC, large unclassified cells.

Table 7.4 Haematology results (Day 28, individual and mean values) following 28 days' continuous intravenous infusion of various concentrations of 0.9% NaCl in CD-1 mice (reprinted with the kind permission of Animal Technology)

Animal	Hb (g/dl)	RBC ($\times10^{12}$/l)	Hct (l/l)	MCH (pg)	MCV (fl)	MCHC (g/dl)	WBC ($\times10^9$/l)	Neut ($\times10^9$/l)	Lymph ($\times10^9$/l)	Mono ($\times10^9$/l)	Eos ($\times10^9$/l)	Baso ($\times10^9$/l)	LUC ($\times10^9$/l)	Plat ($\times10^9$/l)
0.9% NaCl 0 ml/kg/h														
1	14.4	8.81	0.447	16.4	50.8	32.3	5.37	0.84	4.32	0.11	0.18	0.07	0.02	1092
2	14.2	8.95	0.431	15.9	48.1	32.9	8.00	0.63	7.00	0.16	0.09	0.16	0.02	1143
3	13.5	9.04	0.420	15.0	46.5	32.2	4.26	0.37	3.68	0.09	0.24	0.11	0.01	434
4	14.2	8.70	0.421	16.4	48.4	33.8	2.36	0.20	2.00	0.09	0.11	0.08	0.00	1251
5	15.1	9.57	0.463	15.8	48.3	32.6	6.04	0.62	5.08	0.16	0.21	0.13	0.03	976
Mean	14.3	9.01	0.436	15.9	48.4	32.8	5.21	0.53	4.41	0.12	0.17	0.11	0.02	979
0.9% NaCl 4 ml/kg/h														
11	14.2	8.65	0.422	16.5	48.8	33.8	5.81	0.92	4.46	0.24	0.32	0.18	0.00	1148
12	14.3	8.94	0.435	16.0	48.7	32.8	8.11	1.31	6.18	0.24	0.13	0.29	0.07	1546
13	14.0	9.37	0.426	15.0	45.5	32.9	7.87	2.40	4.99	0.35	0.20	0.10	0.01	1468
14	13.7	9.65	0.421	14.2	43.7	32.6	8.51	2.89	4.94	0.35	0.11	0.28	0.04	1267
15	14.8	8.69	0.439	17.0	50.5	33.7	4.27	0.57	3.35	0.16	0.23	0.17	0.01	1476
Mean	14.2	9.06	0.429	15.7	47.4	33.2	6.91	1.62	4.78	0.27	0.20	0.20	0.03	1381
0.9% NaCl 8 ml/kg/h														
21	14.2	9.73	0.446	14.6	45.8	31.8	7.79	1.56	5.69	0.26	0.18	0.12	0.13	1265
22	14.0	9.04	0.423	15.5	46.8	33.1	10.64	2.26	7.81	0.42	0.26	0.08	0.03	1074
23	13.8	9.04	0.414	15.3	45.8	33.4	8.06	1.78	5.65	0.43	0.30	0.14	0.04	1269
24	14.5	9.31	0.450	15.6	48.4	32.3	6.65	1.67	4.49	0.33	0.18	0.13	0.03	1331
25	15.0	9.08	0.453	16.5	49.9	33.0	8.28	2.76	4.73	0.47	0.19	0.27	0.03	1518
Mean	14.3	9.24	0.437	15.0	47.3	32.7	8.28	2.01	5.67	0.38	0.22	0.15	0.05	1291

Having ensured that the optimum design of cannula and tail cuff is used, several procedures can be undertaken to maximise the experimental period. It is important to use as small a size of tail cuff as is required in order to minimise excess movement, but equally the tail cuff must be large enough to accommodate growth during the anticipated dosing period. For a chronic study, the tail cuff is likely to be larger than is ideal, but during the course of the study the wire suture may need to be loosened to allow for growth. A regular check should always be made of tails and tail cuffs during the study, and the need to loosen or tighten the tail cuff should be assessed. In cases where tail lesions become an end-point for an individual animal, the cuff can be removed and the animal retained for post-dose recovery, if appropriate to the experiment (the cannula is cut at the entry to the tail and causes no adverse effects). Ulcerative tail lesions can then be expected to heal.

In addition to the above factors, it is also imperative to the proper conduct of the toxicity study that aseptic techniques are used during dosing procedures such as syringe filling and loading into syringe pumps, as well as any unblocking of cannulae required. The importance of maintaining sterility should be emphasised to all staff involved in the procedures, who should be specifically trained in proper techniques to minimise the possibility of contamination. Ideally, dosing and associated procedures for infusion studies should be carried out by a dedicated team of technical staff.

7.6 Conclusion

The conduct of intravenous infusion in mice has been reported using several different techniques over the years, via different veins. Previously reported techniques have resulted in successful infusion for variable periods of dosing, although the authors have not identified papers that specifically discuss clinical and pathological effects of other cannulation techniques. An understanding of the expected effects of the technical procedures is important when assessing potential adverse effects of test materials following the conduct of multidose toxicity studies.

References

Grindey, G.B., Hoglind Semon, J. and Pavelic, Z.P. (1978) Modulation versus rescue of antimetabolite toxicity by salvage metabolites administered by continuous infusion. *Fundamentals in Cancer Chemotherapy* 23, 295–304.

Lemmel, E. and Good, R.A. (1971) Continuous long-term intravenous infusion in unrestrained mice – method. *Journal of Laboratory Clinical Medicine* 77, 1011–1014.

Paul, M.A. and Dave, C. (1975) A simple method for long-term drug infusion in mice: evaluation of guanazole as a model. *Proceedings of the Society for Experimental Biology and Medicine* 148, 118–122.

Plager, J.E. (1972) Intravenous, long-term infusion in the unrestrained mouse – method. *Journal of Laboratory Clinical Medicine* 79, 669–672.

Van Wijk, H. (1997) A continuous intravenous infusion technique in the unrestrained mouse. *Animal Technology* 48, 115–128.

8 Jugular cannulation and efficacy studies in the mouse

I. Horii

8.1 Introduction

The experimental utilisation of continuous intravenous infusion has been widely recognised as a method applicable not only to toxicological studies (Francis *et al.* 1992; Barrow and Heritier 1995; Salauze and Cave 1995; Barrow *et al.* 1996) but also to pharmacological efficacy studies in disease animal models (Osieka 1984; Langdon *et al.* 1998), and to other pharmacological studies, such as those on self-administration dependence (Jones and Prada 1977; Carney *et al.* 1991; Gold and Balster 1992; Takada and Yanagita 1997; Arroyo *et al.* 1998; Rasmussen and Swedberg 1998) and using telemetry systems (Berger and Phillips 1991; Van den Buuse and Malpas 1997; Balakrishnan *et al.* 1998).

Continuous intravenous infusion has usually been utilised in relatively large animals, such as rabbits, dogs and monkeys, for pharmacological and toxicological studies. There are also many examples of its use in pharmacological and long-term toxicological studies in rats (Steffens 1969; Jones and Hynd 1981; Cave *et al.* 1995; Gregory 1995). Nevertheless, there have been few papers dealing with continuous infusion studies in mice, particularly those involving continuous intravenous infusion (Popovic *et al.* 1968; Lemmel and Good 1971; Plager 1972; Paul and Dave 1975; Hodge and Shalev 1992; Braakhuis *et al.* 1995; Veerman *et al.* 1996; Van Wijk 1997).

For primary efficacy screening studies, mouse continuous intravenous infusion is useful when the compound has low solubility or is quickly metabolised. If a bolus injection were employed for such a compound, the pharmacokinetic parameters would show short half-life and low AUC, indicating a gap between *in vitro* and *in vivo* efficacies. This would mean that efficacy *in vivo* was insufficient even though the compound showed sufficient effects *in vitro*. In such a case, the chance to develop an innovative compound would be lost. The availability of a continuous intravenous infusion technique provides the means to maintain a sufficient blood concentration of the compound to evaluate effectively its *in vivo* efficacy.

This chapter describes procedures for efficient infusion techniques in the mouse, focusing on cannulation via the jugular vein, and gives practical examples of efficacy studies (using femoral cannulation) as a screening system suitable for use with fungal infection mouse models and human xenograft tumour models in nude mice.

Generally, the infusion model provides for three vein routes for surgical implantation of a catheter: jugular vein (Popovic *et al.* 1968; Lemmel and Good 1971; Hodge and Shalev 1992), femoral vein (Van Wijk 1997) and tail vein (Plager 1972; Paul and Dave 1975; Braakhuis *et al.* 1995; Veerman *et al.* 1996). The methods described in this chapter include the improved technologies of a jugular vein system developed at the Nippon Roche Research Center (Inomata *et al.* 1999) and a femoral vein system from Inveresk Research (Van Wijk 1997) as well as the tail vein system.

8.2 Infusion technique through jugular vein in mice

The following technique is generally applicable to efficacy studies with infectious disease mouse models and nude mice bearing human tumour xenografts.

8.2.1 Caging and infusion pump

The animals are individually housed in clear polycarbonate topless cages (Natsume Instruments Manufacturing Co., Ltd, Japan). The animals can move freely and are easy to handle for body weight measurement, blood sampling, etc. A microsyringe pump, KDS220 (KD Scientific Inc., USA) is used. The accuracy of this pump was validated by calculating the coefficient of variation (< 0.1%). The animals are all linked to the same pump to assure that the flow rate is invariant among them throughout the study (Figure 8.1).

8.2.2 Tethering equipment, infusion line and cannulae

A special jacket or implantable disk was designed and custom-made (Figure 8.1a). The jacket is made of cloth and strapped around a rubber tube with surgical tape. The disk is made of silicon and connected to the tube. The end of the tube is attached to a swivel mechanism with vinyl tape. The swivel is held above the cage by a swivel holder, which is held in place by a stand attached to the cage edge (Figure 8.1b). The infusion line from syringe to jugular vein comprises a 20-ml syringe (Terumo Co.,

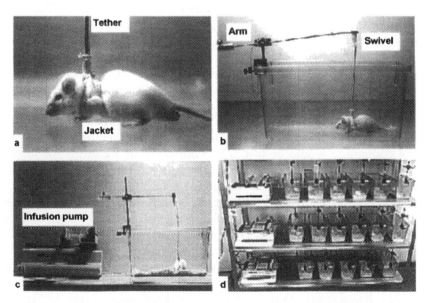

Figure 8.1 Photographs of (a) animal jacket, (b) tethering equipment, (c) infusion pump and (d) caging.

Ltd, Japan), a 23-gauge needle, a 0.22-μm sterile Acrodisc (Gelman Sciences, Germany), clear vinyl tube as a cannula (ID 0.50 mm, OD 0.90 mm) that has been tested for vascular and tissue reaction (Natsume Instruments Manufacturing Co., Ltd, Japan), a plastic swivel (Lomir Biomedical Inc., Canada), and a swivel holder and its stand (Instech, USA) (Figure 8.1b–d).

8.2.3 Animals

Six-week-old male ICR mice are obtained from Japan SLC Inc., Shizuoka, Japan. All animals weighed 25–30 g at the time of surgery.

8.2.4 Surgical procedure for venous cannulation (Figure 8.2)

The animals are anaesthetised with a mixture of 10% Nembutal (Abbott) (50 mg/ml) in physiological saline (Ohtsuka Pharmaceuticals Co., Ltd, Japan) given intraperitoneally. The ventrocervical and interscapular regions of the animal are shaved and swabbed with 70% alcohol. A 5-mm initial incision is made in the skin of the right ventrocervical region, and the connective tissue dissected out to expose the right external jugular vein, which lies below the salivary gland. The external jugular vein is lifted slightly with forceps and separated from the nerves running along the vein (Figure 8.2a). The vein is carefully separated from the nerves and other structures, and secured by the tip of the forceps. The other side of the forceps are placed on a cylindrical supporting rest, at the same height as the thickness of the mouse to avoid compressing the respiratory tract.

Sterilised saline is sprayed on the exposed external jugular vein and over the surgical area to prevent the tissues from drying. Two sutures are then passed under the external jugular vein, with one thread for ligating the vein holding the catheter within (proximal end of opening) and the other thread for ligating the vein and the catheter together (distal end of opening) (Figure 8.2b). Two sutures (Natume Instruments Manufacturing Co., Ltd, Japan) are gently passed below the vein, approximately 3 mm apart. The lifted vein (between the blades of forceps) is incised with precision scissors (less than 0.9 mm, the external diameter of the catheter) and the inserting end of the catheter is checked to confirm that the catheter is filled with saline. The catheter is inserted from the vein opening towards the heart (*c.* 10 mm, depending on the size of animal, leaving approximately 1–2 mm of space between the catheter tip and the clavicle) (Figure 8.2c and g). The cannula is pre-filled with physiological saline in preparation for insertion and physiological saline sprayed again over the operating area to ease the cannula's slide into the inner wall of the vein. An incision is made in the outer wall of the vein between the two sutures with precision scissors (Natsume Instruments Manufacturing Co., Ltd, Japan). The bevelled tip of the cannula is then inserted into the vein and advanced about 10 mm so that the tip is in the vein around the subclavian region. The proximal length of suture thread is tied around the cannulated vein and the distal length of suture used to tie off the vein below the cannulation site and also to anchor the cannula within the vein (Figure 8.2d).

To fix the catheter, one end of the suture used for the distal binding is passed through the subcutaneous tissue and tied with the other end to make a loop. Patency is checked by withdrawing some blood into the syringe. A metal wire is then attached to the catheter, and the wire and the catheter passed together under the skin towards

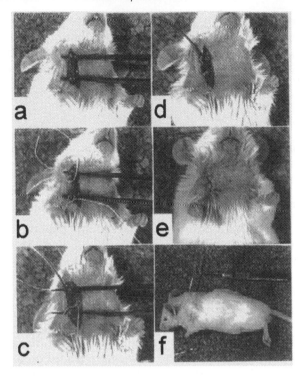

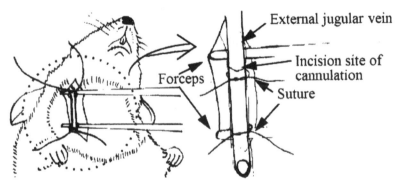

G: Illustration of Photo-C

Figure 8.2 Surgical procedure for venous cannulation through the jugular vein: photographs
of each step and illustration of the cannulating part.

the dorsal area, and then towards the dorsocervical area to the outlet mouth of the
shaved area to prevent the catheter from being pulled out. After the free end of the
cannula is plugged by an arterial clamp, the syringe is removed and a stainless-steel
wire attached to that end. First, the wire is advanced through the subcutaneous tissue
towards the dorsocaudal region of the mouse. Next, the wire is diverted to the shaved
skin in the interscapular region, pushed out through the skin in the area, and the
cannula pulled out with the wire. This loop of the cannula in the mouse's back acts as
a cushion against any accidental twitch. If a silicon disc is used, a 10-mm incision in

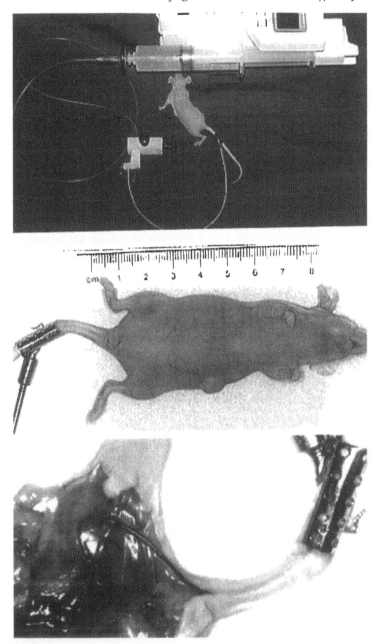

Figure 8.3 Photographs of a nude mice model and continuous infusion system (produced by Hans van Wijk, Inveresk Research).

the interscapular region is made to bury the disc under the interscapular skin prior to pushing through the wire. The cannula is plugged with the clamp again and the wire detached from the cannula. The free end of the cannula is sealed and finally passed into the rubber tube. A jacket or silicon disc is fitted to the animal and the tether attached.

8.2.5 Connection of infusion equipment

After the operation is finished, the mouse is kept warm to prevent hypothermia. When the mouse has aroused completely, it is returned to its home cage. The cannula and its protective rubber tube are connected to the swivel, attached to a swivel holder. The swivel mechanism allows the animal to move freely around the cage. A vinyl tube is connected to the top of the swivel to a syringe containing saline, which is then attached to an infusion pump. Finally, the sealed tip of the implanted cannula is cut, and the cannula attached to the underside of the swivel.

8.2.6 Post-operative care

After surgery, the mouse is weighed with the equipment attached. The body weight of the animal prior to surgery is then subtracted from the weight of the animal after surgery. The initial rate of infusion is 0.3 ml/h for 48 h with saline sterilised through a 0.22-μm sterile Acrodisc, which is connected to the syringe.

8.3 Infusion technique via the femoral vein

The technology for infusion via the femoral vein is an improved one, which could be applicable for long-term continuous intravenous infusion studies in mice, especially for toxicological studies. Details are reported in the preceding chapters and will not be described here. This technology can be also applied to nude mice (Van Wijk *et al.* 1999). The photographs of the nude mouse model for continuous infusion in efficacy studies are shown in Figure 8.3.

8.4 Infusion technique through tail vein

For the efficacy study, Veerman *et al.* (1996) has reported on a conventional intravenous system through the tail vein. Based on Veerman's report, the mice were anaesthetised (i.m.) with Hypnorm (dose 0.02 ml per mouse weighing 20 g) and laid on a water-heated bed. The tail vein was punctured with a 21-gauge needle. A catheter (Intramedic polyethylene tubing, medical formulation PHF; ID 0.58 mm; OD 0.97 mm; Clay Adams) was inserted into the tail vein and flushed with 0.9% NaCl. Both tail and catheter were splinted. The catheter and splint were protected by a plastic tube. Each mouse was put into a separate cage, which was placed on a heated water bed to prevent hypothermia. The catheter was connected to a syringe, which was placed in an infusion pump. Each mouse received 1.2 ml of solution containing compound in 0.9% NaCl per 24 h. After the 24-h infusion the catheter was removed.

8.5 The use of infusion technology for efficacy studies in drug discovery

In the screening stage of drug discovery, the *in vivo* efficacy of a compound has to be estimated by means of intravenous bolus injections to the disease model mice. There may have been some problems, however, with the bolus injection, depending on the characteristics of the compounds. Some compounds have low solubility, poor absorption or are metabolised quickly. Often these characteristics result in short $t_{1/2}$ and low AUC, which may indicate rapid degradation to inactive metabolites. These results may lead to false conclusions, especially when the compound has good efficacy *in vitro*. When bolus repeated injection is applied with high doses to maintain a sufficient blood level, excessively toxic levels in the blood induce acute adverse effects. In contrast, bolus applications at middle/low doses to keep the effective blood level below the non-toxic level may show no pharmacological effect even if the compound has potential efficacy (Figure 8.4). In this example for screening, we need to maintain a sufficiently high plasma concentration of the test compounds to estimate whether a compound is sufficiently active *in vivo* as a lead compound.

In this scenario of an *in vitro/in vivo* gap, the continuous infusion technique is indispensable for finding innovative lead compounds.

8.6 Pharmacological efficacy studies in disease mouse models by using continuous infusion technology

Two practical examples of efficacy infusion studies in mice that have been carried out in the Nippon Roche Research Center and which involved antifungal (Kobayashi *et al.* 1999) and anti-tumour drugs (Kawashima *et al.* 1999) are described below.

8.6.1 Mycology – efficacy infusion study in a fungal infectious disease mouse model

Male ICR mice (4 weeks old, approximately 20 g body weight) infected with *Candida albicans* were used for the antifungal efficacy study. The animals were housed in a temperature-controlled room (22 ± 2 °C) with a 12 h/12 h light/dark cycle (light on 07.00 to 19.00 h). Before surgery, the mice were housed in groups (five per cage) and allowed free access to food and water. After surgical implantation via the jugular vein, each individual mouse was housed in a specialised fitted cage with a swivel. To

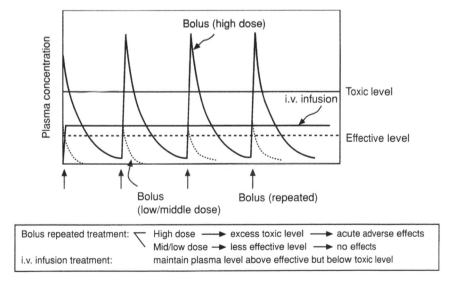

Figure 8.4 Maintenance of effective plasma level with continuous intravenous infusion in a disease animal model.

Table 8.1 Antifungal activity of compounds A and B by continuous intravenous infusion in a fungal infectious mouse model

Compound A			Compound B		
Dose level (mg/kg)	Survival rate (alive/ tested mice)	Day of death	Dose level (mg/kg)	Survival rate (alive/ tested mice)	Day of death
0	0/5	(1–2)	0	0/5	(1–2)
1	1/5	(2–5)	0.3	0/5	(2–5)
3	2/5	(5–7)	0.6	2/5	(3–10)
10	5/5		1.0	5/5	

Notes
Animal: ICR mice, 4 weeks old, body weight ~ 20 g.
Infection: *Candida albicans* (ATCC48130), 6.5×10^6 per animal, i.v. through tail vein 3 h before dosing.
Flow rate: 1–4 ml/kg/h.

establish the infection model, the mice were inoculated with *Candida albicans* (strain ATCC48130, the concentration of *Candida* being approximately 6.5×10^6 per mouse, assuming 25 g body weight) by intravenous injection through the tail vein, up to 1 h before the start of compound infusion. During a 10-day infusion period, survival rate was monitored as anti-fungal activity.

Two anti-fungal compounds, compound A (chitin synthetase inhibitor), which was rapidly metabolised to an inactive metabolite, and compound B (azole antifungal), which has low solubility, were used for the continuous infusion study, because both compounds showed poor efficacy following intravenous bolus injection. As a result, although neither compound had sufficient efficacy when administered by daily repeated bolus injections, intravenous continuous infusion showed dose-dependent antifungal efficacy in both survival rate and prolongation of surviving day (Table 8.1). Their pharmacokinetic profiles revealed that continuous exposure of the compound at sufficient blood level concentrations, which exceeded the effective concentration *in vitro*, exhibited antifungal activity dose-dependently (Figure 8.5).

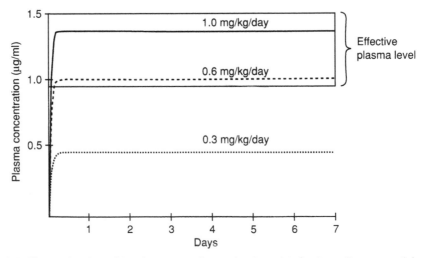

Figure 8.5 Plasma level profile of compound B in the fungal infectious disease model.

Table 8.2 Efficacy study of compound C in MDA-MB-435-bearing CD-1 *nu/nu* mice

Dose (mg/kg/day)	Tumour volume change on Day 42 (mean value, mm³)	Growth inhibition on Day 42 (%)	Survivors on Day 42
Vehicle	112.0	–	6/6
Compound C			
12.5	82.3	26.5	6/6
25	50.6	54.8	5/6 (dead 1/6)*
50	21.7	80.5	6/6
100	18.6	83.4	5/8 (dead 3/8)†

Notes
MDA-MB-435 cells (1.7×10^7) were inoculated (mammary fat pad) into CD-1 nude mice (female) on Day 0.
Drugs were administered (24-h continuous infusion) for 10 days (from Day 33 to Day 42).
*Cause of death not drug related but due to technical problems: leakage of drug solution into thoracic cavity, etc.
†Lethal in prolonged intravenous infusion.
Mean tumour volume on Day 33: 275.0 mm³.

8.6.2 Oncology – efficacy infusion study in nude mice bearing a human tumour xenograft

Nude mice bearing the human MDA-MB-435 breast xenograft were used for this continuous intravenous infusion study. Approximately 4–5 days before treatment, a catheter (vinyl catheter; ID 0.5 mm, OD 0.9 mm) was implanted into the external jugular vein. Animals were maintained on continuous infusion (0.2 ml/h per animal) with physiological saline between implantation and the start of treatment. Especially in the nude mouse study, it was very difficult to maintain the catheter flow over 2 weeks.

As an anti-tumour compound, compound C (cell cycle inhibitor), which has poor solubility and poor absorbability when administered orally, was examined for anti-tumour activity and pharmacokinetic profile by continuous i.v. infusion in nude mice bearing the human MDA-MB-435 breast xenograft. Mice were administered with compound C by continuous i.v. infusion for 10 days. The results were compared with those obtained with 10 days of a twice-daily i.p. administration. Blood samples were taken on Days 1, 3, 5 and 7 during infusion of different doses of compound C. As a result (Table 8.2), over 80% (80–100%) tumour regression was produced by 4 days' infusion at doses of 50–100 mg/kg/day. About 50% (40–100%) tumour regression was also obtained after 7 days' infusion of 50 mg/kg/day to 100 mg/kg/day. However, 60% of animals receiving the 100 mg/kg/day dose died, suggesting that prolonged intravenous infusion of compound C at that dose was lethal. It was also demonstrated that continuous i.v. infusion of compound C was more effective against the MDA-MB-435 xenograft than daily i.p. administration. Pharmacokinetics studies with compound C showed that infusion of 50–100 mg/kg/day produced a steady-state plasma concentration that was, most likely, responsible for the increased efficacy observed. The results indicated that continuous i.v. infusion of compound C was efficacious in nude mice bearing the human tumour xenograft.

Acknowledgements

I wish to thank Hans van Wijk (Inveresk Research) for offering useful information and photographs for infusion technology through the femoral vein, and Akira Inomata, Akira Kawashima, Kazuko Kobayashi and Nobuyuki Shishido (Nippon Roche Research Center) for providing the internal data concerning infusion technology through the jugular vein.

References

Arroyo, M., Markou, A., Robbins, T.W. and Everitt, B.J. (1998) Acquisition, maintenance and reinstatement of intravenous cocaine self-administration under a second schedule of reinforcement in rats: effects of conditioned cues and continuous access to cocaine. *Psychopharmacology*, 140, 331–344.

Balakrishnan, S., Tatchum-Talom, R. and McNeil, J.R. (1998) Radiotelemetric versus externalized catheter monitoring of blood pressure: effect of vasopressin in spontaneous hypertension. *Journal of Pharmacological and Toxicological Methods*, 40, 87–93.

Barrow, P.C. and Heritier, B. (1995) Continuous deep intravenous infusion in rat embryotoxicity studies: the effects of infusion volume and two different infusion fluids on pregnancy. *Toxicology Methods*, 5, 61–67.

Barrow, P.C., Heritier, B. and Marsden, E. (1996) Continuous deep intravenous infusion in rat fertility studies. *Toxicology Methods*, 6, 139–147.

Berger, R.J. and Phillips, N.H. (1991) Suppression of slow wave sleep and circadian rhythms of body temperature in the pigeon by continuous bright light, and their reinstatement by daily melatonin infusions. *Advances in Pineal Research*, 5, 275–277.

Braakhuis, B.J.M., van Haperen, V.W.T.R., Boven, E., Veerman, G. and Peters, G.J. (1995) Schedule-dependent antitumor effect of gemcitabine in in vivo model system. *Seminars in Oncology*, 22, 42–46.

Carney, J.M., Landrum, R.W., Cheng, M.S. and Seale, T.W. (1991) Establishment of chronic intravenous drug self-administration in the C57BL/6J mouse. *NeuroReport*, 2, 477–480.

Cave, D.A., Schoenmakers, A.C.M., van Wijk H.J., Enninga, I.C. and van der Hoeven, J.C.M. (1995) Continuous intravenous infusion in the unrestrained rat – procedures and results. *Human and Experimental Toxicology*, 14, 192–200.

Francis, P.C., Hawkins, P.C., Houchins, J.O., Cross, P.A., Cochran, J.A., Russel, E.L., Johnson, W.D. and Vodicnik, M.J. (1992) Continuous intravenous infusion in Fischer 344 rats for 6 months: a feasibility study. *Toxicology Methods*, 2, 1–13.

Gold, L.H. and Balster, R.L. (1992) Effects of buspirone and gepirone on iv cocaine self-administration in rhesus monkeys. *Psychopharmacology*, 108, 28–294.

Gregory, J. (1995) Practical aspects of continuous infusion in rodents. *Animal Technology*, 46, 115–130.

Hodge, D.E. and Shalev, M. (1992) Dual cannulation: a method for continuous intravenous infusion and repeated blood sampling in unstrained mice. *Laboratory Animal Science*, 42, 320–322.

Inomata, A., Shishido, N., Kawashima, A. and Horii, I. (1999) A continuous intravenous infusion technique for pharmacological efficacy study in mouse disease model. Roche Internal Report.

Jones, B.E. and Prada, J.A. (1977) Effects of methadone and morphine maintenance on drug-seeking behavior in the dog. *Psychopharmacology*, 54, 109–112.

Jones, P.A. and Hynd, J.W. (1981) Continuous long-term intravenous infusion in the unrestrained rat – a novel technique. *Laboratory Animals*, 15, 29–33.

Kawashima, A., Inomata, A. Eda, S. and Horii, I. (1999) Anti-tumor efficacy study with continuous intravenous infusion in nude mice bearing human-tumor xenografts. Roche Internal Report.

Kobayashi, K., Shishido, N., Inomata, A. and Horii, I. (1999) Anti-fungal efficacy study with continuous intravenous infusion in infectious mice. Roche Internal Report.

Langdon, S.P., Ritchie, A.A., Muir, M., Dodds, M., Howie, A.F., Leonard, R.C., Stockman, P.K. and Miller, W.R. (1998) Antitumor activity and schedule dependency of 8-chloroadenosine-3′,5′-monophosphate (8-ClcAMP) against human tumour xenografts. *European Journal of Cancer*, 34, 384–388.

Lemmel, E.M. and Good, R.A. (1971) Continuous long-term intravenous infusion in unrestrained mice – method. *Journal of Laboratory and Clinical Medicine*, 77, 1011–1014.

Osieka, O. (1984) Comparison of a continuous infusion with a daily bolus injection of bleomycin in a heterotransplanted human testicular cancer. *Arzneimittelforschung*, 34, 460 –464.

Paul, M.A. and Dave C. (1975) A simple method for long-term drug infusion in mice: evaluation of guanazole as a model. *Proceedings of the Society for Experimental Biology and Medicine*, 148, 118–122.

Plager, J.E. (1972) Intravenous, long-term infusion in the unrestrained mouse – method. *Journal of Laboratory and Clinical Medicine*, 79, 669–672.

Popovic, P., Sybers, H. and Popovic, V.P. (1968) Permanent cannulation of blood vessels in mice. *Journal of Applied Physiology*, 25, 626–627.

Rasmussen, T. and Swedberg, M.D.B. (1998) Reinforcing effects of nicotinic compounds: intravenous self-administration in drug-naïve mice. *Pharmacology, Biochemistry and Behaviour*, 60, 567–573.

Salauze, D. and Cave, D. (1995) Choice of vehicle for three-month continuous intravenous toxicity studies in the rat: 0.9% saline versus 5% glucose. *Laboratory Animals*, 29, 432–437.

Steffens, A.B. (1969) A method for frequent sampling of blood and continuous infusion of fluids in the rat without disturbing the animal. *Physiology and Behavior*, 4, 833–836.

Takada, K. and Yanagita, T. (1997) Drug dependence study on vigabatrin in rhesus monkeys and rats. *Arzneimittelforschung*, 47, 1087–1092.

Van den Buuse, M. and Malpas, S.C. (1997) 24-hour recordings of blood pressure, heart rate and behavioral activity in rabbits by radio-telemetry: effect of feeding and hypertension. *Physiological Behaviour*, 62, 83 – 89.

Van Wijk, H. (1997) A continuous intravenous infusion technique in the unrestrained mouse. *Animal Technology*, 48, 115–128.

Van Wijk, H., Forsyth, A., Dick, A. and Robb, D. (1999) Continuous intravenous infusion in athymic (nude) mice, an animal model for evaluating the efficacy of anti-cancer agents. Internal Report – Inveresk Research.

Veerman, G.V., van Haperen, V.W.T.R., Vermorken, J.B., Noordhuis, P., Braakhuis, B.J.M., Pinedo, H.M. and Peters, G.J. (1996) Antitumor activity of prolonged as compared with bolus administration of 2′,2′-difluorodeoxycytidine in vivo against murine colon tumors. *Cancer Chemotherapy and Pharmacology*, 32, 335–342.

Part 3

Continuous intravenous infusion in the rabbit

The rabbit is the primary second species model for reproductive continuous infusion studies, although the minipig is emerging as a possible alternative, as described in Part 6.

Within this section, the two different catheterisation techniques via the femoral (Chapter 9) and jugular (Chapter 10) veins are described, as well as the differing methodologies between two laboratories for conducting the in-life phase of the study. The data suggest that neither vessel is shown to be more advantageous than the other. The authors also differ in the method of insemination: both natural (Chapter 10) and artificial (Chapter 9) techniques are detailed. Examination of the data again reveals a similar pregnancy rate between the two methods.

Another fundamental difference highlighted is the use of either exteriorised catheters (Chapter 9) or vascular access ports (VAPs) (Chapter 10). Both techniques have their advantages. For example, if a catheter or external tubing becomes blocked or damaged, it can be easily replaced by disconnecting at the VAP so that the animal is not lost. The use of VAPs also enables the animal to be maintained without infusion until Day 6 post coitum to allow recovery from surgery. The use of an uninterrupted line, on the other hand, could be considered less likely to result in blockages or infection/irritation due to puncturing of the skin when accessing the VAP.

The chapters are consistent, however, in their preference for a tethered approach, rather than using a jacket containing a small pump, because of the excessive weight of the equipment. However, the latter technique should still be explored to alleviate the need for a tethered animal, and as technology improves and pumps become smaller, the weight of the equipment may no longer be a major factor.

The relative merits of each approach are discussed, and new workers will need to identify the method best suited to their own facilities.

Part 2

Contagious infectious diseases Rinderpest
in the cattle

9 Femoral cannulation of the rabbit for reproductive toxicity studies

C. Copeman and K. Robinson

9.1 Introduction

ClinTrials BioResearch Ltd (CTBR) has been performing large-scale GLP pre-clinical intravenous infusion studies, via indwelling catheter, in various rodent and non-rodent species, including rats, mice, dogs, pigs and non-human primates (Enayati *et al.* 1994; Hillebrand-Krallis *et al.* 1994; Armer *et al.* 1995; Badalone *et al.* 1995; Fennell *et al.* 1995), since the early 1980s in response to the demand for pre-clinical safety evaluation studies of biotechnology and pharmaceutical products to mimic the proposed clinical route and treatment regimens. This route of administration has also been used to provide more constant blood levels of the test material where the clinical route provides variable levels in the test species (Clarke 1993). Furthermore, continuous intravenous infusion has been utilised to overcome problems of local irritation or dosing restrictions (e.g. limits of solubility) that may be encountered with intravenous bolus (fast or slow) injections (Perkin and Stejskal 1994). The use of these techniques for dose administration in reproductive toxicity studies to meet the requirements of various regulatory agencies (D'Aguanno 1976; EC 1989; Japanese MOHW 1990), including the recent International Committee on Harmonization (ICH 1993), was established in our laboratories for both rats and rabbits.

The primary method of cannulation utilised at the authors' laboratory in both these species is the vena cava via the femoral vein. Other techniques for intravenous infusion have been developed/validated (e.g. jugular cannulation) and are used as considered appropriate. The transfer of the established femoral cannulation techniques for intravenous infusion to reproductive toxicity studies has yielded results comparable to those obtained in non-infusion studies performed in our laboratories. The feasibility of using this route for embryo-fetal development, fertility and pre- and post-natal studies in the rat to meet the ICH guidelines was established in CD® rats (Robinson *et al.* 1994; 1995a; 1995b). Rodent reproduction toxicology studies using continuous intravenous infusion have been performed in our laboratory since the 1980s, with the first studies in gravid rabbits being conducted in the early 1990s. Definitive Segment II/embryo-fetal development studies in New Zealand White rabbits were initially performed in 1994 (Robinson *et al.* 1995a), and subsequently a number of additional studies have been conducted (Robinson *et al.* 1998).

Procedures to catheterise gravid rats and rabbits to allow for physiological measurements were reviewed in the 1980s (Peeters *et al.* 1984). Subsequently, several authors have described procedures for utilising catheterisation of rabbits for teratological assessments (Robinson 1995a; Barrow and Guyot 1996; McKeon *et al.* 1998; Lynch 1999). In reviewing the use of intravenous infusion via indwelling catheters

Perkin and Stejskal (1994) specified five criteria that should be met for infusion toxicology studies to be acceptable for safety evaluation purposes. These may be summarised as follows:

1. The number of animals per group should be the same as that used for a typical oral study.
2. The procedures for catheterisation should be standardised.
3. The same range of investigations should be performed on infusion studies as are performed on non-infusion studies.
4. There should be consistent daily administration of a known dosage that can be fixed or varied during a study as changes in body weight or any other factor of dose calculation are recorded.
5. Infections due to the presence of a catheter should be as close to non-existent as possible.

In the case of embryo-fetal development studies, a sixth specification should be considered:

6. The fetal data should be comparable to those of non–infusion studies and any differences should be understood in order to appropriately interpret the data from these studies.

The data presented in this chapter were compiled from five definitive embryo-fetal development/teratology studies, representing over 100 control animals. The procedures and results are discussed subsequently. All procedures described here were performed based on CTBR's standard operating procedures and approved by an internal IACUC (International Animal Care and Use Committee) in accordance with *the Guide for the Care and Use of Laboratory Animals* (US Department of Health and Human Services, National Institutes of Health).

9.2 Experimental procedures

Adult New Zealand White female rabbits (4–5 months of age), received from a commercial supplier (Covance Research Products Inc.), are allowed to acclimate to the laboratory environment for 2 weeks prior to surgical implantation of the catheter. Upon arrival, the animals are placed and maintained in a temperature-, humidity- and light-controlled environment (targeted conditions 17 ± 3 °C, $50 \pm 20\%$ relative humidity and 12 h light and 12 h dark). The ventilation is maintained to permit 12–15 air exchanges per hour. The animals are individually housed in stainless-steel cages equipped with a bar-type floor and an automatic watering valve. The animals receive up to 180 g of a standard certified commercial diet (No. 5322, PMI Feeds Inc.) and have *ad libitum* access to municipal tap water (softened, purified by reverse osmosis and sterilised by ultraviolet light).

9.2.1 Surgical catheter implantation

The surgical implantations are performed using aseptic techniques in a surgical suite consisting of pre-operative preparation room, surgical room and post-operative room for post-surgical care. The rabbits are food deprived for a minimum of 15 h prior to surgery in order to reduce the risk of aspiration, bloating and respiratory compromise resulting from abdominal distension. The animals are transported to the pre-operative room, where, in preparation for the surgery, each animal is currently pre-anaesthetised with intramuscular injections of glycopyrrolate (0.01 mg/kg), ketamine (50 mg/kg) and xylazine (5 mg/kg). Previously, Innovar-vet® intramuscular injection (fentanyl citrate, 0.08 mg/kg) was used but this has been discontinued because of the potential for reported residual sedative effects with the use of Innovar-vet®, including reduced respiratory rate and some evidence of respiratory acidosis and hypoxaemia in this species (Guerreiro and Page 1987). The ketamine–xylazine combination is the preferred choice of anaesthetic of many investigators (Green 1979; Lipman *et al.* 1997). It is considered a safe and effective method of inducing light anaesthesia in the rabbit (Lipman *et al.* 1997) despite the fact that hypotension and hypoxaemia are seen to occur. Our experience with ketamine–xylazine combination together with the administration of glycopyrrolate, to prevent bradycardia from vagal reflex and to reduce salivary and bronchial secretions, which can occlude airways (Lipman *et al.* 1997), is that it is an effective pre-anaesthetic that results in a smoother and shortened post-operative recovery in comparison with Innovar-vet®. The mortality rate associated with the surgical procedures for rabbits has decreased from a high of 5% in early pilot teratology studies to essentially 0% with the change in anaesthetic.

While under pre-anaesthesia, the animals are prepared for the surgical implantation by shaving the surgical and exteriorisation sites (the femoral and interscapular regions). The shaved sites are washed with Hibitane® (chlorhexidine gluconate 4%) followed by a liberal application of Betadine® (povidone-iodine 10%). The animals are brought to a surgical room that is equipped with HEPA filtration, and placed in dorsal recumbency on a warmed surgical table. The surgeon applies Betadine® to the surgical sites and a nose cone is placed on the animal to induce complete anaesthesia with isoflurane–oxygen gas, which is maintained for the entire surgical procedure. The limbs are appropriately restrained to permit full extension of the animal. Prior to surgery, while under anaesthesia, bland sterile ophthalmic ointment (Duratears®) is applied to each eye to discourage dryness and/or irritation of the cornea due to the lack of blinking reflex caused by the anaesthesia. A sterile drape is placed over the animal, with a window exposing the groin region to permit the first incision. Surgical instruments and catheters are sterilised by steam autoclaving prior to surgery. Where required, cold sterilisation is performed using 2% or 10% glutaraldehyde solutions followed by rinsing with sterile water. The surgical implantation of the catheter is routinely performed on the right side, but the left side may be used. An incision in the groin region exposes the femoral vein, which is isolated and occluded at two sites, approximately 1 cm apart, using a sterile braid. A small phlebotomy is made in the vein using vein scissors and a medical-grade catheter inserted. The catheter of choice at our laboratory is a medical-grade silicone-based (Silastic®) one, since this is one of the few clinically approved catheter materials that possesses good tensile strength and flexibility together with a smooth yet adhering surface. The standard dimensions of the catheter used for the femoral cannulation in the rabbit are 0.076 mm inner diameter and 0.165 mm outer diameter. The catheters are individually prepared in standard lengths of 1.5 m, with the catheter tip bevelled at an angle of approximately 45° and anchoring bulbs placed at intervals caudal to the tip. The use of other types of catheter (for example Microrenathane® or polyethylene) may occasionally be required in instances of incompatibility of the test material with the catheter (e.g. because of leaching of catheter components, adsorption of the test material to the catheter and/ or potential precipitation of the test material in the catheter). However, in our experience, other types of catheter are generally less acceptable, often resulting in interference with the infusion due to a kink in the catheter line related to lack of flexibility of the tubing and/or increased tissue damage at the infusion site due to the rigidity of the material. It is also important to consider that many of the catheter materials available commercially are not approved clinically.

Entering through the incision in the femoral vein, the catheter is gently inserted into the vena cava to rest approximately at the level of the kidneys. At least two ligatures are made using braided polyester thread (Ticron® 4–0) to secure the catheter in the femoral vein. The patency of the catheter is checked using a syringe containing 0.9% sodium chloride, pH 5.5 (physiological saline), connected to the distal end of the catheter. A loop is made and the catheter secured in place using monofilament thread (Novafil® 4–0). At this point, the animal is placed in a lateral recumbent position. An incision is made in the interscapular region and the catheter directed subcutaneously, with the help of a trocar, to the scapular exteriorisation point in the nape of the neck. The patency of the catheter is again verified and the animal returned to a dorsal recumbent position to permit a gentle irrigation of the femoral site with warmed (approximately 37 °C) physiological saline and Penicillin G Sodium®. The femoral and exteriorisation sites are closed with Novafil® 4–0 suture material and the sutures are removed within 5–10 days following surgery, depending on healing results.

The animal is removed from the surgical table and brought into a post-operative area to permit connection of the catheter to the infusion pump and placement of the animal into a jacket used to hold the tether system, designed for use in our laboratories. The use of post-operative analgesics is not considered necessary for this procedure. The catheter, pre-filled with physiological saline, is fed through the metal tether system and attached to a swivel secured on the outside of the animal cage. A clinical-grade infusion pump is mounted on the outside of each cage and the infusion line attached to the outer portion of a swivel. The swivel–tether system permits free movement of the animal within the cage and the placement of the pump and infusate external to the cage, thus eliminating the potential for the animal to interfere with the pump or infusate. Some laboratories use ambulatory pumps attached to the jacket, which eliminates the need for tether and swivel apparatus. However, in our laboratory the most consistent results are obtained with the external pump system, which prevents the animal interfering with the pump or infusate and avoids the dose volume limitations of the backpack reservoir. The animals are allowed a minimum of 1 week after surgical implantation for recovery prior to insemination, during which time the animals are infused with physiological saline at a rate of 2 ml/h (which is approximately equivalent to the dead volume in the catheter line per hour) to maintain the patency of the catheter. The infusion of physiological saline at this rate is continued post insemination until the initiation of treatment which also provides post-operative fluid replacement that may assist in accelerating the recovery (Peeters *et al.* 1984). The use of anticoagulant(s) to maintain an uninterrupted flow through the catheter, reported by other investigators (Barrow and Guyot 1996; Lynch 1999), has not been required for any of the species routinely used at this laboratory and is considered an unnecessary procedure that can affect interpretation of the data. The infusion volume is monitored on a daily basis by measurement of the weight of the delivery vessel (syringe or bag) prior to and at the end of each daily infusion. This permits calculation of the actual versus theoretical quantities of infusate delivered and allows for the identification of any equipment malfunction or with the condition or placement of the catheter during the post-operative period and during the dosing period. Unlike other species, a pre- and post-surgical regimen of antibiotic injections (benzathine penicillin G + procaine penicillin G administered intramuscularly) is not administered to rabbits because of the susceptibility of this species' natural gut flora to these types of antibiotics. This difference in the procedures has not resulted in any remarkable differences in the incidence (<1%) of infections at the catheterisation/infusion sites. The use of antibiotics in other species is considered prophylactic rather than necessary.

The does are artificially inseminated approximately 1 week after the surgical implantation of the catheter. The technique is identical to that used for non-cannulated animals. All does are luteinised with an intravenous injection of 50 iu of human chorionic gonadotrophin approximately 3 weeks prior to insemination. A second injection of luteinising hormone is administered 2–4 h before insemination. Semen is collected from untreated proven bucks of the same strain and source. The collected semen is pooled and diluted using physiological saline. Each doe is inseminated with a minimum of 0.25 ml of the diluted sample (actual range for the five referenced studies was 0.3– 0.7 ml) to provide a minimum of 12×10^6 spermatozoa/ml. The day of insemination is considered Day 0 of gestation. Following insemination, the animals are randomly assigned to groups using a stratified computer-based randomisation procedure (based on Day 0 of gestation body weights) generally to provide 22 females per group.

On Day 6/7 of gestation, treatment of the vehicle and/or test material is initiated. The duration of the daily infusion may be continuous or intermittent. When the vehicle/ test material is only administered intermittently, the patency of the catheter line is maintained by infusion of physiological saline or vehicle between treatments. Vehicles infused in studies in this laboratory include 0.9% sodium chloride (pH 5.5), 5% dextrose (pH 4), 27 mmol/l sodium citrate (pH 5.45) or sterile citrate-buffered saline (pH 4), infused at rates ranging from 1.7 to 12 ml/kg/h, intermittently or continuously. Note that for two of the five referenced studies, the infusion of the test material was performed intermittently with the infusion of physiological saline at rates of 1–2 ml/h between each treatment, while the remaining three referenced studies were continuously infused with the vehicle or test material (i.e. 24 h/day), throughout the treatment period. In establishing the rate of infusion, various factors have to be taken into consideration: the pH, osmolality and isotonicity of the material, duration of the infusion and duration of the study. In our laboratory, continuous daily infusion with a saline-based solution up to 1–1.5× the total blood volume (considered to be 60– 70 ml/kg for rabbits) can be performed without any appreciable alterations in routine haematological or serum chemistry parameters. This corresponds to rates of 2.5 to 4 ml/kg/h. Higher rates of infusion can be administered but may result in transient or prolonged haemodilution, depending on the duration of the daily dosing. Solutions that are not physiological may result in transient or prolonged changes in haematology and clinical chemistry parameters, and a pilot study is recommended to determine an acceptable infusion rate. For teratology studies, a maximum infusion rate, for continuous 24-h infusion, of 1.5 ml/kg/h for non-isotonic solutions and 3 ml/kg/h for saline-based solutions is usually well tolerated.

Following the completion of dosing, (i.e. gestation Day 19/20), the catheter is cut, tied off and inserted subcutaneously, and the jacket and tether removed.

In-life observations generally include clinical examinations, body weight measurements on gestation Days 0, 6/7, 9, 12, 15, 18, 21, 24 and 29 and food consumption measurements daily during gestation. Blood samples for clinical pathology and/or toxicokinetic evaluations may be collected from specific subgroups of animals before dosing, during dosing and/or at the completion of dosing from the auricular artery (or marginal ear vein, if necessary). Where large numbers of samples are required or the samples are closely timed, a catheter can be placed in the auricular artery (Robinson *et al.* 1996). Typically, blood samples for toxicokinetics are obtained on the first and last days of dosing and additional samples may be collected on specific days of interest during major organogenesis. Fetal blood samples can be collected on gestation Day 19 or 20, while the does are still being infused (Pinsonneault *et al.* 1998). Subsequently, the does in the toxicokinetic subsets are euthanised and, if fetal blood samples have not been collected, given a uterine examination to confirm pregnancy.

A Caesarean section is performed on Day 29 of gestation, the ovaries examined for the numbers of corpora lutea and the uterine contents examined for numbers of live and dead fetuses, resorptions and empty implantation sites. All fetuses are weighed, examined externally and euthanised. Fresh visceral dissections are performed on all fetuses and a proportion of the heads (one-third) are removed for examination by the technique of Wilson (1965). All fetuses are stained with Alizarin Red S for skeletal examination (Dawson 1926). Fetal findings are graded as major malformations, minor anomalies (external/visceral or skeletal) or common skeletal variants (Palmer 1977).

The data generated from intravenous infusion control groups have been compared with data from control animals treated orally, dermally or intravenously (by bolus injection or slow infusion into an ear vein) from the same strain and supplier, and all pregnancies were produced by artificial insemination. Our laboratory control data ranges were derived from over 600 rabbit does used on embryo-fetal development/ segment II (teratology) studies. For maternal and uterine parameters, total numbers or study ranges, as applicable in each subset, have been tabulated. For haematology parameters, the mean and standard deviation (SD) from an infused and a non-infused study are presented. Fetal findings for malformations and anomalies are presented as the percentage of fetuses and litters affected for each subset (infused versus non-infused). Note that for individual malformations/anomalies the control range is shown only for those findings observed among the infusion data set. For common variants the study ranges of the percentage of fetuses affected are given.

9.3 Results

The results of the five intravenous infusion studies included in the data are representative of the findings in our laboratory using this route of administration for teratology studies and are summarised below.

The mortality rate among animals treated by infusion is similar to that obtained in non-infused studies. Among the 105 does inseminated and dosed by infusion, two died during the gestation period, the first on gestation Day 29 with no gross pathological observations and the second on gestation Day 15 in a separate study with findings considered to be secondary to the catheterisation procedure (gross necropsy findings included thickening with pale material at the infusion site and a firm mass at the exteriorisation site). This was the only infused animal of the 105 in the subset showing pathological findings indicative of an inflammatory action and possible infection, supporting our putative infection rate of less than 1%.

The few clinical findings are those commonly seen with intravenous infusion at our laboratories for animals cannulated via the femoral vein. These procedure-related changes are generally related to the presence of the jacket (e.g. redness, thinning fur, scabs or slight lesions in regions in close contact with the jacket). Transient, not biologically significant elevations in total white blood cell counts, occasionally accompanied by an increase in segmented neutrophils, noted on occasion in other species, were not seen among gravid rabbits on gestation Day 20 (Table 9.1).

Changes at the infusion site – seen grossly as firmness/thickening, often correlated with microscopic findings of thrombosis, proliferation of vascular intima and various local inflammatory changes at the infusion site in the vicinity of the catheter tip – are considered normal physiological changes due to the physical irritation of the catheter tip in the vessel. Secondary changes such as pulmonary microemboli or thrombosis and histiocytosis in the hepatic sinusoids, often with erythrophagocytosis, may also be observed (Perkin and Stejskal 1994) in other species. However, these changes were not seen, or seen to a lesser extent, in the rabbit studies probably because the lesions have 9 or 10 days to regress between the end of infusion and necropsy examination.

The body weight progression of the does treated by infusion is similar to those treated by other routes (Figure 9.1). Food consumption is also unaffected (data not shown).

Pregnancy rates among the infused and non-infused does were similar (Table 9.2). The numbers of does aborting appeared to be slightly elevated among the infused control rabbits; however, this is considered to be fortuitous since tabulation of the low doses from these studies yields an abortion rate (2.9%) similar to that of the non-infused rabbits, and for two of the five infused studies the total abortion rates (control and treated groups combined) were 1% and 2% (Robinson *et al.* 1998).

Table 9.1 Mean (SD) haematology data

Parameters	Infusion	Control
Number of pregnant does	17	19
White blood cell count (×10³/mm³)	7.2 (1.81)	7.6 (1.35)
Segmented neutrophils (%)	38.9 (11.06)	37.6 (11.06)
Non-segmented neutrophils(%)	0.0 (0.00)	0.0 (0.00)
Lymphocytes(%)	56.6 (12.75)	54.4 (16.94)
Monocytes(%)	2.1 (1.98)	4.0 (1.86)
Eosinophils(%)	0.7 (1.05)	0.8 (1.00)
Basophils(%)	1.6 (1.94)	3.3 (2.44)
Red blood cell count (×10⁶/mm³)	5.42 (0.38)	5.52 (0.35)
Haemoglobin (g/dl)	12.2 (0.73)	12.0 (0.74)
Haematocrit (%)	35.3 (2.16)	36.0 (2.04)
Mean corpuscular volume (μm³)	65.2 (2.81)	65.3 (1.76)
Mean corpuscular haemoglobin (pg)	22.6 (0.99)	21.8 (0.71)
MCH concentration (g/dl)	34.6 (0.42)	33.3 (0.40)
Red cell size distribution width (%)	14.3 (1.15)	14.3 (1.00)
Platelets (×10³/mm³)	380 (62.9)	390 (94.6)
Mean platelet volume (μm³)	6.6 (0.56)	6.7 (0.61)
Reticulocytes (%)	3.0 (1.53)	4.4 (1.76)

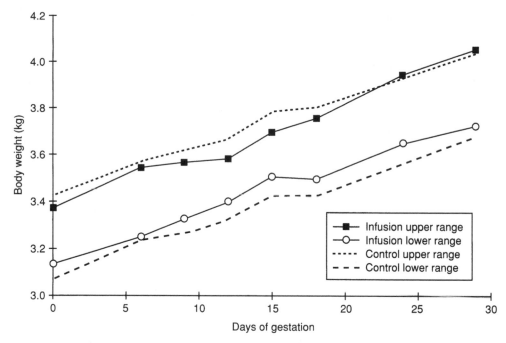

Figure 9.1 Body weights of pregnant females.

Table 9.2 Maternal and uterine findings

Parameter*	Infusion	Control
Total no. of does inseminated	105	385
Total no. of does pregnant	96	349
Pregnancy rate (%)	77.3–100.0	75.0–100.0
Total no. of does with abortion	8	11
Abortion rate (%)	4.8–15.0	0.0–16.7
Total no. of does littering early	0	3
Total no. of does with complete resorption	2	5
Total no. of does dying/euthanised pre-term	2	12
Mortality rate (%)	0.0–5.0	0.0–12.5
Corpora lutea per litter	10.0–12.1	8.5–12.3
Implants/litter	7.3–8.7	5.7–10.6
Pre-implantation loss (%)	14.4–38.5	9.9–33.7
Live fetuses per litter	5.8–7.9	4.9–10.0
Dead fetuses per litter	0.0–0.1	0.0–0.9
Resorptions per litter	0.2–1.4	0.1–1.9
Post-implantation loss (%)	7.1–29.1	1.3–21.7

Notes
*Ranges of study values unless otherwise specified.
Pre-implantation loss (%) = (no. of corpora lutea − no. of implants)/no. of corpora lutea × 100.
Post-implantation loss (%) = (no. of implants − no. of live fetuses)/no. of implants × 100.

The numbers of corpora lutea were similar, and a tendency to slightly higher pre-implantation losses (36% and 39%) noted in initial studies has reduced (21%, 14% and 18%) in more recent experiments (Robinson *et al.* 1995a). The numbers of resorptions (early, middle and late) were similar in infused and non-infused dams. In one study, two litters with total resorption resulted in a slightly elevated post-implantation loss. However, the incidence of total resorption was within the historical control rate and resorption was not seen in any other infused control groups. In consequence, the numbers of live fetuses were unaffected.

Fetal findings showed no increases in the rates of major malformations, minor anomalies or skeletal common variants (Tables 9.3 and 9.4). The types of malformations were similar in the infused and control litters. Only three major

Table 9.3 Summary of fetal findings – fetal weight and external and internal abnormalities

External and internal abnormalities	Infusion		Control	
	% L/E (n = 82)	% F/E (n = 609)	% L/E (n = 312)	2% F/E (n = 2306)
Fetal weight (g) (range)	44.8–48.6		36.9–49.6	
Major malformations (total)	3.66	0.49	7.37	1.08
External				
Craniofacial malformations	1.22	0.16	0.96	0.12
Abdominal herniation/ eventration/gastroschisis	1.22	0.16	1.28	0.17
Limbs – abnormalities	1.22	0.16	0.64	0.13
Internal				
Brain: hydrocephalus	1.22	0.16	1.28	0.22
Minor external and visceral anomalies (total)	23.17	5.42	24.68	4.42
External				
General: subcutaneous haematoma	1.22	0.16	0.32	0.04
Internal				
Intracranial haemorrhage/ oedema/discoloration	1.22	0.16	1.28	0.22
Brain: moderate dilation of lateral/third ventricles	1.22	0.16	0.64	0.09
Eye(s): lens(es) oval	2.44	0.66	2.88	0.48
Heart: minor vessels; variations, carotid/subclavian/innominate arteries	4.88	0.82	3.84	0.87
Gall bladder: absent	2.44	0.33	2.56	0.35
Gall bladder: reduced	7.32	1.81	9.62	1.56
Gall bladder: enlarged	3.66	0.49	1.60	0.22
Liver: variations in lobulation	1.22	0.16	1.28	0.22
Kidney(s): absence/severe/ moderate reduction in renal papillae	1.22	0.16	1.28	0.34

L/E, litters affected; F/E, fetuses affected.

Table 9.4 Summary of fetal findings – skeletal abnormalities

Finding	Infusion		Control	
	% L/E (n = 82)	% F/E (n = 609)	% L/E (n = 312)	% F/E (n = 2306)
Minor skeletal anomalies (total)	46.34	10.02	50.32	12.23
Skull				
Hyoid bone: irregular ossification	3.66	0.66	1.60	0.22
Hyoid bone: reduced ossification	3.66	0.99	13.46	2.52
Hyoid bone: absent	2.44	0.33	0.32	0.04
Hyoid bone: bipartite	1.22	0.16	1.92	0.26
Hyoid bone: semi-bipartite	1.22	0.16	1.28	0.17
Frontal bones: reduced ossification	1.22	0.16	1.60	0.30
Frontal bones: fusion	1.22	0.16	0.32	0.04
Parietal bones: irregular ossification	1.22	0.16	0.64	0.09
Additional suture(s) in frontal/parietal bone(s)	6.10	0.82	3.85	0.65
Vertebral column				
Thoracic centrum: bipartite/semi-bipartite/irregular	18.29	2.96	12.50	2.39
Caudal vertebra(e): reduced number	8.54	1.31	2.88	0.39
Extra pre-sacral vertebrae	2.44	0.33	2.88	0.69
Twenty-five pre-sacral vertebrae	1.22	0.16	0.64	0.09
Ossification centre(s) on first lumbar vertebra	3.66	0.49	4.17	0.56
Ribs				
Absent	1.22	0.16	0.32	0.04
Reduced ossification	1.22	0.16	0.00	0.00
Nodule(s)	1.22	0.16	1.92	0.26
Sternebral column				
Fused	6.1	0.99	8.65	1.34
Extra sternebra(e)/extra ossification centre	2.44	0.16	2.88	0.52
Pelvic girdle				
Reduced ossification of pubic bone(s)	1.22	0.16	2.56	0.82
Limbs				
Reduced no. of bones in pollex(ices)	1.22	0.16	3.21	0.78
Common skeletal variants (%) (range)	*Lower*	*Upper*	*Lower*	*Upper*
Unilateral 13th ribs	9.5	17.0	3.9	24.3
Bilateral 13th ribs	24.3	65.3	23.5	49.4
Total 13th ribs	35.3	74.8	34.0	69.5
Sternebral variants	25.9	45.8	25.1	56.2

L/E, litters affected; F/E, fetuses affected.

malformations were noted in two of the referenced studies from three separate litters. One fetus had hydrocephalus, one multiple craniofacial malformations including micrognathia and the third gastroschisis and limb malformations including ectrodactyly/microdactyly and abnormal limb flexures. The overall rate of fetuses affected was slightly lower among the infused litters than in non-infused animals (0.5 versus 1.1).

In the case of minor anomalies (external/visceral and skeletal), the overall rates of litters and fetuses affected were similar and the incidences of individual findings were also similar. The percentage of fetuses from the infused litters with sternal variants was within the historical control range. A slightly higher percentage of fetuses with bilateral thirteenth ribs (and therefore total thirteenth ribs) was noted in one infused study; however, this difference was probably attributable to biological variation as the findings in the other studies were within the control range.

9.4 Discussion

Success in performing embryo-fetal development (teratology) studies (ICH-3) in rabbits by intravenous infusion via a femorally implanted cannula is measured by the ability to satisfy the six criteria specified in the introduction.

The number of animals per group should be the same as used in a typical oral study

The ICH guidelines (ICH 1993) indicate that a minimum of 15 or 16 litters should be evaluated. In non-infused studies, 20–22 does per group are routinely used. Similarly, for our infusion studies, group sizes are usually set at 22; in all cases this provided a sufficient number of pregnancies to satisfy the guidelines. For both non-infusion and infusion studies, toxicokinetic evaluations are performed on subsets of animals (usually four rabbits). Further, the number of days over which the studies are 'replicated' is 4 or 5 for both non-infusion and infusion-treated study types.

The procedures for catheterisation should be standardised

The authors' laboratory has been performing intravenous infusion studies for almost 20 years, during the first 10 years of which the cannulation procedures evolved to become standardised, to permit within-species and cross-species comparison of results. Certain aspects of the procedures vary slightly between species owing to species-specific anatomical or physiological differences – for example the type of anaesthetic used, the size and length of the catheter, the anchoring procedure, the rates of saline infusion to maintain catheter patency, the use of alcohol to prepare the surgical sites and the use of a prophylactic antibiotic regimen – but, where possible, the techniques have been standardised for all animals.

The same range of investigations should be performed in infusion studies as are performed in non-infusion studies

We have demonstrated that cannulation of the animals for intravenous infusion does not prohibit any of the investigations routinely performed on non-infusion studies, in-life or terminally. For other procedures, results, including the percentage of successful blood samples (maternal and fetal) taken during infusion, are similar to those obtained in non-infused toxicokinetic subsets (Robinson *et al.* 1996; Pinsonneault *et al.* 1998).

There should be consistent daily administration of a known dosage that can be fixed or varied during a study as changes in body weight or other factors of dose calculation are recorded

As each animal is linked to an individual infusion pump, the rate, volume and/or duration of dosing for any given animal can be individually set and monitored. The rate, volume and/or duration of the infusion can be manually changed at any time as required during a study based on body weight changes or any other criterion of dosage adjustment or even within a given daily dose (e.g. administration of a loading dose at a different rate prior to the start of maintenance infusion of a test material). The accuracy of the infusion is measured by recording the weight of the administration vessel (e.g. syringe or bag) before and at the end of the daily infusion. Our expected accuracy of delivery for infusion in any species is generally ± 15% of the theoretical nominal volume on a daily basis and ± 10% of the theoretical nominal volume over the entire treatment period. The accuracy of delivery seen in femorally cannulated rabbits in our laboratories has generally been within ± 10% of the theoretical nominal dose over the period of dosing, with the majority (over 95%) of animals receiving within ± 15% of the daily dosage. As the accuracy of the delivery is monitored and recorded on a daily basis, the impact of any occasional deviation from these ranges can be evaluated and considered in the interpretation of the results for any individual animal.

Infections due to the presence of a catheter should be as close to non-existent as possible

The infection rate recorded in our laboratories is consistently low. For rabbits catheterised femorally, in which no systemic antibiotic regimen is used prior to or shortly after surgery, the level remains less than 1% (actually 0.95%), which is consistent with the incidence occurring in other species infused at this laboratory.

The fetal data should be similar to those of non–infusion studies and any differences should be understood to allow for appropriate interpretation of the studies

As seen from the results presented above, reproductive parameters are similar in infused and non-infused rabbits and the few putative differences are very minor and can be seen only with large, pooled data sets. In particular, the rates of fetal abnormalities are unchanged.

Comparison of our methods with those of other publications reveals differences in several aspects of the procedures, making it difficult for investigators to choose the method most appropriate to their research. To aid this decision, we believe that the following factors should be considered in selecting a method. The choice of anaesthetics commercially available for rabbits and other species is extensive; however, as the surgical procedures involved in cannulation of rabbits are not particularly invasive or prolonged, the use of a shorter-acting anaesthetic(s) is probably more appropriate. In our experience and that of others (Lipman *et al.* 1997), a combination of sedatives for pre-operative preparation and isoflurane–oxygen gas anaesthesia during the course of the surgical procedure provides reliable and efficient anaesthesia for this species. The combined ketamine–xylazine pre-anaesthetic, which can provide sedation for up to 1.5 h (Peeters *et al.* 1984), provides sufficient activity to cover the complete procedure and maintain sedation during the immediate post-operative period.

The site of catheterisation/cannulation provides the most remarkable differences in results. The preferred procedure for cannulation in our laboratories is into the vena cava via the femoral vein as, based on our experience, this results in rapid post-operative recovery. Generally, minimal reactions at the infusion site have been noted in comparison with results obtained in our laboratories from cannulation via the jugular vein. Body weight, food consumption and clinical pathology changes reported by other investigators using the jugular route (Barrow and Guyot 1996) have not been noted in our studies using the femoral route, and increases in the incidence of malformations among fetuses of infused dams implanted with a jugular cannula (Barrow and Guyot 1996) have not been observed in our femorally cannulated rabbits. The duration of post-operative recovery has also been shown to be longer following jugular cannulation, up to 10 days (Lynch 1999), in comparison with our standard period of up to 7 days for femorally catheterised rabbits. Other investigators have published methods for continuous intravenous infusion via the marginal ear vein of the rabbit (Chang *et al.* 1995) but have reported only on infusion of short durations (up to 72 h), which does not represent an effective alternative for embryo-fetal development studies. Also, note that the use of a peripheral vessel has greater potential for local pharmacological effects or irritation because of the relatively smaller blood volume available for dilution of the infused material at the site of administration (Barrow and Guyot 1996).

The use of subcutaneous ports (vascular access ports to which a catheter and tether system is attached) has also been reported by various investigators (McKeon *et al.* 1998; Lynch 1999). Maintenance of infusion was shown by these investigators but our experience with this type of system is that the presence of this much larger foreign body (i.e. vascular access port) in the subcutis generally results in a moderate to severe inflammatory response at the site. In addition, as these set-ups generally require a daily puncture through the skin to insert the dosing needle into the port, there is a constant risk of localised infection, which can exacerbate the inflammatory reaction. Further, the use of non-physiological vehicles frequently results in subcutaneous inflammatory changes owing to unavoidable leakage of the vehicle/test material on withdrawal of the dosing needle from the access port.

The type of histopathological changes observed in both femoral catheterised and jugular catheterised rabbits documented in various publications appears to be similar in nature. However, local infection and irritation at the infusion site, causing thrombophlebitis and occasionally progressing to cutaneous necrosis, as reported by Barrow and Guyot (1996) following jugular cannulation, have not been routinely observed in our femorally catheterised rabbits. The causes of these apparently more severe tissue changes may be related to the site and procedures of implantation or the type of catheter implanted, the material infused or the rate of maintenance infusion.

With respect to embryo-fetal development studies, a difference in the timing of the implantation has also been noted. Some investigators have used time-mated animals and performed cannulation during the gestation period (ranging from Days 0 to 3 of gestation – Barrow and Guyot,1996; McKeon *et al.* 1998) while others (Lynch 1999), including ourselves, have performed the cannulation 1 week to 10 days before mating or insemination. Some of the investigators cannulating during the gestation period have reported an initial adverse impact on body weight and food consumption (Barrow and Guyot 1996). Others (Johnson 1971) have reported an increase in the number of resorptions when surgery was performed as late as 9–10 days post coitum. Whether the cause of these body weight and food consumption reductions is related to the timing of the cannulation or the implantation procedures is uncertain.

9.5 Conclusion

In conclusion, intravenous infusion using an indwelling catheter implanted into the vena cava via the femoral vein under the conditions used in our laboratories is an acceptable route for the conduct of embryo-fetal developmental (teratology) studies in rabbits. These procedures do not adversely affect embryo-fetal development, measurements similar to those taken in non-infused studies can be performed and the studies meet the requirements of the ICH guidelines (ICH 1993). Although cannulation through the femoral vein using the procedures described above is considered to be the most appropriate technique, the conduct of reproductive toxicology studies by intravenous infusion is a comparatively recent innovation and, as further studies are performed both in our laboratories and elsewhere, a greater understanding of the optimum techniques will probably be achieved.

References

Armer, L., Hillebrand-Krallis, P. and Perkin, C. (1994) The effects of continuous intravenous infusion of saline on body weight gain, food intake and clinical pathology in the Beagle dog. *The Toxicologist*, 14, 231.

Badalone, V., Pouliot, L., Lafleur, M. and Perkin C. (1995) An investigative study to assess the suitability of the micro-pig for subchronic intravenous infusion toxicity studies. *The Toxicologist*,15, 110.

Barrow, P.C. and Guyot, J.Y. (1996) Continuous deep intravenous infusion in rabbit embryotoxicity studies. *Studies in Experimental Toxicology*,15, 214–218.

Chang, Y., Fiordalisi, I. and Harris, G.D. (1995) A unique swivel-tether system for continuous I.V. infusions in freely mobile unrestrained rabbits. *Contemporary Topics*, 34, 61–64.

Clarke, D.O. (1993) Pharmacokinetic studies in developmental toxicology: practical considerations and approaches. *Toxicology Methods*, 3, 223–251.

D'Aguanno, W. (1976) *Guidelines for Reproduction Studies for Safety Evaluation of Drugs for Human Use* (based on FDA for Reproduction Studies for Safety Evaluation of Drugs for Human Use, 1966).

Dawson, A.B. (1926) A note on staining of the skeleton of cleared specimens with alizarin red S. *Stain Technology*, 1, 123.

EC (1989) The rules governing medicinal products for human use in the European Community. *Pharmaco-Toxicological Guidelines*, Vol. III, pp. 89–114.

Enayati, S., Pinchuk, L., Marmen, S. and Perkin, C. (1994) A study to determine the feasibility of using mice in intravenous infusion toxicity studies using an indwelling catheter. *The Toxicologist*, 14, 230.

Fennell, S.W., Armer, L. and Perkin, C.J. (1995) The effects of continuous intravenous infusion of saline on body weight gain and clinical pathology in the cynomolgus monkey. *The International Toxicologist*, 7, 85-P-9.

Green, C.J. (1979) *Animal Anesthesia*. Lab Animal Ltd: Colchester.

Guerreiro, D. and Page, C.P. (1987) The effect of neurolepanalgesia on some cardiorespiratory variables in the rabbit. *Laboratory Animal*, 21, 205–209.

Hillebrand-Krallis, P., Armer, L. and Perkin, C. (1994) The effects of continuous intravenous infusion of saline on body weight gain, food intake and clinical pathology in the albino rat. *The Toxicologist*, 14, 229.

International Conference on Harmonisation (ICH) (June 24, 1993) Detection of toxicity to reproduction for medicinal products. In D'Arcy, P.F. and Harron, D.W.G. (eds.) *Proceedings of the Fourth International Conference on Harmonisation, Brussels, 1997*, pp. 1012–1038.

Japanese MOHW (1990) *Guidelines for Toxicology Studies in Japan*. Yakushin 1.24.

Johnson, W.E. (1971) Fetal loss from anesthesia and surgical trauma in the rabbit. *Toxicology and Applied Pharmacology*, 18, 773–779.

Lipman, N.S., Marini, R.P. and Flecknell, P.A. (1997) Anesthesia and analgesia in rabbits. In Kohn, D.F., Wixson, S.K., White, W.J. and Benson, G.J. (eds.) *Anesthesia and Analgesia in Laboratory Animals*, American College of Laboratory Animal Medicine Series. Academic Press: New York, pp. 205–232.

Lynch, S.J. (1999) Rabbit embryo-foetal development studies via continuous intravenous infusion. *Animal Technology*, 50, 51–55.

McKeon, M.E., Walker, M.D., Wakefield, A.E. and Machotka, S.V. (1998) Validation of infusion techniques in nonpregnant and pregnant rabbits using a novel harness system. *Toxicology Sciences*, 42, 292.

Palmer, A.K. (1977) Incidence of sporadic malformations, anomalies and variations in random bred laboratory animals. In Neubert, D., Merker, H.-J. and Kwasigroh, T.E. (eds.) *Methods in Prenatal Toxicology*. Georg Thieme: Stuttgart, pp. 52–71.

Peeters, L.L.H., Mårtensson, L., Gilbert, M. and Penicaud, L. (1984) The pregnant guinea pig, rabbit and rat as unstressed catheterized models. In Nathanielsz, P.W. (ed.) *Animal Models in Fetal Medicine*. Perinatology Press: Ithaca, NY, pp. 75–108.

Perkin, C.J. and Stejskal, R. (1994) Intravenous infusion in dogs and primates. *Journal of the American College of Toxicology*, 13, 40–47.

Pinsonneault, L., Robinson, K., Sey, S. and Kam, M. (1998) A novel sampling for the collection of rat and rabbit fetal blood. *Teratology*, 57, 254.

Robinson, K., Perkin C., Pinsonneault L. and Washer G. (1994). The effect of continuous intravenous infusion on pregnant rats in comparison with historical control data. *The Toxicologist*, 14, 229.

Robinson, K., Pinsonneault, L., Hillebrand-Krallis, P., Washer, G., Mitchell, M.J. and Lafleur, M. (1995a) Continuous intravenous infusion of rabbits for embryo-fetal development studies to ICH guidelines. *Teratology*, 51, 194.

Robinson, K., Pinsonneault, L., Washer, G., Pouliot, L., Mitchell, M.J., Landry, R., Coriveau, L. and Martel, D. (1995b) Continuous intravenous infusion of rats for pre and postnatal studies to ICH guidelines. *Teratology*, 51, 194.

Robinson, K., Pinsonneault, L., Pouliot, L., Hillebrand-Krallis, P., Washer, G., Mitchell, M.J., and Lafleur, M. (1996) Sampling for toxicokinetic evaluations in continuous intravenous infusion embryo-fetal development studies of rats and rabbits. *The Toxicologist*, 30, 190.

Robinson, K., Pouliot, L., Pinsonneault, L. and Washer, G. (1998) Continuous intravenous infusion (CIVI) of rabbits for embryo-fetal development studies – maternal, uterine and fetal data. *Toxicology Sciences*, 42, 259.

Wilson, J.G. (1965) Methods for administering agents and detecting malformations in experimental animals. In Wilson, J.G. and Warkany, J. (eds.) *Teratology, Principles and Techniques*. The University of Chicago Press: Chicago, pp. 262–277.

10 Jugular catheterisation of the rabbit for reproductive toxicity studies

E. Gordon and S. Lynch

10.1 Introduction

In 1993, the International Conference on Harmonisation (ICH 1993) recommended guidelines for reproductive toxicology studies to the regulatory authorities of the European Union, Japan and the USA. These indicated that the form of the substance and route of administration intended for clinical use should also be used, where possible, for reproductive safety assessment studies in animals. This is particularly important for substances with a short half-life, as the parent substance may only exist in the body for seconds. If such substances were administered orally or by intravenous bolus, the duration of exposure would not be adequate during all stages of development in animals. The results obtained therefore may not be predictive for administration to humans.

In the past, the techniques required for continuous intravenous administration of substances to animals did not exist, but this is no longer true. Over the last two decades these techniques have been developed in several species so that formulations can be administered during all stages of reproduction. This includes studies in rats and rabbits, which are the two species most commonly used for reproductive toxicology studies. The use of rats and minipigs, and rabbits using infusion via the femoral vein, for these studies are described elsewhere in this book (Chapters 5, 9 and 19).

Continuous intravenous infusion offers several advantages over more conventional routes of administration. One major advantage of this route is that exposure of the dam to a substance that is rapidly metabolised or eliminated is constant during the dosing period. It is particularly important in reproductive toxicology studies that exposure is continuous as each stage of development of the offspring may only be a few hours in duration. Without continuous exposure, there could potentially be long periods of time where exposure of the dam and offspring is insufficient to produce, and therefore for the investigator to detect, an adverse effect of the substance. This route of administration provides increased exposure in terms of duration with a constant lower plasma concentration of substance. To obtain the same duration of exposure by other routes, it would be necessary to administer much higher doses. Consequently, this may not be tolerated by the pregnant animal owing to high peak plasma concentrations and possible excessive maternal toxicity.

Unfortunately, this method of administration does have disadvantages, such as potential stress to the dam. However, with experience and good animal care, stress can be reduced to acceptable levels and dosing several times a day by other routes, which could be just as stressful, can be avoided. Another disadvantage is the cost in terms of equipment required and labour involved both during surgical preparation and in the animal room.

This chapter describes the procedures developed in house (Lynch 1999) over the past 4 years for both surgical preparation and in the animal room during rabbit embryo-fetal development studies, and includes results obtained from recent studies. In this type of study the dam is treated for 13 days, from the time of uterine implantation on Day 6 post coitum until the morning of Day 19 post coitum, when closure of the hard palate is complete. This is the period of major organogenesis in the rabbit. Day 0 post coitum is the day of mating.

One important factor is that the implantation and litter data obtained from infused animals should mirror the data obtained from animals treated by more conventional routes. A number of improvements are described, all of which have been implemented over time, to improve this model in the authors' laboratory both during surgery and in the animal room. It is essential that aseptic techniques are used for all procedures, both in the surgery and in the animal room, in order to reduce the chance of infection and therefore stress to the animals.

10.2 Animals and housing

In the authors' laboratory, Dutch rabbits from a colony maintained in house are used for reproductive toxicology studies, and therefore background data for this strain are available, which enables comparison between infused and non-infused animals. The rabbits used during investigational and regulatory studies have been between 15 and 38 weeks of age and weighed between 1.7 and 3.0 kg at the time of surgery and the start of dosing.

The rabbits are singly housed in aluminium cages (model RC10DB; W50 × D60 × H45 cm from Aramite, UK) with perforated flooring, which are suspended above trays lined with absorbent tray papers. The tray papers are changed daily during the dosing period to allow changes in urine and faecal production to be monitored. The only modification required to the standard cages used for non-infusion studies is removal of one section of the bar at the top of the door to allow easy attachment of the swivel. The tether, which protects the infusion catheter, is attached to the swivel on the cage door at one end and the infusion jacket on the rabbit at the other end (Figure 10.1). This allows relatively free movement of the rabbit within the cage. Their behaviour is generally unaffected, except that they are unable to groom freely owing to the presence of the infusion jackets and Elizabethan collars. While the animal is in the cage the infusion pump is attached to the outside of the cage (Figure 10.1) so that the animal cannot interfere with it.

The animals have free access to pelleted diet (Teklad TRB, Harlan UK Ltd, Bicester, UK) and tap water. During studies, the water is provided in water bottles so that the amount consumed can be monitored. Each day the animals are also provided with autoclaved hay. As part of an environmental enrichment programme, the animals are now also provided with wooden chew blocks (B & K Universal). Alternative plastic caging (Interlink Rabbit Caging; W72.4 × D65.8 × H50 cm from Tecniplast, UK) is being considered as part of this programme and is currently being used for the stock and breeding animals in the colony. These are provided with shelving and interconnecting doors. However, in order to implement improved caging on infusion studies, alterations to these cages and the infusion equipment would need to be

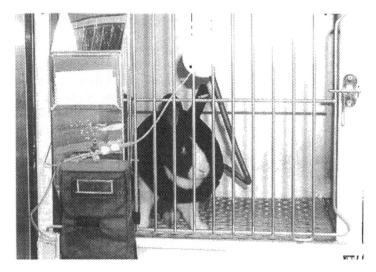

Figure 10.1 Set-up in animal cage with the infusion pump hanging from the food hopper and the swivel attached to the cage door.

considered. This includes increasing the tether length used so that the rabbit can take advantage of the increased space and reach the shelving. Also, the connecting doors would have to be blocked off so that the rabbit would not get trapped in the adjacent cage and the cages modified for attachment of the swivel and pump.

The animals are maintained under artificial light between 06.00 and 18.00 h and in darkness between 18.00 and 06.00 h (GMT). The animal room temperature is maintained between 16 and 20 °C.

10.3 Anaesthesia and analgesia

The current anaesthesia and analgesia regimen used is shown in Table 10.1.

Domitor (SmithKline Beecham) and Vetalar (Pharmacia and Upjohn) are administered together. When previously administered separately, Vetalar caused skin necrosis at the injection site in the flank. This regimen gives sufficient duration of anaesthesia for surgery to be performed. Gaseous anaesthetic (isoflurane) for maintenance of anaesthesia is rarely required. The anaesthesia reversal agent, Antisedan, is administered in the recovery room immediately after surgery.

The analgesic, Vetergesic (Reckitt and Coleman Products Ltd, UK), is given at the same time as anaesthesia induction, with a further dose of analgesia currently being given between 8 and 12 h post surgery. It is considered necessary to administer analgesia to help alleviate post-operative pain and allow the animal to resume normal activity as soon as possible following surgery (Flecknell and Liles 1990; Flecknell 1996). In the future, it is proposed that the analgesia is changed to a single pre-surgery dose of a non-steroidal anti-inflammatory drug such as carprofen (Zenocarp, C-Vet Veterinary Products). Because of the longer duration of action, only one injection will be required. This has been successfully used in other species (Liles and Flecknell 1994) although no data are currently available in the rabbit.

Table 10.1 Anaesthesia and analgesia used

Name	Manufacturer	Dose
Domitor (medetomide hydrochloride 1mg/ml)	SmithKline Beecham	0.5 ml/kg subcutaneously
Vetalar (ketamine hydrochloride 100 mg/ml)	Pharmacia & Upjohn	0.4 ml/kg subcutaneously
Antisedan (atipamezole hydrochloride 5 mg/ml)	SmithKline Beecham	0.2 ml/kg subcutaneously
Vetergesic (buprenorphine hydrochloride 0.3 mg/ml)	Reckitt and Coleman Products Ltd	0.05 ml/kg subcutaneously

Antibiotics are not administered to the surgically prepared rabbits in the author's laboratory. The use of aseptic techniques both during surgical preparation and in the animal room, and the lack of infection observed, indicate that it is not necessary.

10.4 Surgical preparation

The surgical preparation involves insertion of a silicone catheter (Intisil® radio-opaque silicone catheter with a rounded tip; 5 French, 16 gauge, 0.7 mm ID, 1.7 mm OD, 60 cm length; Access Technologies through UNO, The Netherlands) into the jugular vein. The other end of the catheter is attached to a subcutaneously implanted vascular access port (VAP® SLA-AC, polysulphone plastic with silicone rubber septum; Access Technologies through UNO, The Netherlands). A similar method has been used in other laboratories (Perry-Clark and Meunier 1991).

There are five people involved in the surgery: a surgeon, a surgical assistant, two theatre technicians and a recovery technician. The surgery is performed in dedicated facilities, which are thoroughly cleaned prior to each day of surgery. Both the surgeon and the assistant work under sterile conditions throughout the surgical procedures, i.e. scrubbed up, sterile surgical gown and gloves, face mask and hair cover. All instruments are sterile before use for each animal and are used under aseptic conditions during surgical preparation. All of the infusion equipment is supplied pre-sterilised or is sterilised in-house using ethylene oxide prior to surgery. The theatre technician is responsible for administering anaesthesia and analgesia in the preparation area. Once the animal is anaesthetised, the location on the dorsal surface for the VAP and the area above the right jugular vein are shaved. The rabbit is placed on its left side on a draped heated operating table by the theatre technician. The shaved areas are swabbed with diluted antiseptic/detergent solution, e.g. Hibiscrub (chlorhexidine gluconate and detergent solution, ICI Pharmaceuticals, UK), and then 70% ethanol. The surgeon covers the animal with sterile adhesive transparent drape (Dermafilm, Vygon, UK) and autoclaved cotton drape.

Two incisions are made, one above the right jugular vein for the insertion of the catheter and the other on the dorsal surface distal to the scapular region to insert the VAP. Using blunt dissection through the first incision, the jugular vein is cleared (approximately 15 mm length) from the surrounding connective tissue and elevated using curved forceps. Four lengths of non-dissolvable suture (Ethibond, Ethicon) are placed under the cleared vein. The vein is distended and the distal tie is secured. The catheter is filled with sterile saline with a syringe and 21-gauge needle attached at the distal end. Three retention beads are placed approximately 1–2 mm apart at the 4-cm mark on the catheter, so that a tie may be placed between them to secure the catheter in place. The vein is cut using microdissection scissors and the catheter inserted towards the heart for approximately 4 cm. To help insertion of the catheter, the surgical assistant slowly infuses sterile saline from the attached syringe. Blood is aspirated to ensure patency of the system before saline (1–2 ml) is flushed through the catheter to clear the system of blood. The proximal tie is secured. Further ties are secured around the catheter, one between the retention beads and at least one on either side of them. During this procedure, catheter patency is continuously monitored by slowly flushing sterile saline.

The subcutaneous pocket is made for the VAP. A lateral incision (20–30 mm) is made at the chosen site. Blunt dissection is performed to make a subcutaneous pocket large enough to easily accommodate the VAP to one side of the incision. The catheter is clamped close to the vessel, the needle and syringe removed and, using a trocar, the free end of the catheter is tunnelled subcutaneously to the pocket made for the VAP. Air is flushed from the VAP using a 22-gauge Huber needle and syringe of sterile saline. The free end of the catheter is cut to the appropriate length to allow for movement of the animal. The VAP is attached to the catheter and the patency checked. Heparinised saline (0.4 ml of 50 IU/ml) is then infused to fill the VAP and the dead space of the catheter to help maintain patency. The Huber needle is removed from the VAP, while maintaining positive pressure through the syringe to prevent aspiration of blood into the catheter. The base of the VAP is secured to the underlying muscle in the subcutaneous pocket with two Ethibond sutures. Both incision sites are closed using a subcutaneous suture of Vicryl and skin sutures of Ethibond.

The recovery technician transfers the rabbit to a cage in the recovery room, where the anaesthetic reversal agent is administered and the incision area is cleaned with sterile saline. The animal is closely monitored until it regains its righting reflex. The eyes are covered with gauze swabs dampened with saline to prevent drying and the animal is turned over on to alternate sides approximately every 45 min to prevent fluid build-up on the lungs. Once the righting reflex is regained, which is usually within 1–2 h, the animal is returned to its home cage in the study room. An Elizabethan collar (made in house from Plastazote, external diameter approximately 18 cm) is put on to help prevent the animal interfering with the sutures and it is placed on and under a Vetbed to help maintain its body temperature. The covering Vetbed is removed that evening at the time of the second dose of analgesia and the bottom one removed the following morning. The sutures are normally removed after 10 days. If the animal has removed the skin sutures, these may be replaced by skin staples or sutures as considered appropriate, after a local anaesthetic (EMLA, AstraZeneca) has been applied.

Several improvements have been made to the surgical procedure. The use of retention beads on the catheter has reduced the incidence of the accidental removal from the vein. Additionally, while gaining experience with the technique, the position of the catheter was confirmed by radiography to show the catheter tip just above the heart. This is no longer performed routinely.

In the first investigation studies, the VAP was positioned in the intrascapular area. However, in this area there is insufficient muscle to secure the VAP and the sutures did not remain in position, allowing the VAPs to come free and invert in the pocket. In addition, the infusion jacket was found to rub and cause skin abrasion in this area. The VAP is now placed dorsally to the scapular region.

10.5 Comparison of catheters and access ports

An initial investigation study was performed to compare the use of an exteriorised 5-French silicone catheter with Luer lock attachment, and VAPs made of either titanium (VAP® model TI200-AC) or polysulphone plastic (VAP® model SLA-AC) attached to an Intisil® catheter. All of these were positioned in the intrascapular region. In this initial investigation study, retention beads were not used. All catheters and VAPs were obtained from Access Technologies through UNO.

One main concern was that an exteriorised catheter would be pulled out by the male during mating. Therefore, the exteriorised catheter was attached to a female Luer lock attachment approximately 5 cm after exteriorisation so that the system could be fully disconnected during pairing procedures. However, this system was found to be unsuitable for use as three out of five dams pulled the catheter out of the vessel, in some cases even before pairing had taken place. This was probably caused by the lack of retention beads on the catheter. A similar exteriorised system has been used in the past in the author's laboratory and is used successfully in other laboratories where the rabbits are surgically prepared after mating has occurred (Barrow and Guyot 1996).

The use of a titanium VAP caused skin reddening and necrosis above the VAP as it was too big and heavy for the subcutaneous pocket. Three of five rabbits showed signs of infection around the VAP as a result.

The use of the VAPs made from polysulphone plastic was found to be most suitable as there was very little reaction around the VAP area. With increased experience and the use of improved procedures, even the slight reaction seen during early development work very rarely occurs. This system has been used in all subsequent rabbit infusion studies.

The advantages of using the VAP are that it allows recovery from surgery without restriction, the animal is free of the tether during mating, and infusion need not commence until Day 6 post coitum. In addition, if the external catheter becomes damaged or blocked, it can be replaced without losing the animal from the study.

10.6 External infusion equipment

The VAP is accessed using Huber needles, which have a deflected tip to prevent boring holes in the silicone septum of the VAP when the needle is inserted. These are used during surgery and during the recovery period when flushing the infusion system.

The system used during infusion is shown in Figures 10.1 and 10.2. The needle of a right-angled infusion set (22-gauge, 90° needle, 1 m length of Tygon tubing, Access Technologies through UNO) is inserted into the VAP septum, and the distal end connected to a swivel (19-gauge, high-impact plastic single-channel swivel, Lomir Biomedical Inc. through UNO) attached through the cage door. The needle of the right-angled infusion set remains in the VAP septum throughout the dosing period and is only removed if there is found to be a problem with the infusion system, such as catheter blockage or damage, or it is accidentally removed by the animal.

The exterior catheter is protected in the cage by a rabbit infusion jacket and a flexible metal tether (60 cm length; Lomir Biomedical Inc. through UNO). Tubing (19 or 21 gauge, Vygon) then connects the swivel to the remote external infusion pump, via a 3-position Luer stopcock. The stopcock has been included in the system so that the infusion pump can be disconnected, e.g. during body weight recording, while maintaining a closed, sterile system. The infusion pumps (CADD-Plus) are used with sterile medication cassette reservoirs (Sims Deltec Inc., USA). The pump is placed in a pocket hung from the front of the cage. Depending on the infusion rate required to give the required dose and the stability of the formulation, cassette reservoirs (50 or 100 ml) or larger-volume infusion bags (250 or 500 ml) may be connected to the pump.

10.7 Recovery period and pairing procedures

When rabbits are to be mated, they are allowed a post-surgery recovery period of a minimum of 10 days before the start of pairing. When there is no intention to mate the rabbits, they are allowed a recovery period of 7 days prior to the start of infusion. The VAPs are flushed twice weekly during the recovery period and up to Day 6 post coitum, where relevant. For mated rabbits, infusion is started on Day 6 post coitum. Pregnancy rate is similar to that of non-infused animals (Table 10.2). The current procedures were established after several development studies, and by continued assessment and improvements.

After an initial study to establish surgical procedures and equipment to be used, an investigational study was performed to establish the ideal procedure to follow for pregnant rabbits. The animals generally appeared to tolerate the surgery and all study procedures well, with only minimal clinical signs observed in some animals. Administration of saline was started immediately post surgery (0.4 ml/kg/h) for a minimum of 7 days before the mating procedure and started again each day immediately after pairing. With this regimen, the animals were generally unreceptive to mating even after 7 days' recovery and temporary removal of all external infusion equipment. Although the majority of females mated, the pregnancy rate was unacceptably low (53%).

Subsequently, an alternative recovery regimen was investigated (Lynch 1999). The animals were allowed a recovery period of 3–4 days without infusion of saline and

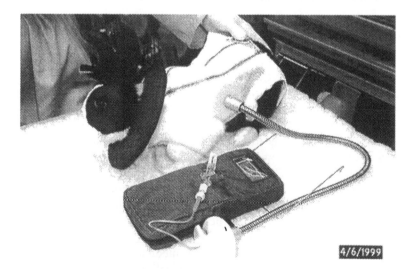

4/6/1999

Figure 10.2 Exterior infusion equipment – rabbit wearing Elizabethan collar and infusion jacket, attached to tether, swivel and tubing with stopcock to infusion pump.

Table 10.2 Implantation parameters

Parameter	Infusion studies			Other	
No. mated	26	22	7	20	20
No. pregnant (%)	20 (78)	20 (91)	7 (100)	17 (85)	18 (90)
No. not pregnant	0	2	0	3	2
No. aborted	1	1	0	0	0
No. with total intrauterine loss	1	0	1	0	0
No. of viable litters Day 29 post coitum	18	19	6	16	18
Mean corpora lutea per rabbit	8.79	8.74	7.29	7.33	7.72
Mean implant sites per rabbit	7.21	6.84	5.43	6.31	6.83
Pre-implantation loss (% per litter)	18.4	20.1	27.6	14.3	12.0
Early intrauterine deaths (% per litter)	11.4	8.4	16.3	4.3	3.1
Late intrauterine deaths (% per litter)	2.2	4.1	0	1.8	0.62
Dead fetuses (% per litter)	0	1.3	0	0	0
Post-implantation loss (% per litter)	13.6	16.4	16.3	6.1	3.7
Mean no. of viable fetuses per litter	6.8	5.8	5.1	5.9	6.6

without any external infusion equipment, apart from Elizabethan collars, until the incisions made during surgery had healed. The infusion jackets were put on the animals for 4 days for acclimatisation, then removed again 2 days prior to mating by natural methods, i.e. pairing did not begin until a total of 10 days after surgery. The longer recovery period allows the animals to recover from surgery completely before acclimatisation starts and then pairing. Instead of continuous infusion with saline during the recovery and pairing periods, the VAPs were flushed twice weekly with sterile saline (1–2 ml) and the dead space filled with heparinised saline (0.4 ml of 50 IU/ml). This is sufficient to keep the infusion system patent; animals have been maintained in this way without infusion for up to 60 days. In an initial study, the VAPs were flushed daily, but this was found to result in the skin around the VAP becoming dry and red, probably because Mediwipes were used to clean the area and because of the greater number of skin punctures. Diluted Hibiscrub is now used to wipe the area prior to insertion of the needle into the VAP. Infusion was started on Day 4 post coitum to ensure that the pumps were delivering the required volume, and to acclimatise the animals to the tether. However, this has since been found to be unnecessary.

10.8 Infusion rates and vehicles

Infusion rates of between 0.4 and 1.0 ml/kg/h have been used at various times in the author's laboratory, resulting in daily administration of between 24 and 60 ml per day for a 2.5-kg rabbit. This range of rates was well tolerated by the animals, with no adverse macroscopic effects being observed. For many infusion studies performed, the limiting factor is the concentration of the formulation that can be prepared. The rate selected for any particular study should be as low as possible so that the volume administered is minimised, but not too low so that the catheter patency is compromised. Using this system, dosing accuracy is within 10% of the nominal dose.

The most commonly used vehicle used in the author's laboratory is 0.9% sodium chloride for injection BP (saline). However, studies have been performed with 20% Intralipid® (Pharmacia and Upjohn) at infusion rates of 0.46 and 0.92 ml/kg/h. The use of Intralipid® at this concentration for continuous intravenous infusion in the Dutch rabbit resulted in catheter blockage and accumulation of lipid in the lung tissue of the rabbits. This led to respiratory problems and death. Intralipid® is therefore not recommended as a vehicle for continuous infusion studies in rabbits.

10.9 Blood sampling procedures

The usual route for obtaining blood samples for clinical pathology and toxicokinetic analysis from rabbits is via the marginal ear vein. However, in the author's laboratory, it was found to be difficult to obtain unhaemolysed samples of sufficient volume consistently from Dutch rabbits while they are being infused. The method was therefore changed and samples are now successfully obtained via the ear artery. When repeated samples are to be taken, the artery is temporarily catheterised to reduce the number of needle punctures. Dosing is maintained during the blood sampling procedures.

10.10 Clinical observations

10.10.1 *Clinical signs*

Animals are examined at least once daily for abnormal signs and to check the integrity of the infusion system.

Any local tissue reactions due to the surgical procedure resolve, and surgical incisions heal, within 2–3 days following surgery. This has been improved by not putting the jackets on the rabbits until 4 days after surgery. During the recovery period, when they are not being infused, the animals generally appear healthy with no abnormal signs.

Minor skin abrasions and hair loss can be observed, caused by chaffing of the jacket or collar. By making appropriate adjustments to this equipment, such reactions usually heal without treatment, although in some cases topical treatment with Dermisol cream (SmithKline Beecham, Animal Health Ltd) can help. Infusion rates of between 0.4 and 1.0 ml/kg/h do not result in any adverse reactions.

Occasionally, the catheter may be pulled out of the vessel, resulting in swelling of the forelimb, neck and thoracic area due to accumulation of fluid. The use of catheters with retention beads reduces the incidence of this. In the most recent study performed, this problem was seen in 4 of 88 animals, only one of which was dosed with saline, and therefore was not associated with formulation problems. The other three may have been related to complications with the formulation.

10.10.2 *Body weight gain*

Body weights are recorded while the animal is still wearing all external infusion equipment up to the stopcock, including the tether and swivel. A sterile cap is attached at the three-way stopcock to maintain a closed system. The actual weight of the animal is obtained by subtracting the weight of the external infusion equipment (nominally 300 g).

Body weight change for several studies is shown in Figure 10.3. For studies A to D, saline was administered by continuous intravenous infusion. For studies E and F, water was administered subcutaneously and orally, respectively. In the most recent

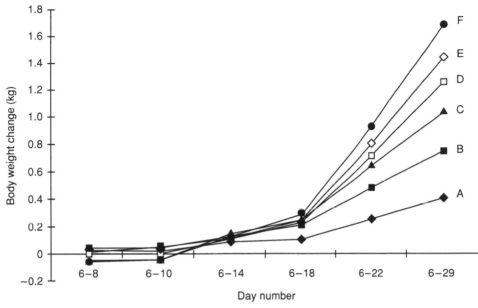

Figure 10.3 Body weight change from Day 6 post coitum.

studies (studies C and D), body weight change during gestation in the rabbit was similar to that obtained following treatment by other conventional routes.

10.10.3 Necropsy findings

In recent studies, animals given saline have exhibited no abnormal macroscopic necropsy observations that were related to the method of administration.

10.11 Reproductive parameters

10.11.1 Implantation and litter parameters

Table 10.2 shows implantation data from a representative sample of studies: three infusion studies, an oral study and a subcutaneous study. The pregnancy rate is similar, although there is an increased tendency for abortion and litter loss in infused animals. There is also a tendency for a slight decrease in the number of implants, indicated by a tendency towards increased pre- and post-implantation loss. This is likely to be caused by the additional stress of the technique. To compensate for this, 22 animals per group are used in an infusion study, compared with 20 per group in studies using drug administration by other routes.

Table 10.3 shows the litter and fetal parameters obtained from the same five studies. The results for all parameters are similar irrespective of the route of administration.

10.11.2 Fetal observations

Generally, there is no difference in the observations seen between infusion and other conventional routes, except for a possible slight reduction in ossification of the skull bones and sternebrae in some studies (data not shown).

Table 10.3 Litter parameters

Parameter	Infusion studies			Other	
Total no. of viable litters	18	19	6	16	18
No. of male fetuses	55	61	16	44	62
No. of female fetuses	68	50	20	50	56
Percentage of males per group	44.7	60.3	44.4	46.8	52.5
Mean no. of fetuses per litter	6.8	5.8	5.1	5.9	6.6
Mean litter weight (g)	265	217	246	239	248
Mean fetal weight combined (g)	39.5	38.0	41.3	40.5	38.4
Mean male weight (g)	39.9	38.7	40.5	41.2	38.7
Mean female weight (g)	39.5	37.1	41.6	40.1	37.4

10.12 Discussion and conclusion

The methodology described and the results presented show that pregnant rabbits may be successfully treated by continuous intravenous infusion via the jugular vein using vascular access ports, with little or no difference observed in the parameters recorded in embryo-fetal development studies. The use of access ports allows the animals to recover completely after surgery without any restriction on normal behaviour due to tethering or the need for immediate infusion of saline to maintain catheter patency. This results in a better mating performance using natural mating procedures in the authors' laboratory. In addition, use of a VAP allows external infusion equipment to be replaced if it is damaged or blocked, without loss of the animal from the study. Where an exteriorised catheter system is used, this is not generally possible. However, this is a relatively new model and continual assessment of methodology will be maintained so as to improve the techniques used and reduce the stress to the animals.

Performing surgery prior to mating, as described here, gives the opportunity for more flexibility in study designs as dosing can start prior to mating for example, but this can affect mating performance. On the other hand, a disadvantage of performing surgery prior to mating is the increase in duration of the study resulting from the 10-day post-surgery recovery and acclimatisation periods. However, this has been shown to produce a model for rabbit embryo-fetal development studies that is comparable to treatment by more conventional routes.

Other laboratories have used femoral vein catheterisation (as described in Chapter 9) as an alternative. This allows the formulation to be delivered to a vessel of larger diameter and blood flow, with less potential for thrombosis and deposition of substance. This will be considered as an option in the authors' laboratory in the future.

References

Barrow, P.C. and Guyot, J.-Y. (1996) Continuous deep intravenous infusion in rabbit embryotoxicity studies. *Human and Experimental Toxicology*, 15, 214–218.

Flecknell, P.A. (1996) *Laboratory Animal Anaesthesia: A Practical Introduction for Research Workers and Technicians*, 2nd edn. Academic Press: London.

Flecknell, P.A. and Liles, J.H. (1990) Assessment of the analgesic action of opioid agonist–antagonists in the rabbit. *Journal of Assisted Veterinary Anaesthesia*, 17, 24–29.

ICH Harmonised Tripartite Guideline (24 June 1993) *Detection of Toxicity to Reproduction for Medicinal Products*.

Liles, J.H. and Flecknell, P.A. (1994) A comparison of the effects of buprenorphine, carprofen and flunixin following laparotomy in rats. *Journal of Veterinary Pharmacology and Therapeutics*, 17, 284– 90.

Lynch, S.J. (1999) Rabbit embryo-foetal development studies via continuous intravenous infusion. *Animal Technology*, 50, 51–55.

Perry-Clark, L.M. and Meunier, L.D. (1991) Vascular access ports for chronic serial infusion and blood sampling in New Zealand White rabbits. *Laboratory Animal Science*, 41, 495– 497.

Part 4

Continuous intravenous infusion in the dog

The dog is currently considered the non-rodent species of choice for pre-clinical testing, although the primate and minipig are increasingly used as alternatives when scientifically justified.

Catheterisation via the jugular vein is described in detail in Chapter 11, but the relative merits of the femoral vein are also discussed in Chapter 13. This section also describes two alternative methods of continuously infusing the dog. The first uses a tethered model with the pump located exterior to the pen (Chapter 12). This method allows mobility within the pen but will not permit exercise or social interaction with other dogs. The second (Chapter 13) describes the jacketed method, whereby the pump is contained within a jacket worn by the dog. This allows social interaction and might be considered better from a welfare angle but might not be suitable under certain circumstances, for example when testing drugs that are temperature sensitive (as they would be in close proximity to the animal's body in a reservoir contained in a pouch in the jacket).

Continuous infusion into the non-surgically prepared animal, through superficial vessels such as the jugular, is feasible and warrants further investigation. At AstraZeneca it has now been shown that animals prepared in this way can be continuously infused for short periods, and this may be the technique of choice for intermittent infusion (up to 4 h per day).

As for other species, new workers will have to balance the relative advantages of each method, from both technical and welfare viewpoints, for themselves.

11 Jugular cannulation in the dog

N. Pickersgill

11.1 Introduction

Techniques for long-term intravenous infusion of dogs have been described since the mid-1960s (Dudrick *et al.* 1966), although the original motivation for the development of these techniques was essentially long-term nutritional support. Subsequently, similar techniques were employed in toxicity studies either involving restraint or short-duration infusion. In recent years, methods that allow longer, subchronic intravenous infusion in unrestrained animals have been developed to enable the pre-clinical testing of new drugs which are intended to be administered by this route. At the Phoenix International site, these methods have existed for about 15 years, and over this period we have had experience with more than 1500 dogs undergoing studies by this route. Initially, a backpack system using a portable infusion pump was used, but currently we use a remote infusion pump connected to an implanted catheter via a tether system and jacket, which allows mobility of the dog within the pen. This chapter provides a brief review of these methods and the capabilities that are required to conduct successful regulatory toxicity studies to allow the prediction of both local and systemic toxicity by this route in man.

11.2 Facilities

11.2.1 Surgery and animal facilities

To enable consistent performance of repeatable regulatory studies, it is important to have dedicated facilities for the implantation of the animals and conduct of the studies. The large animal facility at the author's laboratory comprises a total surface area of over 2000 m², of which approximately 600 m² is dedicated to the housing of dogs. The unit is composed of 18 rooms of various sizes, each housing between 8 and 24 dogs. Infusion studies may be carried out in any room. This facility gives Phoenix International the capability of performing about three full regulatory infusion studies in the dog concurrently (approximately 120 animals on infusion simultaneously).

In a central position within the large animal facility is the surgical block, covering an area of 80 m² and made up of two independent suites, both equipped with operating table(s), surgeons' scrub and changing areas, animal preparation rooms, post-operative recovery rooms, autoclave and storage areas (see Figure 11.1). The unit is also equipped with viewing windows and hatches to reduce passage of personnel and therefore the risk of contamination. Animals always arrive for implantation via the preparation room and leave via the recovery room.

Figure 11.1 Surgical suite.

Before each implantation or session of implantations, the suite is subjected to a rigorous cleaning procedure, which includes the floor and all work surfaces and operating tables. The room is then fogged with formol. Between each animal, the tables and all work surfaces are disinfected and all instruments are cleaned, disinfected and then autoclaved. Operating personnel wear one-use or autoclavable gowns, hats, mask, disposable apron and sterile disposable gloves and scrub between each animal with antiseptic soap and a nail brush.

It is essential that all studies are conducted to current GLP regulations irrespective of the study aims or regulatory status, and consequently all the above procedures are fully documented and open to inspection.

11.2.2 Staff training and ethical issues

It is crucial that all relevant personnel undergo specialised training to attain and maintain the necessary skills in the surgical techniques required for implantation. All technicians undergo a formal, documented internal training course. To date, about 30 selected technicians at the author's laboratory with sufficient aptitude and experience have also undergone additional external surgical training (Introduction to Experimental Surgery) at the Lyon Veterinary School to ensure competence in the procedures required for catheter implantation. In addition, a small number of technicians have undertaken a more advanced course in experimental vascular surgery. Training is reviewed regularly and refresher courses given as appropriate. All training is managed by a dedicated senior technician and conforms to EEC Annex II L358.

Responsibility for the approval of animal research centres in France is undertaken by the Direction des Services Vétérinaires (DSV). The DSV manages the licensing of all laboratories. The principal role is to ensure that appropriate animal welfare procedures are in place and all appropriate regulations are followed. They are concerned with sanitation, ethical issues, veterinary treatment of animals and the processing of all documentation concerning animal supply, transport and importation. There is a national identification system for the accountability of all experimental animals which must be adhered to. The DSV is also responsible for issuing individual experimental licences. These must be obtained by the responsible scientists and require justifications for the type of experiment to be performed and the chosen animal species as well as verification of acceptable facilities, training and experience.

Many research centres now acknowledge the benefits of incorporating an ethical or institutional animal care and use committee into their management systems. The general role of these committees is to ensure the appropriate care and use of the animals and to try to reduce animal use, refine procedures and, where possible, replace the need for the use of animals. Staff training is an important element in the achievement of these goals. The ethics committee at Phoenix International comprises at least one veterinarian, one pathologist, four technicians and a lay person. This committee is responsible for the review of all protocols and standard operating procedures involving animals. They meet on a monthly basis to discuss any up-coming studies that may have ethical issues (studies with extensive blood sampling or with large infusion volumes for example) and issue written communications to the study director, management and other interested parties.

We have experimented with environmental enrichment by the provision of objects for exploration and manipulation, but these were not used by the dogs. The tether allows complete mobility within the pen and, rather than being a hindrance, acts as an additional object for investigation. We feel that, based on the short duration of the study, the animals have good and frequent interactions with the technicians and animal carers and they have adequate opportunity to display a full range of behavioural traits in the environment to which they are exposed.

11.3 Catheter implantation and infusion technique

11.3.1 Catheter implantation

The following describes the current method of catheter implantation into the anterior vena cava via an external jugular vein. Implantation is routinely performed about 1 week before the scheduled start of treatment on the study. The surgical team normally consists of at least three people. One person performing the surgery remains sterile throughout the procedure and usually two others perform the pre- and post-operative procedures and manipulate the animal and any non-sterile materials. The appropriate areas of skin are clipped and cleaned with povidone-iodine soap, wiped with compresses and disinfected with 70% alcohol and then redisinfected with povidone-iodine solution. The dogs are anaesthetised with an intravenous injection of tiletamine and zolazepam at a dose of 10 mg/kg following a subcutaneous injection of atropine sulphate at a dose of 0.1 mg/kg. The animal is placed on a pre-sterilised operating table on its side. The intended incision sites are again disinfected and sterile drapes are used to establish a sterile area at the site of incision. A cut of about 2–3 cm is made in the skin parallel to the jugular vein and the vein isolated by careful dissection. A non-resorbable suture is passed under the vein and the vein lifted slightly to occlude the blood flow. A bent, sterile needle is introduced through the venous wall into the vein and a sterile silicone rubber catheter filled with sterile physiological saline passed into the vein in the direction of the heart via the opening made behind the needle (Figure 11.2). The catheter has an internal diameter of 0.76 mm and an external diameter of 1.65 mm and is thickened at a collar approximately 10 cm from the tip. The collar remains just outside the vein. The catheter is sutured to the vein using surgical thread just above and below the collar. The catheter is then sutured to the musculature a few centimetres from the vein, leaving a small security loop. The tip of the catheter is therefore positioned about 5 cm from the heart (Figure 11.3). The dog is repositioned to give access to the back and side of the neck and the interscapular region of the back again disinfected. A trocar is passed subcutaneously via an incision from the interscapular region to the area of the incision made to introduce the catheter into the vein and the catheter passed back through the trocar to exit at the interscapular region. The end of the catheter is sealed with a sterile Luer lock and the functional integrity of the catheter checked by aspirating a small volume of blood via the catheter using a syringe attached to the Luer lock seal. The catheter is refilled with sterile physiological saline. The tissue overlying the jugular vein is sutured with resorbable surgical thread and the skin closed with non-resorbable thread. The wound is cleaned with oxygenated water and a local application of antibiotic spray. The dog is then transferred to the recovery area and the remainder of the infusion system is put in place. A surgical adhesive dressing is placed on the scapular region. A second loop is made in the catheter (this allows for some movement during routine procedures without stressing the catheter line). The infusion jacket is then placed on the dog. This jacket is custom designed for the dog using heat-malleable plastic and has attached to it a protective metal sleeve or tether through which the catheter passes. The other end of the metal sleeve is attached to a swivel joint which in turn will be attached to a shelf within the dog pen. From here the catheter passes to the tip of a syringe held in a pump on the shelf (Figure 11.4).

The duration of the anaesthesia is about 45 min, and the animal is normally fully recovered after about 2 h. Careful observation of the animal is made during this period. During the post-operative period, the animals are observed frequently and the site of implantation is examined for any signs of discharge or erythema. Local treatment with either antibiotics or disinfectant is performed as necessary. Any indication of

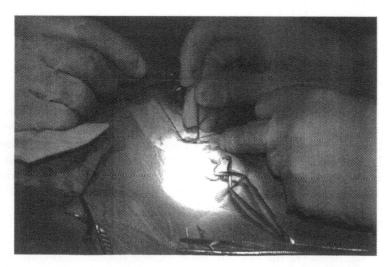

Figure 11.2   Catheter implantation.

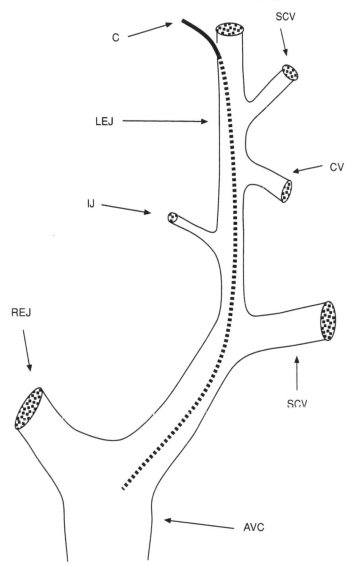

Figure 11.3 Schematic representation of catheter positioning. C, catheter; SCV, superficial cervical vein; LEJ, left external jugular vein; CV, cephalic vein; IJ, internal jugular vein; REJ, right external jugular vein; SCV, subclavian vein; AVC, anterior vena cava.

rubbing of the skin by the jacket is monitored and corrected/treated. Food consumption and body temperature are monitored as necessary. The sutures are removed 10 days after implantation.

If the study design includes a period without treatment, at the end of treatment the catheter is cut and hermetically sealed at the point of exit on the back and the end of the catheter disinfected and allowed to pass subcutaneously. The site is carefully observed for any reaction.

The methods of implantation are continuously under review to take advantage of any new equipment or techniques that may come available and the methods have

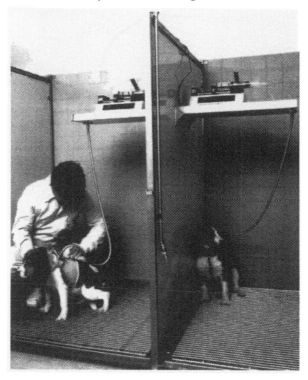

Figure 11.4 Infusion system.

evolved over the 15 years that they have been in place. These improvements are normally of a minor nature and concern changes in details. The most significant modification, however, was in 1994: since then the catheter has been sutured in place. Previously the method had involved the entry of the catheter via a hollow needle and the use of two trocars to implant the catheter. The catheter was glued at the exit site. Since 1994 there has been a significant improvement in the rate of reimplantations (see Figure 11.5), and the addition of the suturing step has probably been the biggest contributor to this. The rate of reimplantation reflects the reliability of the infusion system as well as any undesirable effects of the infused solution. It should be remembered that in the average study 75% of the animals are treated.

11.3.2 Equipment

Among the many types of infusion pump available, we have found the most appropriate type of pump used to be the Harvard type 44 programmable syringe pump. These pumps give a reasonable range of possible flow rates and are reliable and accurate. Another type of syringe pump of more recent design is the Medfusion Medex pump, which allows more sophisticated programming and is of a more compact size. Alternatively, a Vial peristaltic pump can be used for larger dose volumes (above 240 ml per animal per day). These pumps are slightly less reliable and accurate than the Harvard pumps and, consequently, they require more careful monitoring and, in use, calibration. Consequently, an additional flask weighing is performed a few hours

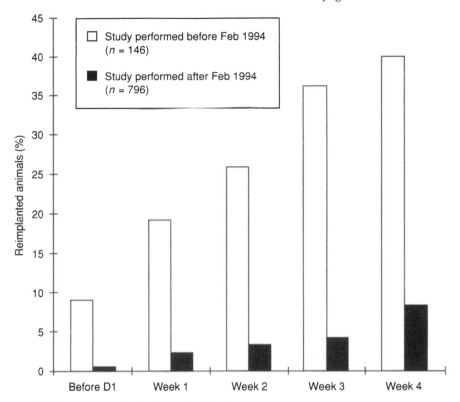

Figure 11.5 Percentage of reimplanted animals.

after each flask change such that any inaccuracies can be identified and accounted for by reprogramming flow rates.

The choice of the type of pump to be used is dependent on the dose volume required and any information available on the compatibility of the test formulation with the infusion system materials. For example, if glass syringes are required this could dictate the use of a particular type of pump over another.

Finally, we still retain a limited capability to perform dog infusion studies with portable Zyklomat pumps for small infusion volumes. All pumps are calibrated before the start of treatment at the flow rate intended in the study. The time of start/end of infusion is recorded and the individual syringes or flasks are weighed before and after being changed so that any anomalies can be quantified. An error rate of $\pm 5\%$ of theoretical infusion volume is considered acceptable and under normal circumstances is routinely achieved. This error rate may rise when problems with the infusion are encountered, such as marked reactions to treatment that result in cessation of infusion or catheter blockage due, for example, to precipitation of test material in the catheters.

11.4 Special considerations

11.4.1 Vehicle selection

In order to achieve acceptably high exposure levels during continuous infusion studies, it is frequently necessary to use a vehicle that provides increased solubility of the test article. It is always important to consider the inclusion of a negative (physiological saline) control group into the study design if a non-standard vehicle is being used as well as a vehicle control group. This will aid in the interpretation of the findings of the study.

For example, propylene glycol is sometimes used in clinical formulations and is generally considered to be innocuous in man, although there is little reported information on the adverse effects of this compound. At Phoenix International we have recently infused small groups of Beagle dogs with propylene glycol at 55% (v/v) over 14 days at 10 ml/kg/day and compared the data with those from a similar group of animals infused with saline. When the animals were sampled for haematology at the end of the 14-day infusion period, there was a slight anaemia in this group of animals which could not be entirely explained by the high rate of infusion. The clinical chemistry data revealed slightly increased serum bilirubin concentrations and elevated aminotransferase and creatine phosphokinase activities when compared with the saline-infused animals. Infusion of high concentrations of propylene glycol in dogs was therefore associated with a haemolytic effect and a possible minor effect on the liver. There were no histopathological lesions relating to these changes with the exception of a very slight increase in severity of the normally expected lesions at the injection site (Pickersgill *et al.* 1994). The use of lipidic vehicles has been associated with an increased risk of infection during the continuous infusion studies performed in the past. This is considered to be because of the favourable conditions this vehicle permits for the growth of micro-organisms. In these cases, antibiotic therapy is now employed routinely before the start and during the conduct of the study; this results in acceptable levels of asepsis.

11.4.2 *Catheter blockage and reimplantation*

Reimplantations become necessary when the infusion stops, either because the pump has stopped or because the animal has torn the flexible sleeve protecting the catheter, occasionally pulling it out. We also have infrequent problems with leaks or kinks in the implanted part of the catheter, which can result in reimplantation becoming necessary. Damage to the implanted part of the catheter would result in subcutaneous administration, which can often provoke a local inflammation reaction with irritant test solutions.

The incidence of reimplantation increases with the duration over which the study is to be performed. It decreases with the technical competence of the technical staff performing the procedure and the expertise gained with repeated performance. The number of reimplantations can also be affected by the solubility of the compound. Depending on the physical nature of the test article, crystallisation may occur in the catheter, stopping or reducing the flow in the system and causing pressure build-up. The daily injection of a small quantity of saline (or vehicle) to ensure patency should be avoided, if possible, since there is the ever-present risk of causing emboli. In addition, if the infusion rate is low, this may result in a significant bolus injection of the test compound that is already in the catheter. Aspiration of blood before injection of saline could be conducted to check catheter patency and function if there is evidence of infusion malfunction. This would reduce the risk of any emboli.

Sometimes the catheter can become definitively blocked and must be changed. The problem here is that this may require the use of the contralateral vein; should a subsequent reimplantation be necessary, this may not be possible and the animal must be removed from the study.

11.5 Comparison of the tether method with other methods

Continuous intravenous infusion in dogs at Phoenix International was first performed in the late 1980s using an ambulatory system consisting of a 'backpack' on the animal which contained the syringe pump and formulation reservoir. Since 1992, our preferred method has been the tether system, which in our opinion has several advantages over the backpack system. The reliability of the portable pumps at the time was relatively poor: they were of the Zyklomat type and were intended for use in man for the control of diabetes. They were susceptible to mechanical damage related to the normal behaviour of the dog within the pen. The weight of the reservoir and pump was prohibitive and considered to alter the normal behaviour of the dog, which could affect the interpretation of any potential test article effects, especially with centrally acting compounds. The portable reservoir had a very limited capacity, severely affecting the maximum achievable infusion rate. The autonomy of the system was limited by the short life of the batteries. The cost of the system was high because of the large numbers of interventions required to change batteries or the infusion reservoir within a 24-h period and there was a range of non-reusable equipment, all of which made this system more costly than the tether system.

In summary, the tethered system provides a cost-effective method of performing continuous intravenous infusion with easy access to the system to monitor and, if necessary, rectify any anomalies. There is a wide range of equipment available commercially to perform such studies (Jacobsen 1998). Although the tethered system precludes socialisation of the dogs outside the pen, the animals can exercise within the pen to the same extent as other dogs within the facility.

Continuous infusion in the rats and monkeys at the author's laboratory is performed via the femoral vein, and this is an alternative method of implantation in the dog also. Antibiotic therapy (500 mg of Clamoxyl) is routinely given about 1 h before implantation with this method. This method could potentially have advantages as it would avoid any complications associated with being close to the heart, which is the case with jugular implantation. However, the advantage of the jugular vein (irrespective of the amount of experience the author's laboratory has now built up with this route) is that it requires much less subcutaneous tunnelling and the vein is much more superficial than the femoral vein, thus making it more accessible and the surgery less invasive.

Phoenix International has also experimented with various types of catheter (polyethylene, rounded tip) but none has shown any clear advantage over the Silastic® type and therefore none is used routinely.

11.6 Conclusion

In recent years, methods have been developed which allow subchronic intravenous infusion in unrestrained animals to enable the pre-clinical testing of new drugs that are intended to be administered by this route. To allow the prediction of both local and systemic toxicity by this route in man, it is important to have dedicated facilities for the implantation of the animals and conduct of the studies and to have a specialised training programme to attain and maintain the necessary skills in the surgical techniques required for implantation. Special consideration should be given to vehicle selection and to the procedures for catheter reimplantation should this become necessary. The tethered system provides a cost-effective method of performing continuous intravenous infusion with easy access and monitoring of the system.

References

Dudrick, S.J., Vars, H.M. and Rhoads, J.E. (1966) Growth of puppies receiving all nutritional requirements by vein. *Fortsch. Parenteral. Ernähung,* 2, 16–18.

Pickersgill, N., Héritier, B., Gregoire, M., Liberge, P., Burnett, R. and Skydsgaard, K. (1994) Effects of continuous intravenous infusion of high concentrations of propylene glycol in the rat and dog. *Toxicology Letters,* Suppl. 1/74 (abstract).

Jacobson, A. (1998) Continuous infusion and chronic catheter access in laboratory animals. *Laboratory Animals,* 27(7), 37–46.

12 The non-ambulatory model in dog multidose infusion toxicity studies

N. Pickersgill and R. Burnett

12.1 Introduction

The aim of pre-clinical studies is to be predictive of possible harmful effects of the test material to humans. It therefore follows that the potential toxicity of any new drug intended to be administered by continuous intravenous infusion is more appropriately assessed by the same route. The principal advantage of continuous intravenous infusion as a route of administration is the ability to maintain a relatively constant blood level of drug within a narrow range. This may be desirable for drugs with a very short half-life or if peak plasma levels are associated with undesirable clinical signs or other effects. Equally, there is a role for continuous intravenous infusion if a drug is very poorly tolerated locally by bolus injection or if there is poor bioavailability by the oral route. As a result, over the last decade, there has been an increasing interest in the development of these types of drug, especially from the booming biotechnology industry, and a corresponding growth in the number and type of techniques which will allow the toxicological evaluation of these new drugs. Phoenix International has been involved in the development and conduct of infusion studies in large and small animal species for the last 15 years and has built up a considerable experience in this area. This chapter describes some of our current designs for the conduct of dog continuous infusion studies to meet the demands of the pharmaceutical industry.

12.2 Test system, husbandry and infusion monitoring

12.2.1 Test system

At the author's laboratory, Beagle dogs are used exclusively from accredited suppliers: either Harlan, France, or Covance Research Products, USA. It is considered advisable to obtain and implant a small number of additional dogs to select the most suitable for the experiment. This is done based on the results of pre-trial clinical pathology assessments and other factors, such as food consumption and body weight gain. It is extremely rare for an animal to be rejected for use on a study owing to changes in behaviour as a result of the wearing of the infusion jacket. The animals are a minimum of 6 months of age at initiation of treatment.

12.2.2 Husbandry

The animals are housed in an air-conditioned building maintained at between 19 and 25 °C and greater than 40% humidity. There are a minimum of eight air changes per hour and a light cycle of 12 h light/12 h dark is routinely maintained. The animals are housed singly in grid-floored pens of approximately 1.44 m². Three hundred grams of a commercially available complete, pelleted diet is offered daily and filtered (0.2 μm) mains water is available *ad libitum*.

12.2.3 Infusion monitoring

From the moment that the animals are infused with the pre-test rate of 1 ml/h saline (or vehicle) they are checked three times per day to ensure the correct functioning of the system. This check includes confirmation of the correct programming of the pump and positioning of the syringe, verification that all taps are in the open position and that there are no leaks/breaks in the infusion line and a check of the catheter to ensure that it is not blocked or kinked. All these checks are documented in the raw data together with any necessary corrective action.

12.3 Study designs

12.3.1 Compatibility test

Before starting the animal phase (or at the same time as performing the preliminary study), it is often useful to carry out a small test to ensure that the test article (or vehicle) is compatible with the material in the infusion system. It is advisable to set up the infusion system in a fashion identical to that of the study conditions. A high concentration of the test article is run through the system, preferably with the catheter placed in a water bath at 37 °C to mimic as closely as possible the conditions in the animal. Samples of the test formulation are taken from the catheter tip for chemical analysis. A visual examination should also be performed to confirm the absence of any crystallisation in the system. This sometimes occurs at joints in the tubing, especially when the bore of the system changes. Here, eddies may be created, allowing high local concentrations of the test article, which are more prone to crystallisation. Based on the results of this test, decisions can be made concerning changing parts of the system at selected intervals, if it is felt that blockage may occur, or on the choice of materials in the system.

12.3.2 Preliminary study

The need for a preliminary study before starting a study with large numbers of animals is paramount. This is indispensable in predicting possible problems on the main study and allows time for these problems to be resolved or alternative strategies to be investigated before committing resources to an expensive study that may not proceed as desired. More and more frequently, preliminary studies of 3 weeks' duration with relatively small numbers of animals are used prior to a 4-week main study to eliminate as many risks as possible. This is considered advisable as many problems encountered with infusion studies (e.g. problems of local tolerance) only become evident after longer than 14 days of infusion. If, once all the parameters have been decided on the preliminary study, it becomes necessary to change any element for the main study then the full implications of doing so should be assessed before starting the main study as this could render all the previous work useless.

Preliminary studies are normally of at least 14 days' duration and include one male and one female. If there is sufficient information on toxicity either by the intravenous (bolus) route or by other routes, it may be sufficient to run a range of appropriate dose levels concurrently with a vehicle control group. This has advantages in ease of conduct, speed of availability of data and concurrent control information. However, the exposure pattern in continuous infusion studies is very different from that in bolus injection studies, and it is difficult to make accurate predictions of dose levels for infusion based on these data. Transient clinical signs, such as inappetance or reduced activity, may be acceptable after bolus administration but can become of concern on infusion studies, when there may be no possibility of recovery between administrations. Alternatively, for test materials without any toxicity data, a dose-escalating phase may be included in the study design before running a single selected dose level for 14 days in a few animals. Special attention should be given to the presence of any signs of local intolerance at the injection site as these may limit the high dose level to be chosen for the main study.

The choice of dose volume will depend to some extent on the solubility of the test material and the chosen vehicle. Our preference is to use a rate of between 3 and 4 ml/kg/h in an attempt to remain within normal physiological ranges and this corresponds to a maximum total daily volume that does not exceed the estimated total blood volume of the animal. These rates are consistent with values reported by other workers, which have been shown to have no effects on haematocrit values and heart rate (Mann and Kinter 1993). Higher infusion rates have been reported in the literature (Perkin and Stejskal 1994) of up to 120 ml/kg/day with no changes with the exception of increased urine output and a slight reticulocytosis.

12.3.3 Main study

All study designs are adapted from OECD 409 and Directive 91/507/EEC. The group size is often four males and four females in each treatment group, with a control group of the same size receiving the vehicle. Consideration is given to the inclusion of a saline control group if a non-standard vehicle is being used which may be associated with changes, especially at the site of injection. This can help distinguish between the effects of the vehicle alone, the vehicle and test article combination (as the vehicle may enhance the effects of the test article) or the effects of the test article alone. The study design often includes a treatment-free period of between 2 and 4 weeks after 4 weeks of treatment. This is normally two males and two females from each of the control and high-dose groups. During this phase, the catheter is left in place but is hermetically sealed and cut at the exit site and allowed to pass subcutaneously. The dose levels are selected based on the results of the preliminary study. The high dose level is selected to be close to the maximum tolerated dose and the low dose level should be a small multiple of the expected human therapeutic dose. The intermediate dose is often the geometric mean of the high and low dose level. If toxicokinetic information is available before the start of the main study, this information can also be used to choose dose levels that provide total exposure (AUC) and peak concentrations (C_{max}) similar to those expected in man. If necessary, a bolus injection of a test material may be given followed by a slow infusion if this is the intended therapeutic regimen in man. This allows a steady state to be achieved more rapidly than by infusion alone.

Clinical observation and care of the animals are vital for the successful conduct of infusion studies. Observations need to be made to monitor the site of implantation for any discharge or hardening of the vein, and around the jacket to confirm the absence of any skin lesions due to rubbing. Additional measurements to monitor the condition of the animals can be used as required, temperature measurement for example. All other evaluations required for a full regulatory toxicity study can be performed on continuous infusion studies without interrupting the infusion, and without having any detrimental effects on the parameters measured. This is done by use of a mobile workstation that can be taken into the animal room and positioned next to each pen. This allows, for example, the measurement of ECG and indirect blood pressure at timed intervals after the start or end of the infusion. Other routine evaluations include ophthalmoscopy, body weight and food consumption and blood sampling for both laboratory determinations (such as haematology and clinical chemistry) and toxicokinetics. Blood samples are normally taken from the cephalic vein. Study designs with bilateral jugular cannulation have been used with success whereby the contralateral side is used for blood sampling. This appears to have advantages as the stress of sampling is reduced. However, if reimplantation becomes necessary, this side may be needed to continue treatment and thus alternative sites of sampling would be required. Sampling for toxicokinetics by this method is not recommended because the site of collection and administration are close together and misleadingly high blood concentrations could be obtained. Urine collection is routinely by direct catheterisation of the bladder because of the complications of putting tethered animals in metabolism cages. However, facilities do exist for taking samples from a limited number of animals using metabolism cages for defined requirements of specific studies.

12.4 Results

12.4.1 Body weight

Table 12.1 compares typical body weight progressions for recently performed continuous infusion studies at two different infusion rates against body weights in an oral (capsule administration) study. The animals infused at 72 ml/kg/day received a buffered saline solution and those at 20 ml/kg/day buffered 4% xylitol. The data reveal no major differences between the infused animals and the orally treated animals. Weight gains during the pre-treatment period are lower in the infusion studies than in the oral study. This is considered to be due to the effects of anaesthesia and surgery for implantation.

12.4.2 Clinical pathology

Using a former method of catheter implantation in which the catheter was not sutured in place, a common effect of continuous intravenous infusion was a slight reduction in the normal red cell parameters (haemoglobin concentration, red blood cell count and packed cell volume) compared with either the pre-implantation values or the background control range for oral studies (Table 12.2). Although the mean white blood cell counts were not elevated, occasional animals did have a slight, transient neutrophilia, which was probably associated with stress. None of these effects was

Table 12.1 Body weight (kg) – males and females

| | Body weight (kg) on Day | | | | Weight gain (kg) over Days | |
	-14	1	15	28	-14 to 1	1 to 28
Males						
Oral (capsule) (*n* = 4)	7.3	7.9	8.2	8.5	0.6	0.6
Infusion (20 ml/kg/day) (*n* = 6)	7.9	7.9	8.1	8.5	0.0	0.6
Infusion (72 ml/kg/day) (*n* = 6)	6.5	6.7	6.8	7.2	0.2	0.5
Females						
Oral (capsule) (*n* = 4)	6.6	7.1	7.3	7.5	0.5	0.4
Infusion (20 ml/kg/day) (*n* = 6)	6.7	6.7	6.9	7.2	0.0	0.5
Infusion (72 ml/kg/day) (*n* = 6)	6.1	6.2	6.5	6.8	0.1	0.6

considered of sufficient magnitude to interfere with the interpretation of the results of the study. It was standard practice at this time to perform a pre- and post-implantation assessment of the clinical pathology parameters in order to monitor this change. Following improvements in the method since 1994 the data sets of infused and non-infused animals are very similar (see Tables 12.3 and 12.4). The post-implantation data are similar to our normal background data and are considered to be sufficient to serve as a baseline before study start. The pre-implantation sampling is therefore no longer performed.

12.4.3 Macroscopic/microscopic histopathology

Many of the lesions seen in experimental animals can also be found following the use of continuous infusion in the clinical situation. These techniques are, therefore, considered as useful predictive models for potential problems in man. These associated lesions are frequently difficult to avoid, particularly with irritant compounds. These lesions must be minimised in order to maintain the sensitivity of the test procedure.

Table 12.2 Mean values of clinical haematology parameters – males and females

	Pre-dose I	Pre-dose II	Week 4	Week 8
Males				
Hb (g/dl)	14.1	12.8	13.2	14.9
PCV (%)	40.7	36.4	37.7	41.3
RBC (10^6/mm³)	6.40	5.78	5.99	6.59
MCV (fl)	64	63	63	63
Platelets (10^3/mm³)	354	268	272	276
WBC (10^3/mm³)	11.9	12.0	10.5	11.5
Females				
Hb (g/dl)	14.8	13.7	13.1	14.0
PCV (%)	42.0	39.1	38.0	40.5
RBC (10^6/mm³)	6.81	6.39	6.18	6.65
MCV (fl)	62	61	61	61
Platelets (10^3/mm³)	404	332	253	407
WBC (10^3/mm³)	11.2	9.6	11.3	9.9

Table 12.3 Comparison between infused and non-infused dogs for selected haematology parameters – males and females

| | Males | | | | | | Females | | | | | |
| | Infused | | | Non-infused | | | Infused | | | Non-infused | | |
Parameter	n	Mean	SD	n	Mean	SD	n	Mean	SD	n	Mean	SD
Hb (g/dl)	43	13.6	0.9	90	13.7	1.7	43	14.1	1	80	13.9	2
PCV (%)	43	38.4	3.1	90	39	4.2	43	39.6	3.1	80	39.6	4.8
RBC (10^6/mm^3)	43	5.87	0.55	90	5.91	0.6	43	6.12	0.49	80	5.96	0.64
MCV (fl)	43	65.5	2	90	66.1	2.4	43	64.5	2.1	80	66.3	2.7
WBC (10^3/mm^3)	43	13.4	4.8	90	12.7	3.4	43	12.1	3.1	80	12.3	3.1
Platelets (10^3/mm^3)	43	305	66	90	391	102	43	301	64	80	360	83

Table 12.4 Comparison between infused and non-infused dogs for selected clinical chemistry parameters – males and females

| | Males | | | | | | Females | | | | | |
| | Infused | | | Non-infused | | | Infused | | | Non-infused | | |
Parameter	n	Mean	SD	n	Mean	SD	n	Mean	SD	n	Mean	SD
Na (mequiv./l)	14	145	2	69	146	2	14	144	1	60	146	2
K (mequiv./l)	14	5.4	0.3	69	5.2	0.6	14	5.1	0.4	60	5	0.5
Cl (mequiv./l)	14	112	2	69	113	2	14	113	3	60	114	2
Glucose (g/l)	14	0.99	0.08	71	1.09	0.09	14	1.03	0.05	62	53	5
Urea (g/l)	14	0.28	0.04	71	0.27	0.08	14	0.28	0.06	62	0.29	0.06
Protein (g/l)	20	56	2	71	54	5	20	55	3	62	53	5
Alkaline phosphatase (IU/l)	14	267	89	62	255	62	14	226	30	62	257	67
Aspartate aminotransferase (IU/l)	14	38	9	62	38	11	14	38	6	62	40	12
Alanine aminotransferase (IU/l)	14	40	20	71	41	25	14	43	9	62	54	82

This is achieved by refining the procedures and continual training of those involved in the surgery.

12.4.4 Histopathological methods

It is our standard procedure to carefully open up the vein from the point of insertion of the catheter down to the heart. Any thrombi are described, measured and frequently photographed. The position of the tip of the catheter is marked with Indian ink. The vessels are pinned onto a cork board and then fixed whole in buffered formalin. Samples are taken from the vessel for histological examination at the level of the tip of the catheter, the injection site and at other appropriate sites. Normally transverse sections are taken at the various levels, though longitudinal sections may be useful to study the extent of a change. A combination of transverse and longitudinal sections of the veins may be the best way to obtain maximum information on local changes. Haematoxylin and eosin staining is used routinely, with Martius, scarlet, blue (MSB) (Bancroft and Stevens 1996) being our preferred special stain to clarify the organisation of any thrombi by the identification of fibrin and connective tissue. Examination of the vein under polarised light should always be performed. This not only shows the connective tissue structure but can reveal the presence of microcrystals. The vein just before and just after the tip of the catheter is the region of the greatest interest in toxicological studies. The changes described are common to the technique and only vary slightly with the different modifications of the procedure (Mesfin *et al.* 1988; Perkin and Stejskal 1994).

12.4.5 External lesions

There are frequent abrasions of the skin related to the chaffing of the restraining jacket; histologically changes are alopecia, slight acanthosis and sometimes mild dermatitis. Where the catheter enters the skin is an obvious portal of entry for infection. Careful management following the surgery limits any problem which may arise at that site. Despite this care, subcutaneous haemorrhage and sometimes abscess formation can occur at this site. Enlargement of the draining lymph nodes due to reactive hyperplasia and histiocytosis can be seen in such cases.

External changes directly associated with this mode of administration are rare. However, in cases where the external jugular becomes occluded, poor venous drainage from the facial region can lead to oedema of the cheeks.

12.4.6 Internal lesions

These can be divided into those associated with the vein and distant lesions.

12.4.6.1 Changes associated with the veins

As discussed (in the description of the method above), the catheter enters the external jugular vein and is passed down past the bifurcation with the internal into the common jugular and rests with the tip in the anterior vena cava (Figure 11.3). The changes in the vein are, therefore, considered in two distinct regions: the external jugular and the common jugular with the anterior vena cava.

EXTERNAL JUGULAR

Where the catheter is inserted into the hole in the external jugular it is secured by sutures to the vein. Reaction to the suture material can occur. These sutures cause some reduction in the blood flow at this point and thrombus formation associated with the traumatised vein is very common. Such thrombi can extend down the length of the catheter (Figure 12.1) and are frequently attached to the vessel wall. These can become well organised, holding the catheter firmly against the wall of the external jugular vein (Figure 12.2). This thrombus can extend virtually to the tip of the catheter (some 10 cm), but typically what is seen is a well-organised thrombus at the point of insertion and patches of mural thrombi along the length of the external jugular. The thrombus is less organised as it approaches the tip and is often only a fibrin sheathing around the catheter in the last few centimetres. In some cases, the external jugular can become occluded. Extension of the thrombus into vessels joining in this region can be seen occasionally. Recannulisation of thrombi can be found in long-term studies. The thrombus formation can give rise to local acute inflammation in the vessel wall. This phlebitis can become extensive, and necrosis of the vessel wall can occur giving rise to a more severe chronic inflammatory reaction. This marked inflammatory reaction may give rise to peri-phlebitis and, in rare cases, can cause exudation and local adhesions. Changes in this region of the vein are clinically important but are of less importance from a toxicological view point.

COMMON JUGULAR AND ANTERIOR VENA CAVA

The tip of the catheter rests in the anterior vena cava where the blood from the other veins in this region (e.g. axillary and azygous) causes turbulence and mixing of the test article. As this is at the point of outflow, this area and that closer to the heart may also show local reaction to the test article and is the area of greatest toxicological importance. A combination of mechanical damage and local irritation can occur even with the vehicle (thus reinforcing the need for a saline control to aid interpretation of

Figure 12.1 Macroscopic view of the external jugular and anterior vena cava opened to display the thrombus around the catheter, with a free part of the catheter at the tip.

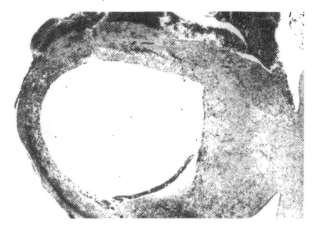

Figure 12.2 Low-power photomicrograph of the histological section of the vessel in Figure 12.1, showing an organised thrombus around the catheter (empty hole) and the attachment of the thrombus to the vessel wall.

the results). The tip of the catheter may move slightly in the bloodstream, and focal loss of a few endothelial cells from the vessel wall is a frequent observation at this point, probably associated with minor mechanical damage. This loss may occur during the euthanasia or even at necropsy as it is seldom accompanied by inflammatory changes. The progression of lesions at this point in the vein depend upon many factors: pH of the test article, flow rate, length of study, etc. Trivial irritation may result in only a slight hypertrophy of a focus of endothelial cells. With some vehicles, where the isotonic balance is poor, swelling of the endothelial cells may become extensive. Prolonged minor irritation of either chemical or mechanical origin results in more pronounced changes to the endothelium. This is seen initially as endothelial damage and loss, which may result in focal hyperplasia of endothelial cells. The border between the normal single endothelial lining and the hyperplasia is often very abrupt (Figure 12.3). This endothelial proliferation can increase and become several cells thick (Figure 12.4). This may progress with irritant compounds to produce florid hyperplasia, which may be thrown into folds (Figure 12.5). This hyperplasia may be accompanied by marked hypertrophy, giving an increase in the thickness of the intima, often with fibrous tissue. Inflammatory cells are not usually a feature of this change. Such changes to the endothelium can promote the formation of thrombi. Endothelial hyperplasia may also be seen in distant sites such as pulmonary vessels. Mural thrombi may give rise to a local inflammation within the vessel wall or around the vessel in adjacent tissue. Thrombi close to the tip may interfere with the local mixing of the blood and test article, leading to high local concentrations, increasing the risk of irritation. In certain cases, this area of high concentration can result in crystal formation with accompanying serious mechanical damage to the vessel, initiating a severe inflammatory response. Thrombi forming in the region of the tip are in a more turbulent bloodstream and are more likely to be dislodged. The influence of products which modify the clotting cascade may, of course, influence the friability of these thrombi. The resultant increase in such thromboemboli is a significant cause of morbidity in long-term infusion studies.

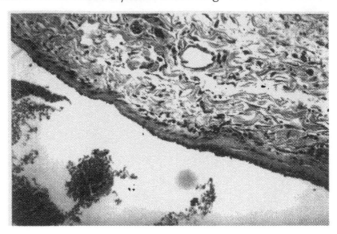

Figure 12.3 Medium-power photomicrograph of the anterior vena cava showing normal single layer of flat endothelium which changes to an area with slight endothelial hyperplasia, in an area corresponding to the tip of the catheter.

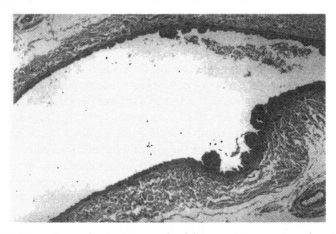

Figure 12.4 Medium-power photomicrograph of the anterior vena cava showing normal single layer of flat endothelium that shows an abrupt change to an area of moderate hyperplasia.

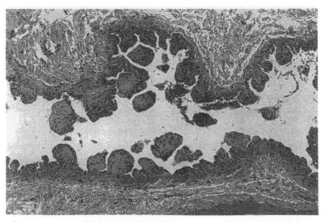

Figure 12.5 Medium-power photomicrograph of the anterior vena cava showing a marked hyperplasia of the endothelium in folds with thickening of the intima.

12.4.6.2 Distant lesions

The majority of thromboemboli from the catheter lodge in vessels in the lungs, and the resultant damage depends on the size of the vessel and upon the collateral circulation. In moderate-sized vessels the embolus may become incorporated into the wall, causing only a minor decrease in flow and little local response (Figure 12.6). In smaller vessels, the embolus leads to occlusion and local ischaemic damage (Figure 12.7). These can be seen macroscopically as minor congested areas to dark red areas which can extend to most of a lobe, and if severe can lead to the sudden death of the animal.

Thromboemboli are rarely seen in other organs, but focal infarction in the kidney related to such thromboemboli has been observed. Small groups of sclerotic glomeruli with adjacent atrophic tubules and some peritubular inflammation can sometimes be found in dogs. These are considered to result from local ischaemia involving a group of glomeruli which have a common blood supply; such lesions are, however, difficult to differentiate from background interstitial nephritis.

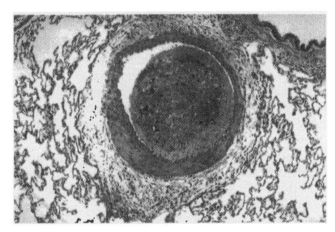

Figure 12.6 Medium-power photomicrograph of a thromboembolus lodged in a moderate-sized vessel in the lung with some thickening of the vessel at the point of attachment but no change to the parenchyma.

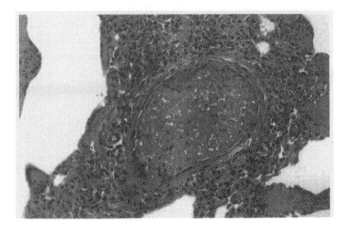

Figure 12.7 Medium-power photomicrograph of a thromboembolus occluding a smaller vessel in the lung with degenerative ischaemic change to the local parenchyma.

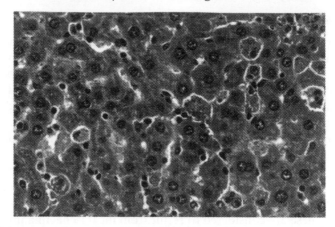

Figure 12.8 High-power photomicrograph of the liver with an increase in histiocytes in the sinusoids and erthyrophagia.

OTHER DISTANT LESIONS

The most frequently observed change is an increase in size of the spleen, which is the result of a moderate to marked increase in extramedullary haematopoiesis. Increased haemosiderin is often seen and is probably a consequence of an increase in erythrocyte fragility. A slight increase in adrenal weights corresponding to increased cortical vacuolation and reduced thymus weights with reduction in the cortex are often noted. These are considered to be a response to non-specific stress which may be enhanced by treatment.

A generalised leucocytosis is invariably seen; this is particularly obvious in the sinusoids of the liver and vessels of the renal glomerulus. In the liver it may be accompanied by histiocytosis with erythrophagia in the sinusoids (Figure 12.8); erythrophagia may also be seen in the lymph nodes.

12.5 Conclusion

The potential toxicity of any new drug intended for continuous intravenous infusion administration should be assessed by pre-clinical studies, which are predictive of possible harmful effects of the test material to humans. Phoenix International has been involved in the development and conduct of infusion studies for many years and has built up considerable experience in this area. An understanding of the limitations and capabilities of the technique has been accumulated, as well as an extensive library of normally occurring background clinical and histopathological findings. These are indispensable in the assessment of any adverse changes due to the drug.

References

Bancroft, J.D. and Stevens, A. (1996) *Theory and Practice of Histological Techniques ED,* 4th edn. Churchill Livingstone: London.

Mann, W.A. and Kinter, B.K. (1993) Characterization of maximal intravenous dose volumes in the dog (*Canis familiaris*). *General Pharmacology,* 24, 357–366.

Mesfin, G.M., Higgins, M.J., Brown, W.P. and Rosnick, D. (1988) Cardiovascular complications of chronic catheterization of the jugular vein in the dog. *Veterinary Pathology,* 25, 492–502.

Perkin, C.J. and Stejskal, R. (1994) Intravenous infusion in dogs and primate. *Journal of the American College of Toxicology,* 13, 40–47.

13 The ambulatory model in dog multidose infusion toxicity studies

T. R. Gleason and C. P. Chengelis

13.1 Background information on the dog

Dogs (*Canis familiaris*) have been used in experimental toxicology since the work of Orfilia, Megandie, and Bernard in the early and late 1800s (Gad and Chengelis 1998). In the twentieth century, dogs were extensively used in physiological and experimental surgery research. They have, in the last 50 years, become the non-rodent of choice for general toxicity testing because of their anatomical and physiological similarities to man; for a more complete review the reader is referred to Haggerty *et al.* (1992). They are a convenient size for surgery and are hardy and tractable, and readily available. In the last 15 years, these two trends, the use of dogs for toxicological testing and in experimental surgery, have merged, with the creation of continuous infusion models in the dog. Dogs also have the advantage in that they are among the few experimental animals that are routinely immunised against their most common diseases; therefore, they require a less rigorous level of care than other experimental models.

In most toxicological research involving dogs, the breed of choice is the Beagle because of its size and temperament. Larger hounds and purpose-bred mongrels are also occasionally used, especially if a larger model is required; for example, if the experiment calls for continuous infusion with telemetric cardiovascular determinations, a larger dog might be required because of the size of the equipment and the amount of surgery.

Depending on supplier, mature Beagle dogs can weigh up to 18 kg, although 9–12 kg is a more normal range. Males tend to be larger than females. Other typical physiological parameters are given in Table 13.1.

13.2 Basic care and husbandry

The husbandry of experimental dogs is relatively straightforward. The applicable National Research Council guidelines (1996) should be followed, as should the applicable animal welfare regulations. Relatively routine animal room conditions (temperature 25 ± 2 °C, humidity at 30–70% and 10–15 fresh air changes per hour) should be maintained. For routine experiments, the animals can be kept in either cages (providing that they meet animal welfare regulations for floor space and head clearance) or pens. The use of cages is still acceptable practice in the US, providing that the animals receive appropriate exercise and socialisation periods. Our interpretation of socialisation is that the dogs not only interact with other dogs, but also receive significant time in contact with human beings. The use of pens, designed

Table 13.1 Normal physiological and reproductive parameters in the dog

Average lifespan	12.5 years
Daily food consumption	25–40 g/kg body weight
Daily water consumption	*Ad libitum*; approximately 600 ml
Glomerular filtration rate	Approximately 5 m/kg/min
Rectal temperature	37.8 ± 0.13 °C (SD)
Respiration rate	20/min (10–30)
Oxygen consumption	0.36 ml/g/h
Tidal volume	24 ml/kg (18–35)
Minute volume	4.5 ± 5.6 l (5 cmH$_2$O)
Functional residual capacity	367 ml (248–540)
Lung compliance	117 ± 5.6 ml (5 cmH$_2$O)
P_{O_2}	73.7 mmHg (61–87)
P_{CO_2}	36 mmHg (29–46)
Heart rate	80–160 beats/min (resting)
Blood pressure (systolic)	110 mmHg
Puberty	6–12 months of age
Breeding age	
Males	10–12 months
Females	9–12 months
Gestation	60–65 days
Oestrus cycle	
Pro-oestrus	5–15 days
Oestrus	5–15 days
Metoestrus	60-65 days
Anoestrus	Length variable
Weaning age	5–8 weeks
Litter size	1–11 pups; 5–6 average
Breeding age (males)	10–12 months
Peak testis development	32 weeks

From Haggerty *et al.* (1992).

for two dogs each, is becoming more common in Europe. This appears to be based on the belief that somehow an 18 square foot enclosure containing two dogs is more humane than a 9 square foot enclosure containing one dog. In reality, the choice is not between cages or pens but between cages and exercise/socialisation or pens. In the case of continuous infusion studies, the options become a little more restrictive. The animals must be housed and maintained in such a manner that their infusion preparation and/or apparatus are not disrupted. Clearly, one cannot keep two tethered dogs together in the same pen. The caging requirements for dogs under the various experimental scenarios will be discussed more specifically below. In general, however, dogs with indwelling catheters will need to be maintained separately to ensure the integrity of the preparations, not to mention the experimental results.

It is also not appropriate to feed dogs *ad libitum*. They will tend to overeat and may become obese if the dietary restriction is not followed. A beagle will eat about 1–3% of body weight in feed per day. In practice, we give a level bowl of about 400 g of a certified feed at a set time each morning. If the treatment regimen is such that the animal's appetite is suppressed, then an extended feeding period is warranted. During an infusion study, a dog's food consumption should be determined by weight daily. Body weights should be taken at least twice per week, and more frequently if appetite suppression occurs. And the dose (amount given) should always be adjusted to the most recently obtained body weight.

For the remainder of this chapter we will focus on the preparation and use of dogs in toxicological experiments that require surgical (or near surgical) implantation of indwelling catheters or cannulae, particularly with regard to those types of preparations that result in untethered dogs. Traditionally, a greater emphasis has been placed on the use of tethered animal preparations (Perkin and Stejskal 1994).

13.3 Types of preparations

The emphasis in this chapter will be on techniques that allow for untethered dogs. This is normally accomplished by using combinations of jackets (with protective collars), lightweight pumps and backpacks. For both tethered and untethered preparations, the jugular vein and, most recently, the femoral vein are the two most commonly used sites of catheter implantation for continuous intravenous infusion studies in the dog. One should be aware that the surgical techniques used for exposing, isolating, cannulating and ligating the target vein are quite similar, regardless of whether or not the animal is ultimately tethered or fitted with a backpack.

We will also make special reference in this chapter to the use of vascular access ports. Examples are shown in Figure 13.1. The advantage of a vascular access port is that, once a vein is cannulated, the cannula is not exteriorised; rather, it is attached to a small reservoir equipped with an injection septum that is placed subcutaneously. The vein is then accessed for infusion via needle puncture of the port. This type of preparation has the advantage of greatly decreasing the chance for infection. It is particularly useful for protocols that require discontinuous infusion, i.e. infusion of several hours' duration per day but not for a continuous 24-h period. When properly used and maintained, a vascular access port can last several months in the dog.

As mentioned, the jugular or femoral vein are the two most widely used sites of cannulation for continuous infusion studies. They each have advantages and disadvantages (Table 13.2). The advantage of the jugular vein is that it is more convenient and requires less invasive surgical procedures to cannulate. The disadvantage is that test materials have less of a chance to mix with blood before coming into contact with the heart. Cannulae have to be carefully trimmed to prevent them penetrating the heart. The femoral vein site provides for better mixing of test materials and avoids potential catheter interactions with the heart.

The jugular vein also provides options with regard to the use of percutaneous introduction of the cannula. Percutaneous introduction is a considerably less invasive and traumatising technique, there is a short recovery period and the animals can be put on study 3–4 days post preparation. Percutaneous introduction uses an introducer needle to puncture through the skin into the vein. The cannula is then introduced through the needle. Once the cannula is threaded well into the vein, the introducer needle is removed from the vein and peeled away (Figure 13.2). The cannula is then secured in place with surface sutures and the site overwrapped for protection. More exact and step-by-step details are given below. Spring or stiffness in the reinforcing material can create problems in that the movement of the animals can cause the cannula to work out of the vein. We have developed a modified procedure in which a light surface incision is made at the implantation site to expose the vein. After the cannula is introduced, the skin is sutured closed over the reinforcing strip. This provides for better protection of the introduction site and keeps the cannula in place.

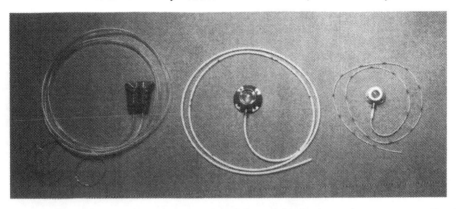

Figure 13.1 Typical vascular access ports. From left to right: dual vascular access port, Davinci Biomedical Research Products; Port-a-Cath access port with titanium portal, SIMS Deltec, Inc.; single-access port, SIMS Deltec, Inc.

Table 13.2 Comparison of percutaneous implantation, traditional cannulation and vascular access methods

	Advantages	*Disadvantages*
Percutaneous	Simpler surgery Better mixing in blood (flow by the tip)	Difficult to use on femoral vein Potential irritation owing to whipping motions
Traditional	Better seating of cannula Less opportunity for tracking	Increased opportunity for tracking infection Increased thrombus formation
Vascular ports	Decreased infection Long-lived preparation	Better suited for intermittent than continuous infusion

13.4 Surgical preparation

A time line for the preparation of dogs for continuous infusion (from receipt of animals to initiation of dosing) is shown in Figure 13.3. The entire process can be divided into the following steps: animal receipt and acclimation, pre-surgical preparation, anaesthesia, surgical implantation, recovery and study initiation.

13.4.1 Animal receipt and acclimation

Animals are acclimated to the facility for approximately 10–14 days before the surgical procedure. During this time, they are acclimated to the feeding schedule, given chew toys and acclimated to the jackets on a progressive schedule. We use old and/or slightly damaged jackets for the acclimation process. The animals are placed in the jacket for approximately 1 h on the first day, 3 h the next, etc. until they are wearing jackets continuously. Some animals accommodate better than others. Animals that continuously struggle or chew at their jackets should be considered for exclusion from study.

(b)

(a)

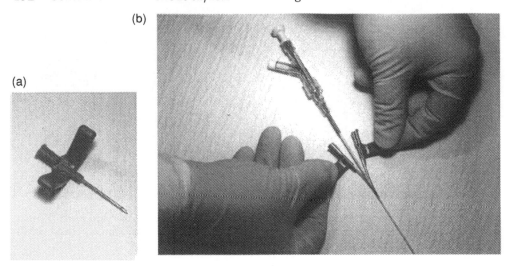

Figure 13.2 (a) Introducer needle. (b) How an introducer needle peels away after insertion of a catheter (SIMS Deltec, Inc.).

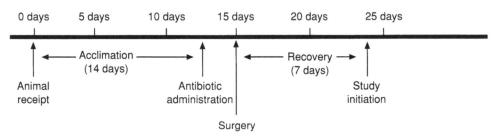

Figure 13.3 Surgical time line for preparation of dogs with implanted cannulae.

Animals must be healthy prior to surgery. Observations for health prior to surgery include: monitoring for normal feed consumption and weigh gain; physical examinations (including at least one veterinary examination); rectal temperature determinations; and clinical pathology evaluation (with emphasis on those parameters that may indicate on-going infection). Because there is a chance that an animal may not appropriately complete the pre-study regimen, dogs in excess of the number actually required for study need to be obtained. We typically obtain one extra dog for every 10 needed to be surgically prepared.

13.4.2 *Pre-surgical preparation*

13.4.2.1 *Aseptic technique*

The most critical aspect of surgical preparation is aseptic technique. The surgery room, including walls, ceilings, surgical equipment and special care equipment, such as monitoring devices and anaesthesia machines, must be disinfected. The air supply should be HEPA filtered. All surgical supplies (e.g. instrument packs, surgeon attire, drapes, cannula, ports) must be sterile. All surgical and support personnel must be properly trained in aseptic technique.

13.4.2.2 Antibiotic treatments

Given the calibre or gauge of the cannulae used on dogs, and the volume of fluid that can be moved through them, maintenance of patency is rarely a problem. Infection can be a problem and we treat dogs prophylactically 2 days before implantation and immediately after the surgical procedure with Naxcel brand of ceftiofur sodium (0.4 ml/kg). It is wise to rotate antibiotics so as to limit the development of resistant strains of bacteria in the laboratory. In addition to systemic antibiotic treatment, we also irrigate the implantation sites with warmed Penicillin G solution (500,000 IU/ ml), and treat the exteriorisation site with Nolvasan antiseptic ointment or McKillip's antiseptic powder.

13.4.2.3 Animal preparation

Food should be withheld overnight prior to surgery to prevent emesis during the surgical or recovery periods and possible aspiration pneumonia after recovery. Once the animal is anaesthetised, the surgical site is shaved. The surgical site is washed with alternating scrubs of betadine and 70% alcohol, after which the animal is draped, leaving only the surgical site exposed.

13.4.2.4 Anaesthesia

For all procedures requiring the implantation of catheters or ports, the dogs have to be anaesthetised. Anaesthetic agents fall into two classes: inhalant or injectable. Most of our procedures have been short and not very invasive, and therefore we tend most often to use injectable agents. Agents such as Telazol provide for short periods of acceptable anaesthesia. For surgical procedures requiring more than 45 min to complete, a volatile inhalant anaesthetic should be used, generally isoflurane. Volatile anaesthetic gases provide for good, well-controlled anaesthesia and analgesia, but the procedures require the use of equipment to administer the gas, ventilate the animals and monitor vital signs. Generally, the animals must be pre-treated with a tranquilliser such as acepromazine, an anticholinergic such as atropine and thiopental or similar ultrashort-acting barbiturate to induce anaesthesia.

13.4.3 Surgical procedures

The procedure for cannulating the jugular vein is detailed in Chapter 11, and so will not be covered here. However, the surgical implantation of vascular access ports (VAPs) will be mentioned.

13.4.4 Vascular access port placement (jugular vein/femoral vein)

A large area is clipped over the vein to be cannulated as well as over the dorsal back in the scapular area. The animal is moved into the dedicated surgery room, placed on a warm air or water blanket and the surgical sites are prepared and draped as described previously.

An incision approximately 4 cm long is made in the skin over the jugular or femoral vein. Using blunt forceps and haemostats, the vein is isolated from the surrounding muscle and interstitial tissue. A proximal and distal ligature of 0 silk is passed around the vein, making one loose throw for each ligature. The clamps at the ends of the ligatures are secured and pulled in the opposite direction to stretch the vessel slightly. The proximal clamp is tightly secured to the drape or towel clamp and the distal ligature tied. A venotomy is made in the vein with microdissecting scissors and the cannula is inserted and advanced to the desired length. The proximal ligature is tightened and secured to the cannula, as is the distal ligature. The incision site is irrigated with 1-2 ml of 500,000 IU/ml warmed Penicillin-G solution. The subcutaneous tissue is closed with 4–0 absorbable suture, and the skin is closed with 9-mm wound clips or 4–0 non-absorbable suture.

At this point, the cannula is threaded subcutaneously with the aid of a trocar and exteriorised in the dorsocervical area (between the shoulder blades). It may be routed through a tether and connected to an infusion pump, or attached to a vascular access port as described below. The length of the cannula and the size of the trocar are dictated by the site of cannulation. A longer cannula and trocar will be required to use the femoral vein.

If a vascular access port is being installed, the vein is first cannulated by the above-mentioned methods and then connected to the port. A subcutaneous pocket is made between the scapula with blunt instruments. The port is filled with saline and connected to the cannula, taking care not to damage the cannula. The cannula is secured to the port via suture or clamps (supplied by the port manufacturer). The port is slid into the pocket and is secured to the underlying musculature with 4–0 non-absorbable suture in such a way that the cannula does not kink. The pocket is irrigated with 1–2 ml of 500,000 IU/ml of warmed Penicillin-G solution. The scapular incision is closed with 9-mm wound clips or 4–0 non-absorbable suture. The port and cannula are flushed with sterile 100 U/ml heparin in saline solution. An injectable antibiotic such as penicillin may be added to the heparin/saline lock. The insertion and scapular incisions are covered with antiseptic cream such as Nolvasan.

The animal is moved to the recovery area and monitored frequently. Once the animal is adequately recovered, the endotracheal tube and i.v. catheter are removed and it is returned to its home cage.

13.4.5 Recovery

Dogs generally recover quickly from the above-described procedures. It is critical to monitor the animal frequently following surgery. Body weight and food consumption are good indications of an animal's health. Haematological parameters should also be monitored. Generally, the animals will be exhibiting normal behaviour and normal appetite the day after the procedure. Analgesics should not be necessary; however, recent literature recommends the administration of pre-operative analgesics, which may decrease sensitivity to pain at the surgical site and reduce the need for post-operative analgesics (Lumb and Jones 1996).

13.5 Study initiation and conduct

The study may be initiated with test article delivery as soon as the animal has recovered from the surgical procedure. Generally, we wait 7 days to ensure that the prophylactic antibiotics have been eliminated from the animal before study initiation and that the animal has fully recovered.

For animals subjected to the percutaneous preparation described above, the cannula is continually flushed with a low steady flow of physiological saline (saline for injection USP). Vascular access ports must be flushed once per week at minimum to maintain patency. The opportunity for infection is decreased if the port site is prepared aseptically before accessing. Further, passing any infusate (for maintenance or dose delivery) through a 0.22-μm filter will decrease the chance of infection. We generally use an extension set with an integrated in-line filter for filtration during dosing.

During the drug delivery portion of the study, recommended flow rates for test article delivery range from 12 to 120 ml/kg/day (0.5–5 ml/kg/h). Both excessively slow and excessively high rates of infusion can be problematic. Low infusion rates may lead to patency problems such as test article precipitation or clotting. High rates of infusion can result in fluid overload, which in turn can result in life-threatening pulmonary oedema. The upper limit of the infusion rate is theoretically set by the glomerular filtration rate, which we have calculated in the dog to be approximately 5 ml/kg/min (300 ml/kg/h). If a 20-fold safety factor is applied to this figure, the upper rate limit for continuous infusion in the dog is 15 ml/kg/h. It would be an extremely unusual study that would require the infusion of so large a volume.

Vehicles for continuous infusion studies must be chosen with considerable care. The pH and tonicity of the infusate should be as close to neutral as possible. As the infusate solution becomes more acidic (or alkaline) or as it become less isotonic, it becomes more irritating, and chronic irritation at the point of infusion can result in a variety of undesirable outcomes such as thrombitis, thrombophlebetis and secondary infections. Chronic use of irritating vehicles can result in sterile abscesses and fibrotic masses that can be quite impressive. We would recommend that any toxicity study conducted by continuous infusion which uses an exotic vehicle should also include a saline control group to control for potential vehicle effects.

Pump accuracy is verified prior to the start of a study. Reservoirs are weighed before and after a suitable use period (e.g. for 24 h for a continuous infusion study). We generally check rates at the low, mid and high end of the rates expected for the up-coming project, using the same vehicle that will be used on the project. Data are tracked by the pump identification number, and the specific pump assigned to a specific animal is documented and maintained in the raw data. We have found the CADD Plus® pumps (Figure 13.4) to be very accurate and reliable over their entire range of delivery speeds. There is a difference, however, between a pump sitting on a laboratory bench and one connected to an implanted catheter. We have found that the rates of delivery during actual usage conditions are generally lower than anticipated based on pump setting and performance during pre-study validation. We hypothesise that this decreased delivery is due to the fact that the pumps are operating against venous pressure. We compensate for this by increasing the pump setting by 10% over that calculated to be appropriate.

Prior to the start of a study, the specific gravity of each infusate is determined. During the study, reservoirs are weighed at each change. We collect weights in our computerised data acquisition system (WTDMS®). During a continuous infusion study, for example, it is advisable to choose an infusate rate such that the reservoir is only changed once every 24 h. However, if the study dictates, the reservoir may be changed

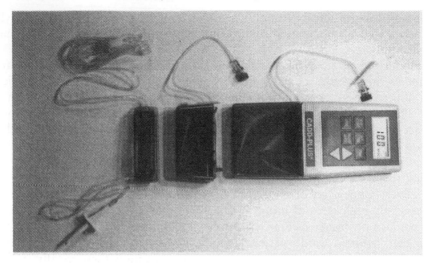

Figure 13.4 CADD Plus® ambulatory infusion pump with 100-ml reservoir attached. Shown with optional 50-ml reservoir and remote reservoir adapter with bag spike. SIMS Deltec, Inc.

several times per day using aseptic technique. For each reservoir, the pre-infusion and post-infusion weights are determined, and the difference is used to calculate the amounts delivered. This is done on a daily basis, and the accuracy of delivery is one of the tools we use to assess system performance while the study is on-going. Data can be presented in terms of ml/h, mg/h or per cent of target.

One of the disadvantages of working with an ambulatory pump system is that the pumps are battery powered. The batteries will have to be changed on a regular basis, with the frequency depending on pump rate. In order to avoid any complications, our procedures dictate that the batteries will be changed with every other reservoir change, as the pump settings are being checked and the backpack system is being accessed. Approximately 500 9-V batteries will be required, for example, for a 28-day study using 32 dogs.

13.6 Different maintenance systems

Once the catheter is in place, a system is required to both protect the catheter and allow for the infusion of test solutions. These fall into two general categories: backpack or tether. In both systems, the dogs must wear protective jackets (Figure 13.5).

Jackets are made by many manufacturers, and it may be necessary to try different varieties to find the one that best suits your needs. In general, we prefer one with canvas-type material on the dorsal aspect and Spandex-type material on the ventral aspect. This type provides for the most protection of the infusion apparatus as well as the greatest comfort for the dog. It is essential that the jacket fits appropriately. A jacket that is too small may cause irritation in the axillary area, but a jacket that is too big may also cause irritation as the weight of the equipment may cause it to shift to one side. Also, an animal may escape from a jacket that is too large. Dusting the axillary area occasionally with an antiseptic astringent powder such as McKillip's may help keep the animal comfortable if it is to wear a jacket for an extended period of time.

Figure 13.5 Dog in ambulatory backpack jacket (Lomir Biomedical) and protective foam collar (Alice King Chatham). Typically, a pump would go in one pocket and infusate reservoir or counterweight in the contralateral pocket.

The backpack jacket is specifically designed for use with an ambulatory system. It usually has a zipper on the midline dorsal area with zippered pockets on the sides. Backpacks with reinforced zippers which can be fastened closed are most desirable to prevent the animal from gaining access to its equipment. The jacket should also be equipped with fasteners to which a protective foam 'C' collar can be attached. The 'C' collar helps to keep the animal from chewing on the pockets of the backpack and from scratching the implantation site of a percutaneous catheter. The pockets should have a reinforced slit on the jacket side to accommodate the extension tubing and bag spike sets. There should also be reinforced areas to run the extension tubing from the pump to the site of catheter access.

With backpack systems, the pump, normally a peristaltic pump designed for human clinical use, is in one pocket of the backpack and the infusate reservoir (generally an infusion bag or a counterbalance weight) is in the other. Typically, the reservoir bag is no larger than 500 ml to minimise the weight on the back of the dog. The pump and infusate bag are attached with bag spike tubing across the back of the animal. Extension tubing then runs through the jacket and to the site of catheter implantation. It is important to keep excess tubing in the pockets so as not to allow the animal access to it. One needs to keep in mind that the 24-h stability of the test article formulation should be determined at normal canine body temperature.

When delivering low volumes of test article (e.g. less than 100 ml), a medication cassette filled with the infusate may be attached to the pump in place of the infusate bag. In this case, a counter weight must be placed in the opposite pocket to balance the jacket. We prefer the CADD pumps supplied by SIMS Deltec. They are microprocessor controlled and provide for a variety of continuous or interrupted flow (e.g. 5 min of delivery every hour). They provide a wide variety of choices in extension sets, bag spike sets and medication cassettes. The original models (CADD Plus®) included high-pressure alarms, but these are being phased out at the time of writing, and will be replaced with pumps that operate in a similar basic (peristaltic) fashion. In addition, these new programmable pumps have more options such as computerised monitoring, pressure and flow alarms that can be routed to a pager or cellular phone to alert staff to problems such as kinked lines or low batteries.

In a tether system, the cannula is run through the jacket and a protective coiled metal protective sleeve to a swivel mounted usually on the cage door. The pump delivering the infusate (normally a syringe pump) is then connected to the other side of the swivel, and is mounted on the outside of the cage. Advantages and disadvantages of the two systems are given in Table 13.3. Of the two systems, we prefer the backpack system because it allows the dog greater freedom of movement.

13.7 Catheter maintenance

Manufactured cannulae and ports are readily available from many companies in various sizes, lengths, materials and anticoagulant coatings. When choosing a cannula it is important to test if the cannula material is compatible with the test article. Silicone is an excellent choice as it is biocompatible, and is very easy to work with; however, polyurethane or Teflon® may be a better choice depending on the test article to be infused. Anticoagulant coating of the catheter is generally not necessary for use in the dog, as patency is rarely an issue with the large-bore cannulae used. For a percutaneous preparation, recommended maintenance (when test article is not being infused) flow rate is approximately 0.5 ml of sterile physiological saline per hour.

When choosing a port, it is best to remember that a low profile will cause the least amount of irritation to the skin. Various materials are available in ports such as titanium, polysulphone and silicone rubber. Ports can be maintained with 0.5 ml of sterile physiological saline per hour or may be flushed and locked with 100 IU/ml sterile heparin/ physiological saline solution at least once per week. It is critical to have only highly trained personnel maintaining vascular access ports. They must aseptically prepare the skin with betadyne and alcohol scrubs prior to access, and ensure that all materials used (e.g. needles and syringes) are sterile prior to use.

Regardless of which catheter or port system is used, it is important to revalidate the system if any change in material takes place. If the manufacturer adds a reinforcing piece, new suture bulbs or any other changes, it is highly advisable to test the new equipment before starting a study. Even something as mundane as a 500-ml reservoir bag provided by a different supplier may cause a problem, as the slight difference in the bags may influence they way they drain and therefore the fashion in which the pump draws on them.

Table 13.3 Comparison of backpack and tether systems

	Advantages	*Disadvantages*
Tether	Pump protected from cage environment Infusate not kept at body temperature No batteries, direct line power Ease to change reservoir	Swivel system more prone to leakage Dog lacks freedom of movement
Backpack	Increased animal mobility	Limited selection on pump type Frequent checking of batteries Animal must be manipulated to change the reservoir

13.8 Pathological complications

Continuous infusion results in the introduction of foreign material into the body of an experimental animal. The response and complications are fairly predictable on that basis.

Localised changes at the injection site may include some minor endothelial damage, thrombus formation, focal hyperplasia/thickening and inflammation, catheter obstruction from fibrin or collagen sheath formation or test article precipitate. Note that high rates of infusion and irritating infusates can exacerbate these effects. Localised response to the catheter also depends on the type of material in the catheter. (See Trooskin and Mikulaschek 1994 for a more complete review on the types of biological response to the different catheter materials.) We have had the best success with medical-grade Silastic® tubing.

Chronic localised inflammatory and thrombogenic changes can lead to predictable sequelae. First, thrombi may break loose and deposit in the lung or kidney. Second, fibrotic response at the injection site can be quite impressive, resulting in masses that could be (in the dog) several centimetres across. Severe localised reactions can make the site more prone to bacterial infections.

Although not as large a problem as venous irritation, reactions on the surface of the animal must also be kept in check. Infection and irritation at the exteriorisation site can be avoided by appropriate preventative maintenance. An ill-fitting jacket can also cause localised reactions. The constant rubbing of a jacket can result in dermal reactions (hair loss, scabbing) at a site remote from the exteriorisation site (e.g. the ventral neck).

13.9 Systemic complications

The most frequent, if not expected, systemic reaction to cannula implantation is fever. Rectal temperature is checked often during the recovery period, and an animal is not assigned to study if the temperature is outside the normal range.

Haematological changes include increases in the total white blood cell and neutrophil counts and often decreases in platelet counts.

General pathological changes may include increased lymphoid hyperplasia in the spleen, increased adrenal cortical vacuolation and generalised increases in white blood cells. In general, the histopathological changes we have observed are consistent with those reported by Perkin and Stejskal (1994): thromboemboli around the implanted canula and then also disseminated to the lungs.

13.10 Conclusion

In summary, the reader should now be aware that there are a variety of tools and techniques available for conducting toxicity studies in the canine via the intravenous route of infusion. The emphasis in this chapter was on those techniques that result in a dog that is untethered, i.e. preparations in which the animals wear a jacket or backpack (with test article and pump deliver system on board). In selecting the most appropriate techniques for a specific study, careful consideration must be given to all aspects of experimental design. For short-term (14-day) continuous infusion studies with relatively non irritating substances, perhaps percutaneous introduction of the cannula via the jugular vein is appropriate. For longer-term studies with a more toxic or irritating material, surgically implanted femoral vein cannulae are perhaps more appropriate. For studies requiring intermittent infusion (e.g. once weekly), a preparation using vascular access ports would be appropriate.

All such studies require acute attention to detail with regard to cannula placement and maintenance and recognition of the limitations of the systems employed. For example, highly irritating materials will be problematic regardless of the techniques used for preparation of the animals. Also, the use of untethered animals may not be appropriate for all situations requiring infusion in dogs, such as those that require the infusions of large volumes (in excess of 500 ml per day) or of heat-sensitive test articles.

References

Gad, S. and Chengelis, C. (1998) Lethality testing. In *Acute Toxicology Testing*, 2nd edn, pp. 157–194.

Haggerty, G., Thomassen, S. and Chengelis, C. (1992). The dog. In Gad, S. and Chengelis, C. (eds.) *Animal Models in Toxicology*. Marcel Dekker: New York, pp. 567–674.

Institute of Laboratory Animal Resources Commission on Life Sciences National Research Council (1996) *Guide for the Care and Use of Laboratory Animals*. National Academy Press: Washington, DC.

Lumb, W.V. and Jones, E.W. (1996) *Veterinary Anesthesia*, 3rd edn, pp. 51–53.

Perkin, C. and Stejskal, R. (1994) Intravenous infusion in dogs and primates. *Journal of the American College of Toxicology* 13, 40–47.

Trooskin, S. and Mikulaschek, A. (1994) Biomaterials used for catheters. In Greco, R. (ed.) *Implantation Biololgy: The Host Response and Biomedical Devises*. CRC Press: Boca Raton, FL, pp. 267–286.

Part 5

Continuous intravenous infusion in the primate

The primate is utilised as an alternative non-rodent species to the dog for pre-clinical testing of pharmaceuticals. The volume of background data available is therefore arguably less than for the dog, but in the laboratories where this species is commonly used this is certainly not the case.

This section describes surgical preparation of the large non-human primate, such as the cynomolgus monkey, via the external iliac vein, and of the smaller marmoset, via the femoral vein. These species are well suited to the use of vascular access ports (VAPs), and the methodology described in these chapters utilises these. The larger primate is an established model for continuous infusion, and the techniques involved in conducting a regulatory general toxicology study are described in Chapters 14 and 15. There is also the choice to be made between tethered and ambulatory models. Although the latter would seem to be preferable from a welfare point of view, in certain situations the tethered model is advisable. For example, if the test compound is heat sensitive, then placing it within a pouch in a jacket close to the animal's body (which can exceed 30 °C within a short period) is not advisable. Other factors to aid the choice of the most suitable model are discussed in Chapter 15.

The marmoset, however, has not, to the editors' knowledge, been used for more than short-term studies but is nevertheless a model that could usefully be developed, not least because of the size of the animals and the reduced compound requirement. Hopefully, the experiences of the author of Chapter 16 will serve to aid any worker who wishes to pursue this model further.

14 Surgical preparation and infusion of the large primate

V. Mendenhall, S. Cornell and M. A. Scalaro

14.1 Introduction

Continuous infusion is the instillation of a fluid compound into a laboratory animal for an extended period of time by any number of routes. The most common application for the associated technology surrounding continuous infusion is to the venous side of the circulatory system; however, long-term access to virtually any other tubular organ in the body is possible today (Jacobson 1998). Chronic access requires the integration of several systems, including surgical placement of appropriate catheters, a means of preventing the animal from disrupting the external infusion lines and an infusion pump. Refinements of these systems over the last 30 years have largely been attempts to eliminate the tether system through the use of programmable-wearable pumps and to decrease thrombotic and infectious events associated with the implanted catheter (Hecker 1981; Peters *et al.* 1982; Winocour *et al.* 1982; Marosok *et al.* 1996). The emphasis in this chapter will be on the perioperative procedures regarding placement of catheters and vascular access ports in non-human primates that we have developed over the last few years. We will also briefly discuss what we believe to be the important technical issues involved for all components of the system. Careful adherence to these procedures has allowed us to successfully perform continuous infusion studies in large numbers of primates for periods as long as 6 months.

14.2 Pumps (tethered and tetherless)

The choice of pump is the most difficult selection decision because there are a myriad of options for a variety of applications and pumps are the most expensive component of infusion systems. Pumps are classified as 'low-flow' or 'high-flow' and as 'tethered' or 'tetherless' (ambulatory) pumps (Jacobson 1998). The various attributes of the pumps currently available vary considerably, and price is commensurate with each one's features and capabilities. As such, the selection of pumps must be based upon the needs of the study and the capabilities of the facility.

14.3 Tether and jacket systems

The jacketed-tether apparatus (Pickens *et al.* 1966) remains the prevailing means for tethering non-human primates intended for studies involving continuous or intermittent infusion (Figure 14.1). Tethers are constructed of stainless steel into a flexible metal coil to act as a conduit for the catheter. A swivel allows the catheter or extension set to extend from the pump to the animal without becoming twisted as it moves about its cage. The jacket should be fitted to the animal based upon the manufacturer's recommendations regarding the ratio of jacket size to body weight. This allows for movement of the animal within the jacket, yet prevents it from having access to the catheter system under the jacket. The areas around the neck and arms are particularly prone to abrasions and lacerations if the jacket is inappropriately adjusted to the animal. It must fit about the limbs and neck and be tight enough to prevent intervention, yet not so tight as to cause discomfort and chafing. As a general rule, the neck region should allow for two fingers to be passed easily between the animal and the jacket, and the arm region should allow for one finger to be easily passed.

14.4 Catheters

The choice of catheter materials has remained relatively unchanged over the past 30 years. Silicone (Silastic®), polyvinylchloride (PVC or Tygon®) and polyethylene (PE) were the materials that had been most commonly used for this purpose (Jacobson 1998). In an effort to decrease thrombotic and infectious events, two notable improvements in catheter access technology have recently been introduced to this field. The use of polyurethane as a catheter material has reduced the incidence of thrombotic events not only when used by itself, but especially so when it is coated with a hydrogel. In addition, the technology to covalently bond heparin to both silicone and polyurethane has recently been developed, thus offering great promise in this area (Arnander *et al.* 1987; Appelgren *et al.* 1996; Jacobson 1998).

14.5 Vascular access ports

Subcutaneously placed vascular access ports (VAPs) have largely replaced percutaneous, externalised catheters, thus greatly reducing the chance of catheter-based infections (Hoeprich *et al.* 1982; Dalton 1985; Gilsdorf *et al.* 1989; Patrick *et al.* 1992; Raad and Bodey 1992; Wojnicki *et al.* 1994; Kinsora *et al.* 1997). The differences between these methods are that externalised catheters exit the animal's skin though a chronic skin opening, whereas the VAP catheter terminates in the subcutaneous tissue in a metal or plastic cylinder with central rubber septum that is accessed percutaneously. The VAP's distal catheter portion is identical to the externalised catheter's distal portion. The externalised catheter's proximal end is connected to the infusion pump or through the jacket/tether/swivel system. VAPs were developed in the early 1980s as a method of delivering chemotherapeutic agents to humans (Jacobson 1998). They were soon used in laboratory animal medicine because of their many perceived advantages over percutaneous catheters (Dalton 1985). As VAPs are catheters that do not exit through the animal's skin, there is no chance of the animal disturbing the device. Although they are best suited for intermittent bolus infusions and periodic sampling or access, they are also especially well suited for use in protracted or continuous infusion studies. The VAP can be accessed with a right angle non-coring needle or catheter that can be subsequently connected to the same pumping apparatus used for continuous infusion through externalised catheters. Their use in this application minimises the chance of infection, since the 'exit' site is so much smaller than that induced by a conventional percutaneous catheter. If the needle's 'exit' site is changed frequently (about every 3 days), the risk of infection is further minimised (Landi *et al.* 1996).

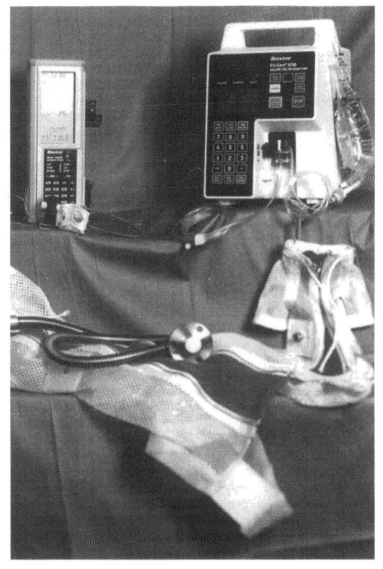

Figure 14.1 The *ex vivo* components of a continuous infusion system: two types of tethered pumps, swivel, small and large jackets and a tether.

VAPs and their associated catheters are foreign bodies, and represent a large surface area in comparison with the size of animals into which they are commonly implanted. It has been estimated that the surface area to body mass of a conventional port placed in a 3.0-kg cynomolgus monkey is nearly equivalent to that of an artificial hip placed in a 75-kg human (Mendenhall *et al.* 1997). Therefore, the successful implantation and later accessing of VAPs is vitally dependent upon an atraumatic and careful surgical implantation procedure, following the strict and conscientious practice of the principles of asepsis at all times. This is true both in their implantation and in their later access.

14.6 General surgical considerations

Everyone involved in a research project that involves surgical manipulation of an animal must embrace the idea that the requirements for successful survival experimental surgery on animals are in fact more rigid than those for clinical surgery, both in animals and in humans. Research studies that involve surgical manipulation of the animal must demonstrate an appreciation of all elements of the perioperative care of the animal, not just the technique itself. The surgeon and his/her team must have an intimate knowledge of the procedures and routines required to ensure aseptic results, as well as possess adequate surgical skills.

Consideration of eight major factors will ensure that all surgical operations will be performed to a uniformly high and reproducible standard. Inattention to any one of these points will invalidate the positive effects of the others. These principal areas of concern include (1) the operating room environment, (2) the necessary equipment, (3) anaesthesia, (4) asepsis, (5) animal preparation, (6) personnel, (7) technique and (8) post-operative care (Mendenhall 1999).

14.6.1 Operating room environment

The specific type and extent of the experimental surgical facilities needed depends to some extent on the animal species and the complexity of the surgery. However, in general, the suite should have only one entrance and be under positive pressure in relation to other areas of the facility. It should have 10–15 non-recirculating air changes per hour, introduced into the suite through HEPA filters. If inhalation anaesthetic agents are to be used, a mechanism for scavenging waste gases must be provided. All furniture, including the operating lights, should be wiped down with antiseptics prior to the animal entering the room (Atkinson 1992a; Brown *et al.* 1993).

14.6.2 Equipment

Simplicity and standardisation are essential in every successful operating room, but the instrument cabinet of the experimental surgical suite should never be the repository of antiquated or discarded instruments. The following is a suggested list of necessary surgical instruments for implantation of an external iliac-based catheter and vascular access port in primates:

   1 small bowl with 4 ×4 gauze sponges
   1 #3 scalpel handle
   1 Brown–Adson thumb forceps
   1 Adson thumb forceps
   2 micro right-angle curved forceps
   2 Harms tying forceps
   1 #5 jeweller's forceps
   4 small towel clamps
   4 mosquito haemostats
   1 small Gelpi self-retaining retractor
   1 medium Gelpi self-retaining retractor
   1 tenotomy scissors
   1 curved Metzenbaum scissors

1 4-mm ID trocar
1 micro-scissors
1 needle holder
1 right-angle clamp
1 Mayo scissors

Instrument packs should be double-wrapped prior to autoclaving, using appropriate indicator tape and expiration dates. Packs wrapped in linen are sterile for only 3–4 weeks, while those wrapped in paper or plastic may retain sterility for 3–6 months (Atkinson 1992b).

14.6.3 Anaesthesia

Anaesthesia is a form of temporary, controlled poisoning, and should, therefore, be treated with the utmost respect. A comprehensive list of specific agents and methods of anaesthesia is beyond the scope of the current discussion; however, there are many excellent texts that deal with this subject (Lumb and Jones 1984; Flecknell 1987). Endotracheal intubation, in conjunction with controlled or assisted respiration utilising an automatic ventilator to deliver inhalant anaesthetic agents, is the best choice for non-human primates.

A number of drugs are currently available to induce what is thought to be the safest method of anaesthesia, commonly termed 'balanced anaesthesia'. In general, this term implies the use of multiple agents. These agents include (a) pre-anaesthetic anticholinergics, narcotics or tranquillisers; (b) barbiturates, dissociative agents or narcotics for induction of anaesthesia; and (c) inhalation anaesthetic agents for maintenance of anaesthesia. Such combinations reduce the required dosage of each agent and in general provide a safer state of anaesthesia. Accepted surgical protocol requires repeated assessment of the physiological status of the animal throughout the surgery. Circulatory and respiratory function, as well as core body temperature, need to be monitored during the anaesthetic and surgical episodes.

14.6.4 Fluid therapy

The purpose of fluid therapy during anaesthesia and surgery is to inhibit antidiuretic hormone levels and induce mild diuresis so as to counterbalance anaesthetic drug-related oliguria and to maintain urinary output during the operative and post-operative period. Attenuation of this antidiuresis can be achieved by replacement fluid therapy during surgery. Tissue desiccation, evaporative losses from the lungs and surgical site, redistribution of extracellular fluid (third space loss) and fluid deficits from the pre-operative fasting period are iso-osmotic losses and therefore can be replaced with the intravenous administration of iso-osmotic solutions. A polyionic fluid, such as lactated Ringer's solution, retards the formation of excess free water that occurs with dextrose-based fluids or hypotonic preparations. It is, therefore, the fluid of choice. An infusion rate of 5–15 ml/kg/h will fulfil all maintenance requirements (Raffe 1985).

14.6.5 Asepsis

Surprisingly little experimental work has been done on the problem of implant infection.

It still appears to be a generally held belief that animal surgery does not require as stringent an adherence to aseptic technique as does human surgery. This erroneous belief may be perpetuated by the lesser importance placed on casualties in animal surgery. Even the belief that rodents are not affected by surgical infections, or that they are more resistant to infection than are other laboratory animals, has not been supported (Dougherty 1986). No species is really any different in its susceptibility to the infective process (Brown *et al.* 1993). In fact, rodents often serve as animal models for studying bacterial infection (Dougherty 1986).

Animal surgery in general, but particularly that involving implanted materials, demands the most rigid aseptic technique. Aseptic routine can, indeed, be both time-consuming and inconvenient, but these considerations are inconsequential when weighed against the cost of compromise to the study. The only treatment for an infected implant is to remove it (Dougherty 1986; Passerini *et al.* 1992; Landi *et al.* 1996; Mendenhall *et al.* 1997). Therefore, an infected implant results in a total waste: waste of the animal's life, of the investigative team's time, of the money spent to place the implant and of the statistical information that would have been provided by the animal. It must be the surgical and post-operative team's very specific goal to perform the implantation and access procedures under the most strictly aseptic conditions possible.

14.6.6 *Animal preparation*

All hair clipping and initial preparation of the animal must be done outside the operating area with an electric clipper and fine blades. Hair removal should be done as near the time of operation as possible, preferably immediately before. Nicks inflicted in the skin several hours before operation have been shown to promote wound infection (Alexander *et al.* 1983; SATS AORN 1992; Shirahatti *et al.* 1993). The hair should be generously removed before draping so that none will become exposed during the operative procedure. Following hair removal, the general area of the operation should be washed thoroughly with antiseptic soap and water before moving the animal to the operating room for the definitive preparation of the skin.

Methods of skin preparation vary and change all the time with the introduction of new products. In general, however, skin preparation is accomplished with an antiseptic detergent applied in a circular motion starting from the proposed line of incision and working out, never going over the same area more than once, to avoid recontamination. The detergent is then wiped off and 70% isopropyl alcohol sprayed on the site. This procedure is repeated three times. The final application of alcohol should be allowed to dry. Finally, a solution of povidone-iodine or similar agent is sprayed on the entire area and also allowed to dry (Polk *et al.* 1983; Guideline for Use of Topical Antimicrobial Agents 1988; Atkinson 1992c; SATS AORN 1992). Alternative methods are now available with the introduction of disinfectant/tackifier solutions. These can be applied to the skin after only one preparation as described above. In addition, they greatly aid in the adherence of adhesive 'incise' drapes to the animal's skin (Gilliam and Nelson 1990; Rochat *et al.* 1993; Roberts *et al.* 1995; Gibson *et al.* 1997).

Draping of the animal is of utmost importance. No area of the animal except the area immediately around the proposed incision site should be visible. All furniture with which operative personnel might come into contact should also be draped. Plastic impermeable drapes are preferred to linen because, once linen gets wet, it no longer functions as a bacterial barrier (Atkinson 1992c). Initially, four small drapes are placed around the incision site, with larger ones then used to cover the animal and furniture. Plastic 'incise' drapes are useful for final draping of the incision site not only for isolating the skin from the tissues beneath, but also for holding the drapes around the operative area in place. If applied after the antiseptic solutions have been allowed to dry, or if 'tackifier' is applied to the skin before their application, they adhere well to animal skin.

14.6.7 Operating room personnel

Before entering the surgical suite, all personnel must don a scrub suit, cap, mask, protective eyewear and shoe covers. No hair should be visible on any person entering the operating room. Members of the team who are to touch the sterile field should scrub their hands and arms to the elbows with an antiseptic detergent solution before each operation. Hand scrubbing by the surgeon and his assistants should be done in essentially the same manner as preparation of the animal's skin for surgery. A sterile gown is donned, and gloves put on using the closed technique or with the assistance of a previously gloved person (Atkinson 1992d).

14.6.8 The operative technique

The majority of technical advances in wound care over the past century have been based on a 'minimal interference' concept: the less 'surgery' the surgeon does, while removing all impediments to normal wound-healing processes, the better the result. Indeed, all technical considerations in surgery may be viewed as methods of minimising the surgeon's interference with normal healing. Extensive tissue injury can prolong the inflammatory phase for months. Large amounts of fibrin, dead tissue fragments, haematomas and fluid collections due to dead space formation act as physical barriers, preventing normal fibroblast penetration and delaying collagen fibre production. The concept of the minimisation of the three Ts – 'time, trash and trauma' – will help lead to a successful outcome.

The maxim of implantation surgery is to leave the animal as a system, and the tissue into which the foreign material is implanted, as normal as possible after the implantation. Maintaining tissue viability is the essence of proper surgical technique. Good surgical technique includes gentle tissue handling, effective haemostasis, maintenance of sufficient blood supply to tissues (which in the case of VAP implantation means ensuring that blood supply to the overlying skin is maintained), asepsis, accurate tissue apposition, proper use of surgical instruments and expeditious performance of the surgical procedure (Mendenhall 1999).

14.6.9 Post-operative care

Three overlapping phases characterise the post-operative period: recovery from anaesthesia and short- and long-term post-operative care. As it is the period of greatest physiological disturbance, recovery from the anaesthesia is the most critical time in post-operative care. The animal should be positioned properly to prevent any impediments to normal respiratory and cardiovascular functioning, the physiological condition should be monitored regularly, and the body temperature maintained.

With respect to the need for post-operative analgesics, an important principle prepared by the Interagency Research Animal Committee and used by the United States Public Health Service (USPHS) is that, 'unless the contrary is established, investigators should consider that procedures that cause pain and distress in human beings may cause pain or distress in other animals' (US Government 1985).

14.7 Implantation of indwelling intravenous catheters and vascular access ports in the non-human primate

14.7.1 Anatomy

Although the tip of indwelling catheters can be placed in almost any location in the body, the most common site for continuous infusion of test articles is into the proper vena cava or right atrium. The choice of which peripheral vein to use as an initial entrance site for the catheter is largely dependent upon the surgeon's experience, preference or simply continuing what he/she was initially taught. However, in our experience of over 500 such implantations over the last 6 years, the best entry site in the non-human primate is the external iliac vein, accessed in its retroperitoneal location immediately ventral to the iliac crest.

The external and internal jugular veins of the primate are of similar diameter, as are their multiple tributaries near the thoracic inlet (innominate, subclavian, superficial cervical and transverse scapular veins). In addition, the superior vena cava is relatively long, being composed of the combined innominates and the azygos vein. In contrast, the vena cava proper is quite short but is formed by many relatively large vessels, including the superior hemiazygos and the veins of the first and second interspaces as they join the left subclavian. On the right side, the first and occasionally the second segmental veins empty into the right innominate vein (Lineback 1971). As a consequence, placing the tip of the catheter into the vena cava proper and into the right atrium can be quite difficult, as it can easily enter any of these other tributaries while being advanced towards the heart from either the right or left internal or external jugular vein. In addition, animals intended for continuous infusion studies must commonly wear a relatively heavy collar around their neck to facilitate conscious capture and restraint, as well as the jacket. If the animal has undergone neck surgery for access to one of the jugular veins, the collar cannot be placed on the animal for a minimum of 2 weeks after the procedure until the skin incision has healed. Even then, its presence, and occasionally that of the jacket, can be quite irritating to the operated area, and can occlude the catheter.

Although the femoral vein is easily accessible, it is relatively small, thus limiting the size of catheter that can be implanted into it. In addition, because of non-human primates' proclivity to sit in a haunched position and pick at anything unusual on their body, the incidence of dehiscence of inguinal incisions can be quite high.

The external iliac vein, on the other hand, is larger than the femoral vein, and its access near the iliac crest is not affected by the stature of the animal. Also, the animal cannot see this incision, and is thus less likely to disrupt it manually. This vein extends cranially from the inguinal ring, being joined by the smaller hypogastric vein at the cranial end of the sacrum. It continues as the common iliac vein until it enters the abdomen at the level of the sixth to seventh lumbar vertebrae. It is covered in this area by the psoas major and minor muscles, the latter of which is easily identified by a broad, flat and glistening tendon. Only the obturator vein joins it in this area, making its mobilisation and subsequent catheterisation relatively easy (Lineback 1971). It does lie within retroperitoneal fat, and the surgeon must remember that the abdominal contents are directly medial to the vessel.

14.7.2 Surgical procedure

The following peri-operative and operative procedures are the current ones that we employ at our institution. They are intended as guidelines only, as the use of specific agents and procedures can vary greatly yet result in the same post-operative result. It is always possible to compromise on specific methods, but one should never compromise on principles.

14.7.3 Analgesia

Buprenorphine (0.01 mg/kg i.m.) is administered to the animal as a pre-anaesthetic agent. It is well documented that analgesics are much more effective if their administration is initiated prior to the surgical procedure (Love 1999). Normally, only the one injection is necessary; however, if the animal elicits any of the classical signs indicative of post-operative pain, this analgesic can be continued as long as necessary, being administered at intervals of 8–12 h each.

14.7.4 Clinical assessment of pain

Wixson (1994) classified the clinical assessment of pain as follows:

> *Behavioural changes*: immobility, reluctance to move, lack of appetite, abnormal vocalisation, abnormal posturing.
> *Physiological signs*: pupillary dilation, blood pressure fluctuations, increased heart rate, increased rate of breathing, increased body temperature.
> *Attitude changes*: unresponsive, depressed, anxious, apprehensive, hypersensitive, and aggressive.

14.7.5 Pre-operative procedures

14.7.5.1 Anaesthesia

Atropine sulphate at 0.04 mg/kg, is administered subcutaneously (s.c.) along with the buprenorphine approximately 15 min before initially anaesthetising the animal with ketamine hydrochloride at 10 mg/kg, administered intramuscularly (i.m.). Endotracheal intubation is then performed as soon as possible in order to maintain an open airway and to allow for anaesthetic maintenance through the use of isoflurane. A ventilator is used to control respiration and maintain a slightly alkalotic state. An intravenous catheter is placed in a peripheral vein for administration of lactated Ringer's solution at a rate of 10 ml/kg/h. Any additional drugs for appropriate anaesthetic management should be readily available for administration if indicated.

14.7.5.2 Surgical preparation

An ophthalmic lubricant is applied to each eye to prevent corneal desiccation. The hair is removed from the abdominal and femoral region, and from over the entire pelvis. A preliminary wash of the area is performed with a suitable disinfectant and detergent solution, and the animal is then taken to the operating room. It is placed in lateral recumbency, according to the preference of the surgeon. In general, if the surgeon is right-handed, the animal should be placed on its left side to allow access to the right iliac vein. As the procedure is performed with the surgeon facing the animal's back, introduction of the catheter into the vessel is facilitated if this can be done with the dominant hand. The surgical site is then prepared and draped for aseptic surgery as described earlier. The area is then draped appropriately for strict aseptic surgery.

14.7.6 Surgical procedure

The skin incision is made approximately 1 cm ventral to the iliac crest, and is about 4–5 cm long. The underlying muscles are exposed, and the wound retracted with two self-retaining retractors. The external iliac artery and vein are exposed by incising through the gluteal fascia, and by performing a longitudinal myotomy of the fleshy portion of the gluteus maximus lateral to the sacrum. The psoas minor can then be identified by its characteristic broad, flat tendon (Figure 14.2). The self-retaining retractors are then repositioned to retract the psoas minor dorsally, and the retroperitoneal fat ventrally.

The surgeon must remember that an incision into this fat, or placement of a retractor too deep into it, will enter the peritoneal cavity directly. The artery overlies the vein, and therefore must be retracted separately, using a temporarily placed long ligature. The external iliac vein is then mobilised a distance of about 1–2 cm cranial to its exit from the femoral ring, but prior to its entrance into the abdominal cavity. Two encircling ligatures are placed around the vein, approximately 1 cm apart, and the distal one is

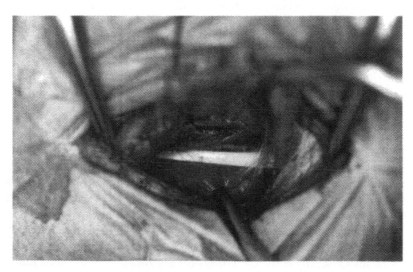

Figure 14.2 Exposure of the external iliac artery ventrally and the tendon of the psoas minor muscle dorsally.

tied (Figure 14.3). A 5-Fr catheter is then attached to the VAP and filled with irrigation solution. While elevating the cranial ligature, a small phlebotomy is made between the two ligatures and the tip of the catheter is introduced through it into the vein a distance of about 10 cm. To ensure proper placement of the tip, blood should be aspirated through the system and flushed back into the animal. The ligatures are tied around the catheter to secure it in place (Figure 14.4). A trocar is passed subcutaneously from the ilial incision to exit the skin at a point overlying the last few ribs, and about 1 cm ventral to the dorsal midline. The catheter is temporarily occluded with a 'guarded' haemostat. Its distal end is removed from the VAP so that it can be passed through the trocar to exit the skin at this location. The exit site is lengthened and the underlying subcutaneous tissue elevated from the muscle a sufficient amount to accommodate the presence of the VAP. It is very important to ensure that its septum comes to lie a minimum of 1 cm dorsal to this skin incision (Figure 14.5). At least two sutures of non-absorbable material are then pre-placed in the underlying muscle and through the 'stay' holes of the VAP, so as to facilitate their placement. These sutures prevent the VAP from rotating or even turning over completely within the subcutaneous pocket. The VAP is placed within the pocket, and the 'stay' sutures tied. The cranial portion of the catheter is cut to an appropriate length so that it can be reattached to the VAP in such a manner as to prevent any kinking of the catheter as it enters the VAP. The system is filled with a 'locking' solution designed to help prevent the build-up of thrombus at the tip of the catheter. The preferred 'locking' solution is 50% dextrose, containing 100 U/ml heparin, and 1 mg/ml vancomycin (Schwartz *et al.* 1990; Cowan 1992). This hypertonic solution acts as an osmotic pump, as interstitial fluid can enter the catheter, maintaining a slightly positive pressure within it. In addition, such a highly hypertonic solution is bacteriostatic.

The wound is closed in at least two layers in such a way as to eliminate as much 'dead space' from around the VAP as possible, utilising appropriately sized absorbable

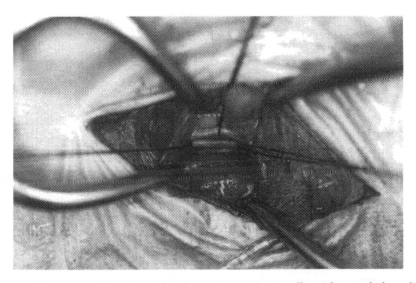

Figure 14.3 The external iliac artery has been retracted ventrally with a single long ligature. The external iliac vein has been ligated distally, and is being elevated cranially with an untied ligature.

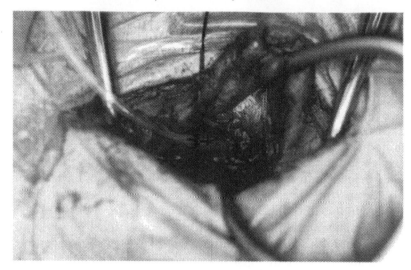

Figure 14.4 The catheter has been introduced into the external iliac vein, and is held in place with the two previously placed ligatures. The distal one is tied caudal to a 'suture bulb' on the catheter.

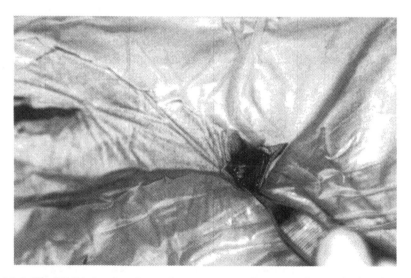

Figure 14.5 The VAP is in place in a subcutaneous pocket created over the last few ribs. Note that the septum of the VAP is a minimum of 1 cm away from the skin incision.

suture material. The VAP should not be visible at all once the first layer of closure is completed. The skin is closed with appropriately sized absorbable suture material, placed in a subcuticular pattern. The ilial incision is likewise closed in at least two layers, taking care not to involve the retroperitoneal fat. The ureter lies within this fat, and it can be inadvertently ligated should the closure sutures be placed too deep within it. Again, the skin should be closed with small absorbable suture material placed in a subcuticular pattern. The animal can then be allowed to recover from anaesthesia.

14.7.7 VAP maintenance

As many different methods to maintain patency of the VAP and catheter system are published as there are groups performing this work. Apart from ensuring that access to the port system is conducted under strict adherence to the principles of asepsis, the general principle is to minimise the number of different people performing the designated tasks (Coatney and Kissinger 1997). Accessing the system is a technical skill, developed over time. If the system becomes permanently impatent or infected, the only treatment is to either reoperate on the animal or euthanise it. These animals are extremely valuable and should be handled only by personnel that have experience in caring for them.

The reason for maintenance of the system is to lengthen the amount of time that the port and catheter is functional. Early on, the objective is to prevent the build-up of fibrin and thrombus over the tip of the catheter. Later, encapsulating fibrous tissue can completely cover the catheter and occlude its lumen (O'Farrell *et al.* 1996). The following procedures have worked well in our institution to maintain patency and system function in a large percentage of cases for periods longer than 1 year. Of course, when the system is to be used for continuous infusion studies, these issues are of less importance, since patency of the system will be ensured while the animal is under study.

Maintenance should be conducted under the strictest adherence to the principles of asepsis as possible. It should be performed daily for the first 3 days after implantation, and then approximately monthly. The skin over the VAP is first cleansed with a disinfectant/detergent solution, and then wiped with 70% isopropyl alcohol. This procedure should be repeated at least three times. Finally, the site is covered with a disinfectant solution and allowed to dry. Alternatively, the site over the VAP can be washed clean once with the disinfectant/detergent solution and 70% isopropyl alcohol and, when dry, covered with a disinfectant/tackifier solution. Wearing sterile gloves, a sterile non-coring needle attached to an injection cap is inserted into the septum of the VAP. Using a 3- to 5-ml syringe with attached cannula or needle, the syringe is aspirated to withdraw the previous 'locking' solution. The syringe is then removed and replaced with a 3- to 5-ml syringe filled with sterile saline or similar solution. The VAP and the catheter are then vigorously flushed, using a pulsatile motion while rotating the syringe to clear the port of any residual material. The second syringe is then removed and replaced with a third one containing a volume of 'locking' solution equal to the 'dead space' of the VAP and catheter + approximately 50% or more. As the last few tenths of a millilitre are injected into the VAP/catheter, the needle is withdrawn while still injecting in order to minimise the chance of backflow of blood into the tip of the catheter. Care should be taken to secure the VAP to prevent its lifting while withdrawing the needle. Should the system become occluded with thrombus, an attempt can be made to clear the clots with 0.1 N hydrochloric acid or 5000 U/ml urokinase (Schulman *et al.* 1988; Tschirhart and Minoo 1988; Fraschini 1991; Lawson 1991; Holcombe *et al.* 1992; Stone 1996). Urokinase is most effective in the first few days after implantation. Later on, hydrochloric acid is more effective. Further attempts at clearing the catheter can then be instituted after at least 30 min, although more success is usually gained if the hydrochloric acid or urokinase is left in place overnight.

14.7.8 Continuous infusion

Before initiation of the study, the specific types of catheters and filters to be used should be infused with the actual test article to ensure catheter/filter/test article compatibility. The points at which the tether is attached to the swivel system on the cage must be checked to ensure that they are sufficiently strong to preclude the system from becoming detached upon movement of the animal. These points can be subjected to a great deal of pressure and, as such, should be checked daily during the study to ensure that the points of attachment remain secure and that the swivel is operating free of any leakage. Leakage is evidenced by the presence of droplets or crystals at the swivel base. If present, the swivel should be changed as soon as possible. The distal portion of the swivel is then attached to the pump system, which must be matched to the duration, rate and volume required for the period of infusion.

On the day continuous infusion of a test article through the system is to begin, the jacket used to hold the tether system is placed on the animal while it is temporarily immobilised with ketamine. The catheter leading from the pump is securely anchored to the swivel system attached to the front of the animal's cage and then fed through the tether system to the jacket from the pump and swivel. This catheter is then connected to a right-angle non-coring needle or catheter, and introduced into the VAP. Prior to accessing the VAP, a similar preparation of the skin is done as for VAP maintenance. The location of the needle should be changed every 3–5 days, after similarly preparing the skin. The skin under the needle is examined daily to ensure that it is long enough, and not causing any necrosis of the skin. If the skin is observed to be erythematous, the access needle should be replaced with a longer one, or if not previously used, a non-coring catheter. During the study, the infusate should be changed at least every 24 h and should be pumped through a 0.2-μm non-protein-binding filter in order to maximise aseptic conditions. Finally, to optimise success in the delivery of the test articles, daily evaluation of the entire system should take place and replacement of any worn materials should occur as soon as possible.

The rate of infusion used at this laboratory to maintain patency is 2 ml/h. It is recommended that the rate of infusion of a drug does not exceed 5 ml/kg/h for a long-term study. For periods of 24 h or less, the rate should not exceed 10 ml/kg/h.

14.8 Conclusion

Long-term recovery and utilisation of animals implanted and maintained for the purpose of participating in continuous infusion studies is vital to the success of the study. This period will be greatly facilitated by a management programme that is as well organised as the implantation procedure itself. The key to post-operative management and conduct of the study is good communication among the members of the team and careful observation by trained and caring personnel. The surgical wound and the entry sites over the VAP should be observed daily for signs of infection, dehiscence, necrosis or self-inflicted trauma. If percutaneous cannulae and catheters are used rather than VAPs, they should be cleaned and disinfected daily.

The institution and the implantation and study team must always be flexible enough to re-evaluate all aspects of the study, including the pre-operative, operative, post-operative and study conduct plans, as well as their implementation, at any time during the study. In a new study, some changes often need to be made for subsequent

Figure 14.6 A fully instrumented animal, undergoing continuous infusion.

procedures, as the specifics of the study may never have been done before. This evaluation requires the input of the surgical team, attending veterinarian, research technicians and animal care staff. The study environment must reflect an attitude in which all constructive suggestions, from all members of the team (regardless of their position), are taken seriously, and are given due consideration. Changes to the procedures must arise from group discussion, not authoritarian mandate. Obviously, modifications that do result from these meetings need to be reviewed with all other personnel involved and, when significant changes are to be made, with the IACUC.

The post-operative care programme to be followed during the study must be tailored to each individual study, and may be individualised for the well-being of each animal. The study in general, and the plan specifically, must be characterised by prospective thought involving all of the appropriate personnel. Of paramount importance in this activity is thorough and accurate documentation of all aspects of the study, using appropriate medical records and logs, with regular evaluation of the entire programme, carried out by trained personnel during all phases of the work.

Utilising these principles, the practice and art of continuous infusion of test articles to non-human primates can be accomplished over long periods of time with minimal stress to the animal, and few complications (Figure 14.6). If the study was well designed, based on extensive literature review, early toxicity testing, and sufficient prospective discussion involving all pertinent personnel; if the protocol was well thought out and the experiment performed accordingly; and if the operation was well performed and the access system properly maintained, then the results of the experiment will be useful. Significant data will be obtained and the animal will have served the purpose of improving health care to both mankind and animals.

Acknowledgements

The authors would like to acknowledge the invaluable assistance of the surgical and infusion technicians for their suggestions and ideas, offered openly and freely, to accomplish our goal to constantly improve all methods of animal experimentation. Without their input, improvements in all areas would not occur. Likewise, the authors would like to acknowledge the support of the upper management of Primedica Corporation for its enthusiastic support of new techniques.

References

Alexander, J.W., Fischer, J.E. and Boyajian, M. (1983) The influence of hair-removal methods on wound infections. *Archives of Surgery*, 118, 347.

Appelgren, P. Ransjo, U., Bindslev, L., Espersen, F. and Larm, O. (1996) Surface heparinization of central venous catheters reduces microbial colonization *in vitro* and *in vivo*: results from a prospective, randomized trial. *Critical Care Medicine*, 24(9), 1482–1489.

Arnander, C., Bagger-Sjoback, D., Frebelius, S., Larsson, R. and Swedenborg, J. (1987) Long-term stability *in vivo* of a thromboresistant heparinized surface. *Biomaterials*, 8, 496–499.

Atkinson, L.J. (1992a) Physical facilities. In *Berry and Kohn's Operating Room Technique*, 7th edn. St. Louis: Mosby, pp. 36–45.

Atkinson, L.J. (1992b) Sterilization and disinfection. In *Berry and Kohn's Operating Room Technique,* 7th edn. St. Louis: Mosby, pp. 126–153.

Atkinson, L.J. (1992c) Preoperative preparation of surgical patients. In *Berry & Kohn's Operating Room Technique,* 7th edn. St. Louis: Mosby, pp. 265–272.

Atkinson, L.J. (1992d) Attire, surgical scrub, gowning and gloving. *In Berry & Kohn's Operating Room Technique*, 7th edn. St. Louis: Mosby, pp. 192–205.

Brown, M.J., Pearson, P.T. and Tomson, F.N. (1993) Guidelines for animal surgery in research and teaching. *American Journal of Veterinary Research*, 54, 1544–1559.

Coatney, R.W. and Kissinger, J. (1997) Survey of VAP infections in cynomolgus monkeys and dogs. Paper presented at the 1997 Laboratory Animal Long-Term Access Roundtable, San Antonio, 4 September 1997.

Cowan, C.E. (1992) Antibiotic lock technique. *Journal of Intravenous Nursing*, 15(5), 283–287.

Dalton, M.J. (1985) The vascular port, a subcutaneously implanted drug delivery depot. *Laboratory Animals*, 14(7), 21–30.

Dougherty, S.H. (1986) Implant infections. In von Recum, A.F (ed.) *Handbook of Biomaterials Evaluation*. New York: Macmillan, pp. 276–289.

Flecknell, P.A. (1987) *Laboratory Animal Anaesthesia: An Introduction for Research Workers and Technicians*. London: Academic Press, Inc.

Fraschini, G. (1991) Urokinase prophylaxis of central venous ports reduces infections and thrombotic complications. Paper presented at the NAVAN Fifth Annual Conference Syllabus, September.

Gibson, K.L., Donald A.W., Hariharan, H. and McCarville, C. (1997) Comparison of two pre-surgical skin preparation techniques. *Canadian Journal of Veterinary Research*, 61(2), 154–156.

Gilliam, D.L. and Nelson, C.L. (1990) Comparison of a one-step iodophor skin preparation versus traditional preparation in total joint surgery. *Clinical Orthopedics*, 250, 258–260.

Gilsdorf, J.R., Wilson, K. and Beals, T.F. (1989) Bacterial colonization of intravenous catheter materials *in vitro* and *in vivo*. *Surgery*, 106(1), 37–44.

Guideline For Use of Topical Antimicrobial Agents (1988) *American Journal of Infection Control*, 16(6), 253–266.

Hecker, J.F. (1981) Thrombogenicity of tips of umbilical catheters. *Pediatrics*, 57(4), 467–471.

Hoeprich, P.D., Wolfe, B.M., Jerome, C., Olson, D.A. and Huston, A.C. (1982) Long-term venous access in rhesus monkeys. *Antimicrobial Agents and Therapeutics*, 21(6), 976–978.

Holcombe, B.J., Forloines-Lynn, S. and Garmhausen, L.W. (1992) Restoring patency of long-term central venous access devices. *Journal of Intravenous Nursing*, 15(1), 36–41.

Jacobson, A. (1998) Continuous infusion and chronic catheter access in laboratory animals. *Laboratory Animals*, 27(7), 37–46.

Kinsora, J.J. Jr., Christoffersen, C.L., Swalec, J.M. and Juneau, P.L. (1997) The novel use of vascular access ports for intravenous self-administration and blood withdrawal studies in squirrel monkeys. *Journal of Neuroscience Methods*, 75, 59–68.

Landi, M., Schantz, J., Jenkins, E., Warnick, C. and Kissinger, J. (1996) A survey of vascular access port infection in cynomolgus monkeys. *Scandinavian Journal of Laboratory Animal Science*, 23(1), 441–448.

Lawson, M. (1991) Partial occlusion of indwelling central venous catheters. *Journal of Intravenous Nursing*, 14(3), 157–159.

Lineback, P. (1971) The vascular system. In Hartman, C.G. and Straus, W.L. (eds.) *The Anatomy of The Rhesus Money (Macaca mulatta)*. New York: Hafner Publishing Company, pp. 248–265.

Love, J.A. (1999) Analgesia in experimental animals. *Journal of Investigative Surgery*, 12, 63–64.

Lumb, W.V. and Jones E.W. (1984) *Veterinary Anesthesia*, 2nd edn. Philadelphia: Lea & Febiger.

Marosok, R., Washburn, R., Indorf, A., Solomon, D. and Sherertz, R. (1996) Contribution of vascular catheter material to the pathogenesis of infection: depletion of complement by silicone elastomer *in vitro*. *Journal of Biomedical Materials Research*, 30, 245–250.

Mendenhall, H.V. (1999) Surgical procedures. In von Recum, A.F. (ed.) *Handbook of Biomaterials Evaluation*. Philadelphia: Taylor & Francis, pp. 481–492.

Mendenhall, H.V., Piechowiak, M. and Raikowski, D. (1997) A long term study on the use of vascular access ports in multiple species. Paper presented at the 1997 Laboratory Animal Long-Term Access Roundtable, San Antonio, 4 September.

O'Farrell, L., Griffith, J.W. and Lang, C.M. (1996) Histologic development of the sheath that forms around long-term implanted central venous catheters. *Journal of Parenteral and Enteral Nutrition*, 20, 156–158.

Passerini, L., Lam, K., Costerton, J.W. and King, E.G. (1992) Biofilms on indwelling vascular catheters. *Critical Care Medicine*, 20(5), 665–673.

Patrick, C.C., Plaunt, M.R., Hetherington, S.V. and May, S.M. (1992) Role of the *Staphylococcus epidermidis* slime layer in experimental tunnel tract infections. *Infection and Immunity*, 60(4), 1363–1367.

Peters, G., Locci, R. and Pulverer, G. (1982) Adherence and growth of coagulase-negative *Staphylococci* on surfaces of intravenous catheters. *Journal of Infectious Diseases*, 146(4), 479–482.

Pickens, R., Hauck, R. and Bloom, W. (1966) A catheter-protection system for use with monkeys. *Journal of Experimental Analytical Behaviour*, 9(6), 701–702.

Polk, H.C., Simpson, C.J. and Simmons, B.P. (1983) Guidelines for prevention of surgical wound infection. *Archives of Surgery*, 118, 1213.

Raad, I.I. and Bodey, G.P. (1992) Infectious complications of indwelling vascular catheters. *Clinical Infectious Diseases*, 15, 197–208.

Raffe, M.R. (1985) Fluid therapy in the surgical patient. In Slatter, D.H. (ed.) *Textbook of Small Animal Surgery*. Philadelphia: W. B. Saunders, pp. 90–102.

Roberts, A.J., Wilcox, K., Devineni, R., Harris, R.B. and Osevala, M.A. (1995) Skin preparations in CABG surgery: a prospective randomized trial. *Comparative Surgery*, 14(6), 741–748.

Rochat, M.C., Mann, F.A. and Berg, J.N. (1993) Evaluation of a one-step surgical preparation technique in dogs. *Journal of the American Veterinary Medical Association*, 203(3), 392–395.

SATS AORN (1992) Recommended Practices. Skin preparation of patients. *AORN Journal*, 56(5), 937–941.

Schwartz, C., Henrickson, K.J., Roghmann, K. and Powell, K. (1990) Prevention of bacteremia attributed to lumenal colonization of tunneled central venous catheters with vancomycin-susceptible organisms. *Journal of Clinical Oncology,* 8(9), 1591–1597.

Shirahatti, R.G., Joshi, R.M., Vishwanath, Y.K., Shinkre, N., Rao, S., Sankpal, J.S. and Govindrajulu, N.K. (1993) Effect of pre-operative skin preparation on post-operative wound infection. *Journal of Postgraduate Medicine,* 39(3), 134–136.

Shulman, R.J., Reed, T., Pitre, D. and Laine, L. (1988) Use of hydrochloric acid to clear obstructed central venous catheters. *Journal of Parenteral and Enteral Nutrition,* 12(5), 509–510.

Stone, J. (1996) Treatment of central venous catheter occlusions with urokinase and hydrochloric acid in rhesus macaques. Paper presented at the 1996 Laboratory Animal Long-Term Access Roundtable, Philadelphia, 10 May.

Tschirhart, J.M. and Minoo, K.R. (1988) Mechanism and management of persistent withdrawal occlusion. *American Surgeon,* 54, 326–328.

US Government (1985) *Principles for Utilization and Care of Vertebrate Animals Used in Testing, Research, and Training* . Office of Science and Technology Policy, Federal Register, 50, pp. 20864–20865.

Winocour, P.D., Cattaneo, M., Somers, D., Richardson, M., Kinlough-Rathbone, R.L. and Mustard, J.F. (1982) Platelet survival and thrombosis. *Arteriosclerosis,* 2(6), 458–466.

Wixson, S.K. (1994) Rabbits and rodents: anesthesia and analgesia. In Smith, A.C. and Swindle, M.M. (eds.) *Research Animal Anesthesia, Analgesia and Surgery.* Greenbelt, MD: Scientists Center for Animal Welfare, pp. 59–92.

Wojnicki, F.H., Bacher, J.D. and Glowa, J.R. (1994) Use of subcutaneous vascular access ports in rhesus monkeys. *Laboratory Animal Science,* 44(5), 491–494.

15 Multidose infusion toxicity studies in the large primate

M. D. Walker

15.1 Introduction

Non-human primates represent an increasingly important animal model for the safety evaluation of investigational drugs, as well as for the development of novel therapeutic modalities. The rapid growth in the biotechnology industry has resulted in an increased demand for primates in research, primarily because of the tissue and species specificity of most recombinant proteins. Pre-clinical study designs have become increasingly complex, with dosing strategies that often mirror clinical protocols. Ancillary tests for pre-clinical infusion studies (such as electrocardiographic evaluations, haemodynamic monitoring and toxicokinetic studies) present their own unique challenges when using this species. The tethered system for continuous intravenous infusion in the large primate is well established (Pickens *et al.* 1966; McNamee *et al.* 1984) and has remained the method of choice for many laboratories. Ambulatory infusion has gained favour in recent years, as a result of the advent of improvements in technology, as well as aesthetic, ethical and animal welfare considerations. This chapter will review the use of ambulatory infusion technology, describe the necessary equipment, methods and materials, discuss pertinent issues related to novel intravenous formulations and briefly review technical considerations. A common protocol and study design will be outlined, and typical clinical pathology, gross necropsy and histomorphological findings related to the infusion procedure will be discussed. The information provided in this chapter is based on the experiences of the author at his laboratory, and reflects the results from validation efforts and the conduct of multidose infusion studies in monkeys during the last 7 years.

15.2 Infusion technology

Mendenhall *et al.* have reviewed the standard tether and jacket infusion system in detail in the previous chapter. This is an established and reliable technology; however, there has been an increasing demand for an alternative system that allows the animal free movement in the cage and decreases its perceived physiological and psychogenic stress, while providing for a more efficient and cost-effective study conduct. Ambulatory infusion employs the use of a custom-fit infusion jacket, a battery-powered, programmable infusion pump contained within a dorsal thoracic pouch, a self-contained dosing reservoir, an external infusion access line and a subcutaneously implanted vascular access port and intravenous catheter (Wojnicki *et al.* 1994). As such, the system is entirely self-contained and allows the animal full mobility, minimising technique-related stress.

When making the choice between tethered and ambulatory infusion (if the option is available), there are technical limitations to each system, as well as advantages, that should be considered. Solutions, emulsions or fine-particle suspensions that have limited room temperature stability may render ambulatory technology inappropriate. The temperature reached within the dorsal pouch, where the pump and dosing reservoir are contained, has been measured at 30–35 °C within 2 h of initiation of infusion in the male and female cynomolgus monkey (M. Walker, unpublished data). The conductive transfer of heat from the animal will, of course, vary according to species, body size, activity and ambient room temperature. In addition, dose volumes that exceed 250 ml (intermittent daily dosing) or 3 ml/kg/h (for continuous infusion studies) will require at least twice-daily reservoir changes, increasing the labour requirements as well as the potential for secondary animal stress.

Infusion lines and reservoirs require that a volume of liquid be wasted, as a result of priming requirements, material remaining in the infusion lines and necessary residual within the dosing reservoir to prevent air embolism. Ambulatory infusion holds a distinct advantage when the test material is in limited supply or quite costly, as the dead-space fluid volume for the infusion lines and reservoir can be as little as 10 ml, depending on the specific components utilised. In contrast, tethered infusion systems may require as much at 30–50 ml of residual fluid volume to accommodate the lengthy externalised infusion lines, reservoirs and priming needs for the fluid path. When calculating test material requirements for a repeat dose infusion study, the type of infusion system must be considered.

15.3 Caging and equipment

Ambulatory infusion employs standard animal caging that requires neither modification nor unique features. Each monkey is individually housed in a stainless-steel cage measuring approximately $70 \times 60 \times 76$ cm (d × w × h); this cage is of adequate size for cynomolgus and rhesus monkeys up to 8 kg in body weight (Figure 15.1; Allentown Caging Company, Allentown, PA, USA). The cages include a squeeze-back to facilitate catching and restraint of the animals. The cages are arranged in the room so that the animals have constant visual, olfactory and auditory contact (Figure 15.2). All monkeys are hand-caught at our facility; however, for safety reasons, monkeys weighing less than 5 kg are used most of the time. Each animal is identified by a chest tattoo during quarantine and acclimation; following randomisation, each animal is assigned a permanent animal number mapped to an implantable microchip device (Biomedic Data Systems, Inc.). Prior to assignment to study, the animals are surgically implanted with a subcutaneous vascular access port and venous polyurethane catheter (Port-a-Cath® Low Profile, SIMS Deltec, Minneapolis, MN, USA). Although jugular venous catheterisation is most common, the femoral route is used approximately 20% of the time. The route of infusion is typically chosen by the sponsor and is based on previous experience, physical characteristics of the test material formulation, potential for vascular irritation or the known toxicity profile. Polyurethane has been found to be a superior material when implanted, resulting in decreased incidence of thrombosis and complement degradation (Marosok *et al.* 1996). The many advantages of vascular access ports, as well as surgical technique, post-operative care and maintenance of the device, have been reviewed in the previous chapter. While many such devices are available for pre-clinical use, it has been the author's experience that titanium ports

Figure 15.1 Stainless-steel monkey caging. Each rack is composed of four cages; adjacent cages are fitted with dividing doors that may be opened to allow for commingling (when appropriate). Each cage is designed with a perch, a hanging mirror and at least one toy.

with minimal residual volume, in conjunction with Silastic® or polyurethane catheters, are superior for repeated-dose studies because of their relative durability and biocompatibility.

There are many available sources for ambulatory infusion jackets. Of paramount importance is proper jacket fit and appropriate padding around the shoulders, neck and chest and around the abdominal region. In addition, all system components must be out of reach of the monkey, a species that can be quite limber and tenacious. As seen in Figures 15.3–15.7, the Covance infusion jacket (Lomir® Biomedical, Montreal, Canada) contains many features that have effectively reduced the incidence of dermal ulceration while providing greater comfort for the animals and, ultimately, better individual animal compliance. The dorsal pocket contains the infusion pump, reservoir and externalised infusion lines, and it is constructed of black canvas to protect the dosing solutions from light.

Figure 15.2 Positioning of multiple racks of cages within a typical room. Banks of cages are oriented so as to face each other; if that is not possible, mirrors are mounted to the opposing wall.

Ambulatory infusion pumps have been used in clinical medicine for many years, and the advancements in this technology have enabled us to use many of these devices in the pre-clinical setting. The pumps of choice at Covance, based on validation studies, knowledge of the manufacturing techniques, operative characteristics and experience, are the CADD-Plus® model 5400 and the CADD-Prizm® (SIMS Deltec®; Figure 15.8). The pumps are shock and moisture resistant, are programmable for daily intermittent, pulsatile or continuous infusion modes, and (in the case of the CADD-Prizm®) can provide flow rates from 0.1 to 350 ml/h. A lock-out code programmed into the pump prevents the animal from modifying dose administration. Each unit is powered by single 9-V alkaline battery, but it may be used with an AC adapter as pole-mounted infusion pumps. Alarms include high pressure (> 25 psi), low residual dose volume (< 5 ml), completion of dosing and low battery. The CADD-Prizm® maintains a log of activity and alarm data that may be downloaded from resident memory and reviewed on a personal computer. In addition, a remote monitoring system is being developed by the manufacturer, which will provide for 24-h response to dosing interruptions and technical problems.

Figure 15.3 Cynomolgus monkey fitted with a Covance ambulatory infusion jacket. Significant padding is provided in the cervical, ventral thoracic and abdominal regions. While of no practical use, the sleeves provide a more even fit and may increase animal compliance.

The externalised infusion lines are connected to the infusion pump via one of two primary set-ups. For dosing volumes of 100 ml or less, a 50-ml or 100-ml dosing cassette and winged infusion set (½-inch Huber-point needle, right angle) are used. If the dose volume is greater than 100 ml, a remote reservoir set-up is employed. All infusion reservoirs and lines are available in medical-grade polyvinyl chloride, and select components are available in low-phthalate ester content plastics, such as trioctyl trimellitate (TOTM). TOTM and polyolefin components are frequently used when the dosing formulations contain relatively high concentrations of alcohols and solubilising agents, such as Cremophor® EL or Solutol®HS (BASF Corporation). During infusion, the dosing reservoirs are changed once daily, or when the reservoir is depleted. The frequency of reservoir change, however, may be dictated by dosing formulation stability data. The externalised infusion lines are changed approximately once weekly if the lines are disconnected and contaminated or if any components are damaged. It should be re-emphasised that strict aseptic technique must be adhered to at all times to prevent iatrogenic infections.

Figure 15.4 The ambulatory infusion pump and fluid reservoir are contained within a dorsal pouch constructed of black canvas. The location of the pocket minimises problems related to inadvertent animal access to the infusion system.

15.4 System validation and monitoring

Infusion pumps are bench validated twice yearly to confirm that the units are functioning normally in both intermittent and continuous modes and to confirm accuracy of dose delivery. All pumps are tested at three rates (typically 1, 10 and 25 ml/h) in both programmes, and pump accuracy must be ± 3% of target volume. Any pump that fails validation is returned to the manufacturer for repair because, as FDA-regulated therapeutic devices, they cannot be adjusted and the user cannot modify the resident programmes. All infusion pumps are uniquely identified and are assigned to specific monkeys; pump assignment and copies of validation records (for the previous year) are maintained in the study files.

For intermittent infusion studies, daily dose accuracy is monitored during the conduct of an infusion study by calculating a percentage of target volume based on pre-dose and post-dose reservoir weight (in grams, with or without adjustment for density, as appropriate) for each animal. Continuous infusion dosing requires that both the time of reservoir use and target volume/time be factored into dose delivery

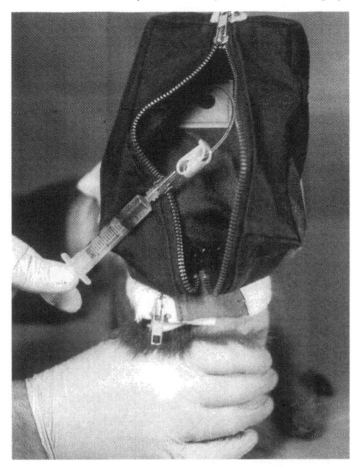

Figure 15.5 The jacket and winged infusion set (which contains a right angle, ¹/₂-inch Huber point needle for port access) are fitted at least 24 h prior to initiation of infusion. The port and catheter are flushed with heparinised saline and the externalised line is mechanically locked.

calculations. An alternative approach has been to designate that usage is assumed to be 24 h if the reservoir is changed within ± 1 h of the previous day. It is recommended that the procedures be defined in the facility standard operating procedures (SOPs).

15.5 Husbandry and enrichment

All monkeys are fed a certified primate diet (8726C Harlan Teklad® or PMI® Certified Primate Diet® #5048 are common choices) twice daily. Tap water is provided *ad libitum* via an automatic watering system. Each cage contains a perch, and at least one toy is provided within the cage. Toys are alternated on a regular basis to decrease boredom and complacency. In addition, each cage has a movable mirror mounted on a chain outside the cage, and this has proven to be a highly successful manipulada for non-human primates. A variety of fruits, vegetables, popcorn and certified primate treats are offered at least three times weekly, and a multivitamin is provided once weekly. Institutional programmes for nutritional enrichment vary substantially but

Figure 15.6   The dorsal pocket may be reflected and/or removed, if necessary. A dorsal strip of
Spandex ensures a snug and comfortable fit.

are usually based on providing a variety of foodstuffs to stimulate natural eating behaviours (Rosenblum and Andrews 1995). Foraging boards are used with some success but are not uniformly accepted. It is suggested that many treats, toys, and activity-stimulating devices be offered in a rotating fashion. Commingling, a highly successful practice of allowing monkeys of the same sex and treatment group to spend defined periods in a common cage (typically by removing solid dividers between adjacent cages), is used routinely for toxicology studies at Covance. This practice is not acceptable for infusion studies because of obvious concerns that the infusion systems may be damaged during play, endangering the health of the animals and potentially invalidating the study. As a result, additional efforts should be made to provide an enhanced environment and supplemental dietary enrichment to these animals.

15.6 Selection and evaluation of intravenous formulations

The selection of an intravenous formulation is based on many factors but primarily on the relative solubility of a test material, the stability of the final product and *in*

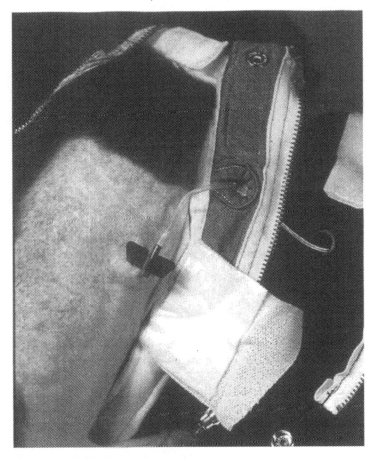

Figure 15.7 View of the Huber point needle inserted into the vascular access port, located in the dorsolateral mid-thoracic region. Placing a Telfa® pad and elastic adhesive bandage over the access site protects the port and minimises contamination. The port site is monitored on a daily basis for development of infections, drainage or poor seating of the needle.

vitro haemocompatibility of the test material and the resulting excipients at the proposed concentrations. In general, our experience is that solutions in a pH range of 3.0–9.0, with an osmolality between 100 and 600 mosmol, have been found to be readily tolerated. Electrolyte solutions, such as isotonic (0.9%) saline and lactated Ringer's solution, are preferred and (in moderate volumes) produce minimal effects on the cardiovascular and renal systems. Infusion of saline solutions will result in a mild diuretic effect and natriuresis, but this is a compensatory effect that is of no toxicological significance. Isotonic (5%) dextrose solutions function as mild osmotic diuretics and should be used with some caution, as they may produce electrolyte loss and mild dehydration at higher volumes. Lactated Ringer's is a buffered electrolyte solution of some popularity, but metabolic acidosis may be exacerbated (due to the need to metabolise sodium lactate) in instances where hepatic metabolism is affected by the test material.

Although it is preferred that isotonic electrolyte solutions be used for intravenous infusion, it is frequently impractical because of the poor solubility of the test material

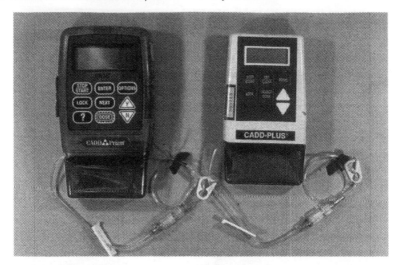

Figure 15.8 Photograph of the Cadd-Prizm® and Cadd-Plus® ambulatory infusion pumps with cassette dosing reservoirs and winged infusion sets attached. A similar model will replace the Cadd-Plus® during 2000.

in aqueous solutions. Stock solutions of the test material formulation are frequently prepared using a known combination of one or more solubilising agents, and the stability of the diluted material (both chemically and physically) is then assessed. Cremophor® EL (BASF Corporation, Mt. Olive, NJ, USA), a well-known fatty acid ester surfactant, and ethanol, at a 10:1 ratio (typically 10% Cremophor® and 1% ethanol), have been used for a variety of compounds, most notably in the development of cyclosporin A. The stock formulation is then diluted with dextrose (2.5% or 5%) or sterile water, at a minimum of a 1:3 ratio (stock to diluent) prior to dosing; electrolyte solutions are to be avoided, as the salt components may cause the test material to precipitate.

Solutol® HS15 (BASF Corporation), another ester surfactant used to improve aqueous solubility, is becoming a popular alternative to Cremophor®; it is prepared as a simple aqueous solution or in conjunction with PEG50 and ethanol, but final concentrations of Solutol® may safely reach 7.5%. It should be noted, however, that the intravenous administration of these agents might result in adverse reactions due to the allergenic potential of Cremophor® and, to a lesser degree, Solutol® (Liau-Chu *et al.* 1997). The systemic responses are probably due to a hapten-like effect, and the intravenous administration of Cremophor® and Solutol® in dogs produces a sudden histaminic reaction, heralded by agitation, tachycardia, hypotension, urticaria, generalised erythema and maxillofacial oedema and emesis. The severity of effect is dose dependent, but dogs are highly sensitive to systemic allergens, partly because of a robust population of tissue-specific and circulating histiocytes.

In the non-human primate, mild cardiovascular effects (decreased blood pressure, tachycardia, increased capillary refill time) may occur initially, but acclimation occurs over time. Pre-treatment with diphenhydramine hydrochloride (or a similar antihistaminic agent) has been shown to substantially decrease the severity and duration of adverse clinical signs in dogs. The inherent toxicity of each component of the proposed formulation, as well as potential pharmacological effects for the species utilised, must also be considered. The intravenous administration of formulations containing Solutol® has resulted in decreased serum triglyceride levels in rats, dogs and cynomolgus monkeys (Woodburn *et al.* 1995); however, no evidence of impaired hepatic function or hepatotoxicity has been found. Continuous infusion of formulations containing Dextran®70, a synthetically derived colloid used as a volume expansion agent in hypovolaemic shock, will produce a mild non-regenerative anaemia and histiocytic infiltrates in multiple tissues, with no treatment-related clinical observations or tissue toxicity (Peterson *et al.* 1997; Maddison 2000). It is essential, therefore, to evaluate the advantages of the characteristics of the formulation and determine if the toxicology of the product in the target species is acceptable and allows for a clearly defined safety assessment of the drug.

The interaction between the test material formulations and the infusion apparatus must be evaluated in a prospective manner. Effluent testing is a simple procedure that evaluates, in an *ex vivo* manner, the test material concentration(s) after preparation and over the course of a typical infusion dose by collecting aliquots of the fluid from the infusion apparatus at defined time points. Decreases in test material concentration may indicate adsorption to the inner surfaces of the reservoir, tubing or catheter materials. The appearance of unexpected analytical peaks, or an increase in test material concentration, may indicate that substances are leaching from the plastics from components in the infusion system. Of particular concern is the leaching of phthalates from polyvinyl chloride tubing and reservoirs when alcohols or organic solvents are used; phthalate esters have been found to be mutagenic and possibly carcinogenic in rats (David *et al.* 1999). The overall impact of one or more of these phenomena must be considered when selecting an infusion system and the individual components of the fluid path.

15.7 Study design considerations

The design and planning phases of an infusion study require that one address many technical and scientific issues that ultimately affect the ability to successfully initiate and complete the project. While the number of animals assigned to study and the primary battery of in-life and post-mortem evaluations are frequently dictated by regulatory requirements, each study proposes unique challenges to the toxicologist. This section will review the issues that should be addressed and will briefly review the aspects of study design and conduct that are common to pre-clinical multidose toxicity studies in monkeys, regardless of the route of administration.

15.7.1 *Instrumentation pool and animal selection*

The number of animals implanted with intravenous catheters and vascular access ports will vary according to study as well as laboratory; however, it has been the experience at this laboratory that 10–15% more (approximately two of each sex) is sufficient. Three to four animals of each sex per group are ultimately assigned to study; if required, additional animals are assigned for post-treatment recovery. Selection of animals is based on catheter patency, physical examinations, clinical pathology (evaluated at least 7 days post-operatively) and any additional pre-treatment evaluations (as specified by protocol).

15.7.2 Post-operative monitoring

Post-operative care of the instrumented monkey has been addressed in the previous chapter. The port and catheter should be flushed with an appropriate intravenous solution, such as heparinised saline (10–100 IU/ml), a minimum of once weekly (Tschirhart and Minoo 1988). If animals are to remain in the colony for an extended period, the port and catheter should be flushed a minimum of once monthly after the initial 1-month healing period. The flushing process should include an initial attempt to extract fluid that remains in the catheter, followed by a fresh syringe of heparinised saline. It has been our experience that it is difficult to aspirate blood from up to 50% of monkeys with chronic intravenous catheters, most likely because of margination of the catheter and/or the formation of fibrin tags at the catheter tip. As a result, a catheter is considered patent if the flushing process is performed with no evidence of back-pressure and if no swelling is found at the catheterisation site.

15.7.3 Group designations and dosing regimen

The number of treatment groups may be determined by the sponsor; however, three groups treated with the test material (low-, mid- and high-dose) are common. The dose levels reflect multiples of the proposed clinical doses, and the dosing regimen should mimic the clinical plan. If the vehicle is not a standard electrolyte solution, it is important to include a vehicle diluent (such as saline) and a vehicle control group. The toxicological assessment of the vehicle controls (when compared statistically with the saline control and the treated groups) will assist in determining the contribution of the vehicle to clinical, gross pathology and histomorphological changes.

The monkeys are fitted with an ambulatory infusion jacket, containing a partially filled infusion reservoir, at least 3 days before the initiation of dosing. All necessary infusion lines, as well as the winged infusion set, are placed on each monkey the day prior to initiation of dosing. This gradual process of acclimation greatly decreases the incidence of dosing errors on Day 1 that are attributed to mechanical or behavioural problems.

15.7.4 Infusion volumes

The maximum infusion volumes, for both continuous and intermittent delivery, are dependent upon the duration of exposure, the length of the dosing period, the chemical characteristics of the formulation and the toxicity of the test material. The requirements for maintaining normal fluid balance in the dog and cat have been estimated at 60 ml/kg/day (Maddison 2000). Acute changes to the volume within the circulatory fluid compartment have been associated with pulmonary hypertension and oedema; however, compensatory responses (such as initial vasodilatation and, ultimately, increased renal excretion of fluid and solutes) enable one to infuse much larger volumes, if the duration of treatment is extended. It has also been found that very low infusion rates may lead to insufficient mixing of the infusate with intravascular fluids (Blacklock *et al.* 1986). Recommendations at each laboratory are based on experience and typically are conservative in nature. Potential changes to clinical pathology, as well as secondary effects to the cardiovascular and renal systems, may occur at higher volumes; therefore, conservative guidelines are typically provided. In general, our laboratory recommends dose rates of ≤ 15 ml/kg/h for intermittent infusion dosing (6 h of duration or less), and a maximum infusion rate of 4 ml/kg/h (100 ml/kg/24 h) for continuous infusion studies of 2 weeks' duration or longer.

15.7.5 In-life observations

Standard observations include twice-daily mortality and morbidity checks and a minimum of once-daily detailed observations for indications of toxic or pharmacological effects of treatment. In addition, the integrity of the infusion system is monitored at least once daily (continuous infusion) or at the beginning and end of the infusion period (intermittent infusion). The infusion pump, dosing reservoir, dose administration tubing and port site are carefully evaluated, problems are addressed and corrected, and all information is documented appropriately.

Ophthalmological and electrocardiographic evaluations are conducted once prior to initiation of dosing (post-surgery) and, at a minimum, before terminal sacrifice. Ophthalmological evaluations are strongly recommended to evaluate fully the potential effects of treatment on the ocular vasculature (choroid) and the macular body. Agents that increase vascular permeability and/or fragility may produce changes that can only be detected via indirect ophthalmoscopy (Rubin 1974). Electrocardiography is a valuable tool in pre-clinical safety assessment, and one may detect the arrhythmogenic potential and possible myocardial toxicity of the test material (Tilley 1992). Electrocardiographic examinations should be performed with reference to the time of dose administration (intermittent infusion) or the projected time of maximum systemic drug concentration (T_{max}).

Serial blood samples for toxicokinetic studies are collected at least twice during the course of treatment. The number of time points per interval is dependent upon previous experiments with the compound in monkeys, in which the metabolic pathway has been examined. Blood sample collections for intermittent infusion studies explore the increase in drug concentration during infusion, as well as the decrease over time after completion of the dosing period. For continuous infusion studies, samples are collected at a minimum of five time points during the initial phase of exposure and, depending on the total duration of infusion, at a minimum of 3 days during the course of treatment, to determine steady-state drug concentrations (C_{ss}). Sample collection subsequent to Day 1 is performed at the same time of day and accounts for interruption in dosing during reservoir changes (as necessary). Samples are collected after completion of treatment period to evaluate the rate of metabolic degradation, conversion, and/or systemic elimination ($t_{1/2}$).

Clinical pathology sample collections (haematology, coagulation, serum and urine chemistry, and urinalysis) occur at least once before initiation of dosing and at appropriate intervals during the course of, and after completion of, the infusion phase. For studies of 2 weeks' duration or longer, it is recommended that sample collections occur at least once during treatment. Evaluations should also be performed after completion of a post-treatment recovery phase (if included) to monitor the resolution of potential treatment effects and/or delayed toxicity. While the requirements for specific tests may vary according to the regulatory guidelines used in the study design, the following can be used as a basic reference for global registration of pharmaceuticals:

Haematology
Bone marrow
Erythrocyte count
Haemoglobin
Haematocrit
Platelet count
Mean platelet volume
White blood cell (leucocyte) count
Differential blood cell count
Blood cell morphology
Reticulocyte count
Mean cell volume
Mean cell haemoglobin
Mean cell haemoglobin concentration

Decreases in red cell mass can be associated with increased loss of erythrocytes (thrombosis, treatment-related lysis or decreased lifespan) or bone marrow suppression. Decreased platelet counts are associated with increased thrombosis and should be monitored carefully. As expected, increased leucocyte counts, along with an increase in relative (%) and absolute neutrophil counts, are diagnostic for bacterial infections. Abnormal cell morphology may indicate a toxicosis related to either treatment or sepsis.

Coagulation
Prothrombin time
Activated partial thromboplastin time

Changes to coagulation parameters may indicate liver toxicity, increased rate of thrombosis, a toxic effect of the test material or impaired liver function.

Clinical chemistry
Glucose
Urea nitrogen
Creatinine
Total protein
Albumin
Globulin
Albumin–globulin (AG) ratio
Cholesterol
Triglycerides
Total bile acids
Total bilirubin
Alanine aminotransferase (ALT)
Alkaline phosphatase
Gamma-glutamyltransferase (GGT)
Aspartate aminotransferase (AST)
Calcium
Inorganic phosphorus

Sodium
Potassium
Chloride

Serum chemistry evaluations include a standard battery of tests that assess the health and metabolic state of major organ systems. However, there may be instances where additional parameters assist in determining the physiological and toxic effects of continuous infusion of test material formulations, as well as effects secondary to administering large volumes.

Urinalysis
Appearance/colour
Specific gravity
pH
Protein
Urobilinogen
Glucose
Ketones
Bilirubin
Blood
Microscopic examination of sediment

Urine chemistry
Sodium
Potassium
Chloride
Sodium excretion
Potassium excretion
Chloride excretion
Volume

Urinalyses are conducted to monitor the renal function, as well as the potential toxicity of the test article formulation. Evaluating the concentration of solutes, the excretion of electrolytes and microscopic presence of cellular debris monitors glomerular and tubular functional integrity.

15.7.6 Post-mortem evaluations

After completion of the designated treatment or recovery periods, a necropsy and macroscopic evaluation of tissues are conducted. It is of particular importance that the tissues surrounding the vascular access port, subcutaneous catheter tract, the entrance site of the catheter into the vein and the infusion site (described as tissues surrounding the tip of the catheter) be critically examined. The heart, liver and lungs should be examined to determine if excessive thrombosis, infarcts or scarring are present. The presence of a pleural or pericardial effusion may be associated with cardiac toxicity, pulmonary hypertension or pneumonia. Evidence of infection, extensive development of granulation tissue or deterioration of tissues at the cannulation and infusion sites may indicate that direct intravascular delivery of the drug was not occurring prior to sacrifice.

Organ weights are recorded for at least the following tissues: heart, liver, spleen, kidneys, adrenals, thymus and brain. Additional organ weights may be collected according to the regulatory requirements and knowledge of potential target tissues.

Histomorphological evaluations are performed on a comprehensive list of preserved tissues. Particular attention should be paid to the adrenals, brain and spinal cord, catheterisation and infusion sites, eyes, bone marrow, heart, kidneys, liver, lungs, lymph nodes, pituitary gland, spleen, thymus and thyroid/parathyroid glands.

15.8 Clinical findings related to infusion procedures

Clinical observations related to the infusion procedure are primarily due to poorly fitting jackets and resulting dermal ulcerations and sores in the cervical, axillary and dorsal thoracic regions. All jackets should be washed and dried prior to use to soften the material and to ensure a more comfortable fit; readily available detergents and automatic laundering are preferred. Daily monitoring of jacket fit should be performed at the time of infusion system checks. Jackets should be replaced at least every 2 weeks during treatment or if excessive wear or heavy soiling warrant a more frequent schedule of change. Local antibiotic ointments (such as Neosporin®) or combination products containing an anti-inflammatory medication (Panalog® ointment or cream) are to be applied promptly to developing sores to prevent infection and minimise the incidence of inflammatory ulcerations. If lesions persist, the jacket may need to be modified and additional padding provided on a case-by-case basis.

Food consumption may decrease during the immediate post-operative period (due to post-anaesthetic suppression of appetite), but there are generally no effects that are related to the infusion procedure. Observations of poor appetite not attributed to treatment with the subject test material may be indicative of infection, dehydration or possibly an inability to acclimate to laboratory procedures. As with all studies using non-human primates, monitoring of body weights and food consumption are critical barometers of animal health. It is recommended that nutritional support (additional treats, liquid primate diets, fruit juices, etc.) be considered for those monkeys that lose 10% of their previous body weight in 1 week (or less), or a total of 25% over the course of an assessment period.

There are no ophthalmological changes that have been attributed to the infusion procedure. Retinal or choroid plexus abnormalities have been associated with systemic infections, toxicosis, increased rates of thrombosis, prolonged clotting times, increased vascular permeability and primary cytotoxicity, all of which can be related to test material toxicity or iatrogenic infections (Hackett and Stein 1991). Electro-cardiographic changes are rare and may be noted in the immediate post-operative period. The abnormalities (noted as atrial premature complexes) are related to intracardial catheter placement and mechanical irritation (Bolton 1975); these abnormalities usually resolve within the first 2 weeks post surgery. Changes during the treatment period that are indicative of direct cardiac toxicity include prolongation of the Q-T interval, changes in the T- or P-wave polarity, premature ventricular contractions (usually detected in moribund animals) or qualitative changes to the QRS complex (Tilley 1992).

Saline control group reference ranges for select haematology and serum chemistry parameters are included in Tables 15.1 and 15.2. Mild decreases in red blood cell count (RBC), haematocrit (HCT) and haemoblogin (HGB) are observed during the

Table 15.1 Haematology parameters

		Males					Females				
		Pre-treatment/ post-surgical	Week 1	Week 2	Weeks 4–5	Recovery	Pre-treatment/ post-surgical	Week 1	Week 2	Weeks 4–6	Recovery
RBC (10^6/µl)	Mean (SD)	6.47 (0.637)	5.67 (0.552)	5.54 (0.803)	5.76 (0.813)	6 (0.935)	5.91 (0.423)	5.34 (0.557)	5.5 (0.66)	5.76 (0.564)	5.72 (0.77)
	n	23	11	17	21	5	23	11	17	21	6
HCt (%)	Mean (SD)	41 (2.32)	35.1 (2.39)	34.9 (3.53)	37.1 (2.99)	38.4 (3.85)	38.9 (3.2)	34.5 (2.11)	35 (2.79)	37.5 (2.33)	42 (1.69)
	n	23	11	17	21	5	23	11	17	21	6
HGB (g/dl)	Mean (SD)	12 (0.71)	10.2 (1.06)	10.1 (1.18)	11.7 (0.61)	10.5 (1.41)	11.9 (0.91)	10.1 (0.67)	10.1 (0.92)	12 (1.3)	12.3 (1.29)
	n	17	11	11	9	3	17	11	11	9	4
Reticulocytes (% RBC)	Mean (SD)	0.19 (0.4)	0.13 (0.179)	0.24 (0.331)	0.16 (0.221)	0.03 (0.036)	0.11 (0.321)	0.21 (0.362)	0.18 (0.34)	0.2 (0.39)	0.03 (0.031)
	n	17	11	11	9	3	17	11	10	9	4
Platelets (10^3/µl)	Mean (SD)	406 (79.1)	353 (86.7)	427 (161.9)	379 (93)	411 (58.8)	377 (106.5)	406 (157.5)	427 (139.8)	380 (112.3)	354 (62.7)
	n	23	11	17	21	5	23	11	17	21	6
WBC (10^3/µl)	Mean (SD)	13.3 (5.52)	11.1 (4)	12.2 (4.82)	17.1 (8.42)	16.7 (3.59)	12.7 (4.14)	10.6 (3.02)	11.2 (3.36)	14.8 (3.78)	13.8 (4.73)
	n	23	11	17	21	5	23	11	17	21	6

Table 15.2 Chemistry parameters

	Males					Females				
	Pre-treatment/ post-surgical	Week 1	Week 2	Weeks 4–6	Recovery	Pre-treatment/ post-surgical	Week 1	Week 2	Weeks 4–6	Recovery
ALT (u/l) Mean (SD) / n	44 (18.3) / 23	48 (21.5) / 11	48 (21.6) / 17	45 (17.6) / 21	39 (9.8) / 5	52 (26.2) / 23	63 (24.4) / 11	55 (21.7) / 17	46 (11.3) / 21	44 (18) / 6
AST (u/l) Mean (SD) / n	39.6 (12.88) / 23	43.5 (14.25) / 11	40.8 (12.72) / 17	36.5 (10.81) / 21	30.2 (9.98) / 5	40.3 (13.19) / 23	45.5 (23.74) / 11	37.9 (21.11) / 17	37 (28.43) / 21	33.2 (9.47) / 6
Alkaline phosphatase (u/l) Mean (SD) / n	560 (213.1) / 23	643 (268) / 11	531 (143.1) / 17	524 (268.5) / 21	463 (161.5) / 5	344 (176.5) / 23	390 (227.9) / 11	382 (220.8) / 17	333 (150.4) / 21	289 (109.5) / 6
GGT (u/l) Mean (SD) / n	99 (33) / 15	83 (13) / 3	86 (24.5) / 3	90 (33) / 18	115 (38.4) / 4	53 (15.3) / 15	49 (27) / 3	51 (16.1) / 3	51 (13.2) / 18	47 (17.8) / 4
Urea N (mg/dl) Mean (SD) / n	21 (2.7) / 23	19 (4.5) / 14	21 (5.5) / 17	20 (3.5) / 21	23 (4) / 5	22 (5.5) / 23	21 (6.2) / 14	23 (4.3) / 17	21 (3.1) / 21	23 (5.5) / 125
Creatinine (mg/dl) Mean (SD) / n	1 (0.14) / 23	0.9 (0.09) / 14	0.9 (0.14) / 17	1 (0.12) / 21	1 (0.19) / 5	0.9 (0.14) / 23	0.9 (0.12) / 14	0.8 (0.11) / 17	0.9 (0.13) / 21	0.9 (0.08) / 6
T protein (g/dl) Mean (SD) / n	8.4 (0.64) / 23	7.5 (0.6) / 11	7.8 (0.54) / 17	8.3 (0.77) / 21	8.2 (0.86) / 5	8 (0.48) / 23	7.5 (0.55) / 11	7.9 (0.49) / 17	8.3 (0.54) / 21	7.4 (0.29) / 6
Albumin (g/dl) Mean (SD) / n	4.5 (0.35) / 23	4.1 (0.37) / 11	4 (0.42) / 17	4.1 (0.34) / 21	4.2 (0.39) / 5	4.2 (0.32) / 23	4.1 (0.3) / 11	4.1 (0.35) / 17	4.1 (0.43) / 21	4.1 (0.35) / 6
Globulin (g/dl) Mean (SD) / n	3.9 (0.6) / 23	3.4 (0.42) / 11	3.8 (0.61) / 17	4.2 (0.85) / 21	4 (0.91) / 5	3.8 (0.53) / 23	3.4 (0.41) / 11	3.8 (0.63) / 17	4.2 (0.64) / 21	3.3 (0.12) / 6
AG ratio Mean (SD) / n	1.17 (0.207) / 23	1.24 (0.168) / 11	1.1 (0.255) / 17	1.2 (0.284) / 21	1.14 (0.375) / 5	1.14 (0.213) / 23	1.23 (0.158) / 11	1.12 (0.251) / 17	1 (0.241) / 21	1.24 (0.133) / 6
Inorganic phosphate (mg/dl) Mean (SD) / n	7 (0.9) / 23	6 (1.3) / 11	6 (1.3) / 17	6 (1.2) / 21	7 (0.5) / 5	6 (0.7) / 23	5 (1.1) / 11	6 (1.3) / 17	6 (1) / 21	5 (1.5) / 6
Sodium (mequiv./l) Mean (SD) / n	154 (5) / 23	151 (3.4) / 11	151 (3.7) / 17	152 (6.2) / 21	154 (4.4) / 5	152 (3.3) / 23	151 (3) / 11	152 (4.3) / 17	151 (3.4) / 21	151 (4.2) / 6
Potassium (mequiv./l) Mean (SD) / n	5.3 (0.69) / 23	4.6 (0.6) / 11	4.8 (0.46) / 17	4.8 (0.58) / 1	5.1 (0.25) / 5	5.1 (0.61) / 23	4.6 (0.65) / 11	5 (0.69) / 17	5.1 (0.35) / 21	4.8 (0.33) / 6
Chloride (mequiv./l) Mean (SD) / n	112 (4.7) / 23	109 (4.8) / 11	110 (4.7) / 17	111 (4.5) / 21	108 (1.5) / 5	111 (3.2) / 23	108 (3.7) / 11	111 (3.9) / 17	112 (3.7) / 21	110 (2.7) / 6

initial week of infusion, probably due to the frequency of blood sample collections for clinical pathology and toxicokinetic studies, as well as the likely haemodilution effects of treatment. Reticulocyte counts (as a percentage of RBC) are elevated by week 1 (females) or 2 (males), and indices of erythrocyte mass are similar to pre-treatment values by week 4. There is no clear evidence of an effect on platelet counts. Mean total leucocyte counts (WBC) generally increase with the duration of treatment, suggesting that a chronic inflammatory response remains present. The differential leucocyte data were unremarkable (data not presented).

There were no changes noted in the reviewed urinalysis data, which were considered to be biologically significant. As expected, increased urine volumes, with lower solute concentrations, were found but were considered normal compensatory responses to intravenous infusion.

Selected gross pathology findings for saline control group monkeys are presented in Table 15.3. Gross necropsy observations were infrequent and restricted to the lungs, catheter site, spleen, kidney and thymus. Dark or mottled areas on the lungs may be indicative of thromboembolism, and adhesions are associated with parasitic migration. Thickening at the catheterisation site may occur as part of the normal healing process; however, excessive scarring may be indicative of chronic infections. The spleen may be enlarged as a post-mortem artefact (secondary to barbiturate anaesthesia) or possibly in response to infection. Discoloured areas on the cut surface of the kidney have been associated with thromboembolism and infarcts, but histopathology is required to confirm this diagnosis. The finding of a dark, small or gelatinous thymus is non-specific and may be due to the collection technique, route of catheterisation and secondary changes at the infusion site, or chronic stress.

Histomorphological findings in the lungs, liver, infusion site, catheterisation site, spleen, kidney, thymus, heart, lymph nodes and adrenal gland are presented in Table 15.4. In the lungs, perivascular and peribronchial macrophages, as well as the presence of parasitic pigments, are commonly noted in this species and are unrelated to the infusion procedure. Inflammation of the interstitium and pleural tissues may be associated with minimal to slight fibrosis and/or haemorrhage, but these findings are infrequent and are of no toxicological significance. The presence of thrombi in small arterioles is uncommon, but increases in frequency, distribution and severity in chronic inflammatory conditions. In the liver, chronic inflammation and pigment in the reticuloendothelial cells are slightly more common in animals that have chronic indwelling catheters. Changes at the infusion site (tissues adjacent to the tip of the catheter) include thrombus formation, chronic inflammation, fibrosis and haemorrhage, as well as fibrotic and hypertrophic changes to the vessel wall. The formation of large thrombi will result in a showering of microthrombi into the lungs, particularly in cases of sepsis. The catheterisation site reactions (inflammation, fibrosis, perivascular arteritis and oedema) are common and represent a normal healing response. In the heart, chronic myocardial inflammation and the presence of occasional thrombi in small vessels may occur, but the significance of these changes remains unclear. Findings in the spleen, kidney, thymus, adrenal gland and lymph nodes are of a character and frequency common for this species and are considered unrelated to the infusion procedure.

Table 15.3 Control data for surgically catheterised primates – gross pathology

Diagnoses	Unscheduled deaths		Terminal sacrifice		Recovery sacrifice	
	Males	Females	Males	Females	Males	Females
Lungs						
Not remarkable	1/1	N/A	15/17	14/16	4/4	4/4
Failure to collapse	–		–	1/16	–	–
Dark area	–		1/17	–	–	–
Interlobar adhesion	–		–	1/16	–	–
Mottled	–		1/17	–	–	–
Liver						
Not remarkable	N/A	N/A	17/17	16/16	4/4	4/4
Infusion site						
Not remarkable	N/A	N/A	17/17	16/16	4/4	4/4
Catheter site						
Not remarkable	1/1	N/A	15/16	16/16	3/3	4/4
Thickened	–	–	1/16	–	–	–
Spleen						
Not remarkable	0/1	N/A	3/13	1/12	4/5	5/6
Enlarged	1/1	–	1/16	–	1/5	–
Small	–	–	–	–	–	1/6

Kidney					
Not remarkable	–	N/A	17/17	14/16	4/4	4/4
Unequally sized	–	–	–	1/16	–	–
Pale area	–	–	–	1/16	–	–
Thymus					
Not remarkable	1/1	N/A	16/17	15/16	4/4	4/4
Dark	–	–	–	1/16	–	–
Small	–	–	1/17	–	–	–
Heart					
Not remarkable	1/1	N/A	14/14	13/13	4/4	4/4
Mesenteric lymph nodes					
Not remarkable	–	N/A	14/14	13/13	4/4	4/4
Mandibular lymph nodes					
Not remarkable	–	N/A	14/14	13/13	4/4	4/4
Inguinal lymph nodes					
Not remarkable	–	N/A	14/14	13/13	4/4	4/4
Adrenal cortex					
Not remarkable	1/1	N/A	14/14	13/13	4/4	4/4
Adrenal medulla					
Not remarkable	1/1	N/A	14/14	12/13	4/4	4/4
Small	–	N/A	–	1/13	–	–

N/A, not applicable.

Table 15.4 Control data for surgically catheterised primates – histopathology

Diagnoses	Unscheduled deaths		Terminal sacrifice		Recovery sacrifice	
	Males	Females	Males	Females	Males	Females
Lungs						
Not remarkable	0/1	N/A	8/17	6/16	1/5	2/6
Perivascular/peribronchial infiltrate, macrophage						
Minimal to slight	1/1	–	6/17	3/16	1/5	–
Parasitic pigment						
Present to slight	1/1	–	11/17	7/16	2/5	4/6
Inflammation						
Minimal to slight	–	–	2/17	–	1/5	–
Fibrosis, pleura						
Minimal to slight	–	–	1/17	1/16	–	–
Thrombus, pulmonary arteriole						
Slight	–	–	–	1/17	–	–
Haemorrhage						
Minimal	–	–	1/17	–	–	–
Inflammation, chronic pleura						
Minimal	–	–	–	1/16	–	–
Inflammation, granulomatous						
Minimal	–	–	1/17	1/16	–	–
Focal pneumonitis						
Minimal	–	–	–	1/16	1/5	1/6
Pleura, fibrous adhesions						
Present	–	–	–	1/16	–	–
Liver						
Not remarkable	0/1	N/A	4/14	5/13	2/5	1/6
Inflammation						
Minimal	–	–	3/14	3/13	–	–
Chronic inflammation						
Minimal to slight	1/1	–	7/14	4/13	2/14	4/13
Reticuloendothelial cells, pigment						
Minimal	–	–	2/14	1/13	–	–

Focal necrosis	–	–	–	–	–	–
Minimal	–	–	–	1/13	–	–
Necrosis, individual cell	–	–	–	–	–	–
Slight	–	–	–	1/13	–	–
Pigment	–	–	–	–	–	–
Minimal	–	–	1/14	–	–	–
Lipoidosis, focal	–	–	–	–	–	–
Slight	–	–	–	1/13	–	–
Focal vacuolisation	–	–	–	–	–	–
Minimal	–	–	1/14	–	–	–
Infusion site[a]	–	N/A	–	–	–	–
Not remarkable	–	–	1/14	1/13	1/4	0/4
Thrombus	–	–	–	–	–	–
Moderate to moderately severe	–	–	–	2/13	–	–
Organised thrombus	–	–	–	–	–	–
Moderate to moderately severe	–	–	3/14	1/13	–	–
Soft thrombus	–	–	–	–	–	–
Slight	–	–	–	1/13	–	–
Inflammation	–	–	–	–	–	–
Minimal to moderate	–	–	6/14	5/13	–	–
Chronic inflammation	–	–	–	–	–	–
Minimal to moderate	–	–	6/14	7/13	2/4	3/4
Perivascular tissue, arteritis	–	–	–	–	–	–
Present	–	–	–	1/13	–	–
Fibrosis	–	–	–	–	–	–
Minimal to severe	–	–	8/14	9/13	2/4	2/4
Haemorrhage	–	–	–	–	–	–
Slight to moderately severe	–	–	3/14	3/13	–	–
Suture material	–	–	–	–	–	–
Present	–	–	–	2/13	–	–
Pigmented macrophages	–	–	–	–	–	–
Minimal	–	–	1/14	–	–	–
Fibrous capsule	–	–	–	–	–	–
Moderate	–	–	1/14	2/13	–	2/4
Intimal reaction	–	–	–	–	–	–
Slight	–	–	1/14	1/13	2/4	1/4
Medial hypertrophy	–	–	–	–	–	–
Moderate	–	–	–	–	1/4	–

Table 15.4 Control data for surgically catheterised primates – histopathology (continued)

Diagnoses	Unscheduled deaths		Terminal sacrifice		Recovery sacrifice	
	Males	Females	Males	Females	Males	Females
Catheter site[b]						
Not remarkable	0/1	N/A	0/17	1/16	0/5	0/6
Inflammation						
Minimal to moderate	–	–	4/17	3/16	–	1/6
Fibrosis						
Minimal to moderately severe	–	–	8/17	7/16	2/5	2/6
Chronic-active inflammation						
Exit site						
Minimal to slight	1/1	–	3/17	3/16	1/5	1/6
Entrance site						
Minimal to slight	1/1	–	3/17	2/16	1/5	2/6
Above the tip						
Slight to moderate	–	–	2/17	2/16	1/5	2/6
Tip						
Minimal to slight	1/1	–	3/17	3/16	1/5	2/6
Below the tip						
Minimal to slight	1/1	–	1/17	1/16	–	2/6
Chronic inflammation						
Minimal to moderate	–	–	7/17	7/16	3/5	3/6
Suture material						
Present	–	–	2/17	1/16	–	–
Thrombus						
Present to moderate	–	–	5/17	5/16	2/5	3/6
Perivascular tissue, arteritis						
Present	–	–	–	1/16	–	–
Haemorrhage						
Minimal	1/1	–	–	1/16	–	–
Hypertrophy						
Minimal	–	–	–	1/16	–	–
Fibrous capsule						
Slight to moderate	–	–	4/17	4/16	2/5	2/6
Foreign body material						
Slight to moderate	–	–	3/17	4/16	2/5	2/6

Oedema						
Present	–	–	–	–	1/5	–
Spleen						
Not remarkable	0/1	N/A	15/20	17/19	5/5	6/6
Hyperplasia, lymphoid	1/1	–	–	–	–	–
Minimal						
Increased pigment	1/1	–	–	–	–	–
Minimal						
Chronic inflammation	–	–	3/20	2/19	–	–
Minimal						
Protein precipitate, follicular centres		–	–	1/20	–	—
Slight						
Germinal centres, eosinophilic material	–	–	1/20	–	–	–
Minimal						
Kidney						
Not remarkable	0/1	N/A	1/17	7/16	0/5	1/6
Infiltrate, mononuclear cell	1/1	–	1/17	3/16	–	1/6
Minimal to slight						
Mineralisation, papilla	1/1	–	1/17	1/16	1/5	2/6
Minimal						
Tubule dilatation	1/1	–	–	–	–	–
Minimal						
Lymphocytic infiltrate, focal, cortex	–	–	3/17	1/16	1/5	–
Minimal to slight						
Inflammation	–	–	4/17	1/16	–	–
Minimal to slight						
Chronic inflammation	–	–	7/17	4/16	2/5	2/6
Minimal						
Tubule regeneration	–	–	3/17	1/16	–	–
Minimal to slight						
Arteriosis	–	–	–	1/16	–	–
Present						
Microcalculi, papilla	–	–	2/17	2/16	2/5	1/6
Minimal						
Cortical scar	–	–	1/17	–	–	–
Minimal						

Table 15.4 Control data for surgically catheterised primates – histopathology (continued)

Diagnoses	Unscheduled deaths		Terminal sacrifice		Recovery sacrifice	
	Males	*Females*	*Males*	*Females*	*Males*	*Females*
Thymus						
Not remarkable	–	N/A	12/14	12/13	3/4	4/4
Cyst						
Present	–	–	1/14	1/13	–	–
Lymphoid depletion						
Slight to moderate	–	–	1/14	–	1/4	–
Heart						
Not remarkable	1/1	N/A	11/17	11/16	4/5	4/6
Inflammation						
Minimal	–	–	–	1/16	–	–
Chronic inflammation						
Minimal to slight	–	–	4/17	1/16	1/5	1/6
Chronic-active inflammation						
Minimal	–	–	–	1/16	–	1/6
Arteritis						
Present	–	–	–	1/16	–	–
Thrombus						
Minimal to slight	–	–	2/17	1/16	–	–
Mesenteric lymph node						
Not remarkable	1/1	N/A	16/17	15/16	5/5	6/6
Granuloma, foreign body						
Moderate	–	–	–	1/16	–	–
Pigment						
Moderate	–	–	1/17	–	–	–
Mandibular lymph node						
Not remarkable	1/1	N/A	17/17	16/16	5/5	6/6
Irginual lymph node						
Not remarkable	–	N/A	4/6	5/6	–	–

Pigmented macrophages						
Minimal to slight	–	–	2/6	–	–	–
Haemorrhage						
Slight	–	–	–	–	1/6	1/6
Adrenal cortex						
Not remarkable	1/1	N/A	16/17	16/16	5/5	6/6
Mineralisation						
Present to minimal	–	–	1/17	–	–	–
Adrenal medulla						
Not remarkable	0/1	N/A	16/17	15/16	4/5	6/6
Mineralisation						
Present to minimal	–	–	1/17	–	–	–
Pigment						
Minimal	–	–	–	1/16	–	–
Mononuclear cell infiltrate						
Minimal	1/1	–	–	–	–	–

Notes
[a]Tissues directly adjacent to intravascular catheter tip.
[b]Intravascular and extravascular tissue at the site of catheter insertion (femora or jugular vein).
N/A, not applicable.

15.9 Conclusion

The successful conduct and completion of an infusion study in non-human primates relies heavily on adequate planning and careful experimental design. The choice of an infusion system may depend on the preferences and experiences of the laboratory, but one should exercise flexibility and use the technology best suited for the species, test material formulation being evaluated, experience of the current staff and previous experiments that are referenced. While the caging used is standard and requires little to no modification for infusion studies, the pumping systems are quite costly and will require dedicated technical support to provide routine validation for operational accuracy. Training of the technical staff must be considered for surgical support, post-operative monitoring, infusion system operation and maintenance and in-life evaluations. Proper evaluation of the chemical and physical interactions between the infusion system components should be conducted in a prospective manner, and the results considered.

The selection of an acceptable intravenous formulation typically depends upon the physical and chemical characteristics of the test material; however, biocompatibility and haemocompatibility must be assessed to determine if the mixture is suitable for intravenous administration. A number of study design considerations have been reviewed, with the ultimate aim of providing a basic guideline for animal selection, in-life evaluations, clinical pathology and post-mortem procedures. Selected findings for control animals were presented so that an investigator can critically evaluate study results and provide sound judgement as to the toxicity of a proposed therapeutic material. The cumulative experience of the subject laboratory will increase the likelihood of a successful infusion study in this important species.

Acknowledgements

The author would like to acknowledge the invaluable assistance of Ms Stephene Cyrek in the preparation of this manuscript. Ms Cyrek is an associate toxicologist who specialises in primate and infusion studies in mammalian toxicology at our Vienna laboratory. Her dedication, technical skills and scientific abilities are an asset to our company. I would also like to thank the technical staff who conduct the studies with the highest level of skill and integrity; without them, all scientists would be helpless. Lastly, I would like to thank my many colleagues within the company and in other laboratories for their willingness to share experiences and technical improvements, as well as provide critical feedback on similar experiences. It is through cooperation that these efforts ultimately benefit all of us in the field of pre-clinical toxicology.

References

Blacklock, J.B., Wright, D.C., Dedrick, R.L., Blasberg, R.G., Lutz, R.J., Doppman, J.L. and Oldfield, E.H. (1986) Drug streaming during intra-arterial chemotherapy. *Journal of Neurosurgery*, 64(2), 284–291.
Bolton, G.R. (1975) The arrhythmias. In Edwards, J.N. (ed.) *Handbook of Canine and Feline Electrocardiography*. Philadelphia: W.B. Saunders, pp. 60–87.
David, R.M., Moore, M.R., Cifone, M.A., Finney, D.C. and Guest, D. (1999) Chronic peroxisome proliferation and hepatomegaly associated with the hepatocellular tumorigenesis of di(2-ethylhexyl) phthalate and the effects of recovery. *Toxicological Sciences* 50(2), 195–205.

Hackett, R.B. and Stein, M.E. (1991) Preclinical toxicology/safety consideration in the development of ophthalmic drugs and devices. In Hobson, D.W. (ed.) *Dermal and Ocular Toxicity*. Boca Raton, FL: CRC Memorial Press, pp. 607–626.

Liau-Chu, M., Theis, J.G. and Koren, G. (1997) Mechanism of anaphylactoid reactions: improper preparation of high-dose intravenous cyclosporine leads to bolus infusion of Cremophor EL and cyclosporine. *Annals of Pharmacotherapeutics*, 31(11), 1287–1291.

Maddison, J. (2000) Adverse drug reactions. In Ettinger, J.T. and Feldman, E.C. (eds.) *Textbook of Veterinary Internal Medicine*, 5th edn. Philadelphia: W.B. Saunders, pp. 321–347.

Marosok, R., Washburn, R. Indorf, A., Solomon, D. and Sherertz, R. (1996) Contribution of vascular catheter material to the pathogenesis of infection: depletion of complement by silicone elastomer in vitro. *Journal of Biomedical Materials Research*, 30, 245–250.

McNamee, A., Wannemacker, R.W., Dinterman, R.E., Rozmiarek, H. and Montrey, R.D. (1984) A surgical procedure and tethering system for chronic blood sampling, infusion, and temperature monitoring in caged nonhuman primates. *Laboratory Animal Science*, 34(3), 303–307.

Peterson, T.V., Carter, A.B. and Miller, R.A. (1997) Nitric oxide and renal effects of volume expansion in conscious monkeys. *American Journal of Physiology: Heart and Circulatory Physiology*, 41(4), R1033–R1038.

Pickens, R., Hauck, R. and Bloom, W. (1966) A catheter-protection system for use with monkeys. *Journal of Experimental and Analytical Behaviour*, 9(6), 701–702.

Rosenblum, L.A. and Andrews, M.W. (1995) Environmental enrichment and psychological well-being of nonhuman primates. In Bennett, B.T., Abee, C.R. and Henrickson, R. (eds.) *Nonhuman Primates in Biomedical Research*. Philadelphia: Academic Press, pp. 101–112.

Rubin, L.F. (1974) *Atlas of Veterinary Ophthalmoscopy*. Philadelphia: Lea & Febiger.

Tilley, L.P. (1992) *Essentials of Canine and Feline Electrocardiography*, 3rd edn. Philadelphia: Lea & Febiger.

Tschirhart, J.M. and Minoo, K.R. (1988) Mechanism and management of persistent withdrawal occlusion. *American Surgeon*, 54, 326 328.

Wojnicki, F.H., Bacher, J.D. and Glowa, J.R. (1994) Use of subcutaneous vascular access ports in rhesus monkeys. *Laboratory Animal Science*, 44(5), 491–494.

Woodburn, K., Sykes, E. and Kessel, D. (1995) Interactions of Solutol HS15 and Cremophor EL with plasma lipoproteins. *International Journal of Cell Biology*, 27(7), 693–699.

16 Surgical preparation and multidose infusion toxicity studies in the marmoset

C. R. Schnell

16.1 Introduction

Marmosets are New World monkeys coming originally from Brazil. Their species-typical characteristics, notably their small size, fecundity and relative ease of maintenance in laboratory situations, have contributed to their increasing popularity in many areas of fundamental and applied research. The common marmoset (*Callithrix jacchus*) is the species of marmoset which has been used most extensively in physiological, pharmacological and toxicological studies, and it has contributed to significant advances in the field of neuroscience, reproductive biology and cardiovascular research.

They provide a useful primate model to perform pharmacodynamics and pharmacokinetics studies involving absorption, distribution, metabolism and elimination of xenobiotics. In the last two decades, the marmoset has become more widely accepted as an alternative non-rodent (or second) species in pre-clinical safety evaluation and other regulatory toxicity studies. In drug development research, their small size is advantageous in that it enables evaluation of relatively small amounts of test material, which helps in the early development of biotechnology products or of molecules that are difficult to synthesise. Compounds that are not orally active can be delivered by intravenous infusion to test their safety and bioactivity pattern. Infusions can be most conveniently delivered via indwelling devices allowing chronic access.

With the increasing use of marmosets in biomedical research, especially in pharmaceutical development and safety testing laboratories, safe and reliable techniques for chronic catheter access have become necessary. However, there are relatively few published data about pre-clinical chronic infusion in the marmoset. This chapter reviews and describes various methods for implanting indwelling intravenous catheters in the marmoset.

Most of the existing techniques presently used in conscious marmosets are techniques that were originally designed for use in rats. The different chronic continuous infusion models presented in this chapter relate to two main categories: catheter access models for short-term infusion (i.e. less than 1 day) and infusion models for long-term infusion (i.e. more than 1 day). Some of the techniques described in detail are based on our own experiences. These techniques were developed and applied in our marmoset facilities over the past 15 years for pharmacokinetic and pharmacodynamic studies with antihypertensive compounds. Other techniques are reviewed from the literature and/or based on personal communications with other laboratories engaged in marmoset research.

16.2 Animals

The common marmosets (*Callithrix jacchus*, Ciba-Geigy, Sisseln, Switzerland) used in all studies were normotensive adults (4–7 years old) of both sexes weighing approximately 350 g (Figure 16.1). All the animals were born in captivity. They were kept in a room ventilated with air maintained at a temperature of 24–27 °C and a humidity of 50–80%. A 12-h light/dark cycle from 06.00 to 18.00 h and from 18.00 to 06.00 h was used. The marmosets were maintained in family groups of 2–4 animals per cage. They were fed a normal-salt pellet diet (Na$^+$ 100 mmol/kg, K$^+$ 250 mmol/kg, NAFAG, Gossau, Switzerland) supplemented with fruits, eggs, meat and milk. Water was given *ad libitum*.

16.3 Chronic catheter access models

The most common application for chronic catheters is access to the venous side of the

Figure 16.1 A couple of common marmosets (*Callithrix jacchus*).

circulatory system for fluid infusion and blood withdrawal. Two types of catheter devices are used: indwelling externalised catheters and, more recently, subcutaneous access port catheters (Jacobson 1998).

16.3.1 *Surgical implantation of indwelling catheters*

Marmosets were anaesthetised with a combination of 18 mg/kg alphaxolone-alphadone (Saffan; Glaxovet, Uxbridge, UK), 0.15 mg/kg atropine (Atropin; Sigfried, Zofingen, Switzerland) and 0.75 mg/kg diazepam (Valium; Roche, Basle, Switzerland) given intramuscularly on two different sites. The marmosets were positioned in dorsal recumbency and shaved on the lateral side of one leg (thigh) area and placed on a warmed sterile operative field (Aquamatic, K module, American Medical Systems, Cincinnati, OH, USA) maintained at a temperature of 39 °C. The shaved skin was scrubbed with polividone-iodine and care was taken to maintain sterility throughout the operative procedure. A skin incision of 1–1.5 cm was made over the femoral triangle in the inguinal region. The femoral vein was exposed by blunt dissection. This was performed with the use of a surgical microscope and care was taken not to traumatise either the nerve or the artery accompanying the femoral vein. Tissues were kept moist by frequent flushing with sterile, isotonic saline. The vein was cleared of connective tissue and two ligatures (3–0 silk) passed beneath it to facilitate lifting and occluding during puncture and initial catheter insertion. The caudal end of the vein was ligated and the upper surface of the exposed vein was cut transversely proximal to the ligature with spring-type iris scissors. A custom-made catheter (or cannula) was filled with sterile physiological saline and inserted 3 cm into the femoral vein in the direction of blood flow so that its tip lay in the inferior vena cava caudal to the kidney. The final position of the catheter was adjusted until the maximal blood flow was obtained.

The catheter was then tied into the vein and the free end tunnelled subcutaneously and exited through a small skin incision made at the middle of the tail (Figure 16.2). The catheter was filled with a sterile heparin solution (250 U/ml in isotonic saline) and sealed by heating. The catheter was constructed from a 20-mm piece of polyethylene (PE) 10 tubing (ID 0.28 mm; OD 0.61 mm; Laubscher, Basle, Switzerland), inserted into a 200-mm length of PE 50 tubing (ID 0.58 mm; OD 0.96 mm; Laubscher, Basle, Switzerland) and heat sealed. A 10-mm length of Silicone (Silastic® 602–105) tubing (ID 0.3 mm; OD 0.63 mm; Ulrich AG, St. Gallen, Switzerland) was welded at the end of the catheter to ensure a soft tip. As a general rule, the softer tubing materials are less traumatic to the intimal lining of the vessels and therefore induce a less severe host defence response. The tip of the silicone rubber catheter was not cut to a bevel of about 45° to facilitate venous insertion as often used in other species, but left in its original shape. The reasons for doing this are as follows. Firstly, the convoluted route of the femoral vein high in the groin makes cannulation with a bevelled catheter very difficult and dangerous as it can result in severe damage or even to perforation of the blood vessel. Secondly, blood withdrawal can be severely compromised in these animals as a result of total or partial obstruction of the bevel tip of the catheter by direct aspiration of the vascular wall. The constructed catheter was then sterilised with ethylene oxide and a minimum of 10 days' aeration was allowed before use. To prevent destruction by the marmosets when returned to their home cages, the end of the vascular catheter was twisted around the base of the tail and covered with soft tape, a metal mesh band and several further layers of strong tape. The skin was closed with polyester thread 4–0 (Dagrofil; Braun, Neuhausen, Germany).

Figure 16.2 Exit site of intravenous catheters located at the tail of the marmoset.

The animals were administered penicillin (5000 IU per animal, Duplocilline LA, Veterinaria AG, Zurich, Switzerland) and piroxicam (2 mg per animal, Piroxicam-Mepha, Mepha Pharma AG, Aesch/BL, Switzerland) intramuscularly immediately after surgery to prevent infection and pain, and skin incisions were sprayed lightly with Nobecutan® (Astra, Pharmaceutica AG, Dietikon, Switzerland). The implanted marmoset was allowed to fully recover in its sleeping box placed over a commercially available electric heating-pillow (approximately 2 h) before returning to its home cage (1.55 × 1.23 × 0.86 m) together with its family group. We found that this social environment is pivotal for the well-being of the marmoset immediately after surgery. Because marmosets exhibit highly developed social systems, particular care must be taken, especially in highly applied research situations, to ensure that the social needs of the animals are addressed within the constraints of the experimental protocol. At least a 20-h period was allowed for total recovery from surgery until the beginning of any experimental protocol.

During an experiment, the marmoset was placed in the horizontal position in a restraining tube in various sizes to accommodate animals of different ages and sizes

(Figure 16.3). In our experience, the maximum time period a marmoset could be continuously restrained under this condition was 4–5 h. The venous infusion catheter was connected to a syringe secured on a syringe infusion pump (Precidor, Typ 5003, Infors HT®, Bottmingen, Switzerland) for controlled drug delivery. Using this technique, we could follow the hypotensive action of a specific renin inhibitor, CGP 29287, after both intravenous infusion and intravenous bolus application in the marmoset (Figure 16.4). Although the animals were previously trained to be accustomed to this procedure, restraint resulted in stress, which could clearly be demonstrated by comparing heart rate data obtained via telemetry with those from unrestrained animals (Schnell and Gerber 1997). In addition, because of the dextrous nature of the marmosets, it was found not to be possible to maintain chronic catheterised animals for periods longer than 24 h. For this reason, the day after the experiment, catheters were taken out under aseptic conditions and light anaesthesia induced by ketamine (Ketalar, Parke-Davis, 10 mg/kg i.m.) supplemented with diazepam 0.75 mg/kg i.m. (Valium, Roche) and atropine 0.15 mg/kg (Atropine, Siegfried s.c.). Bleeding was prevented by a temporary ligature proximal to the leak, which was then closed with acrylic tissue glue (histoacryl blue; Braun-Melsungen, Melsungen, Germany). After removal of the temporary ligature and the application of tissue glue, blood flow in the vessel was restored, allowing the distal recannulation at a later date. The skin was closed with absorbable 4–0 sutures (Ethicon). As an alternative to the femoral vein for chronic cannulation, the lateral tail vein of the marmoset was also used with success in our laboratories. However, because of their small size and their susceptibility to collapse and thrombosis, it was only possible to cannulate the lateral tail vein of the marmoset at the base of the tail. After a recovery period of 4–6 weeks, the marmoset could be operated on again, but the number of times this could be repeated was limited to 6–8 (three times for each femoral and twice for each tail vein).

The drawbacks we encountered using this method over a period of 10 years were mainly related to thrombotic catheter occlusion as a result of the host's defensive clotting cascade, or the removal or damaging of the catheter by the animal itself (or

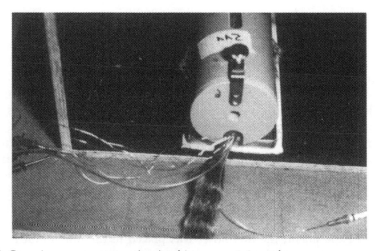

Figure 16.3 Conscious marmoset maintained in a restraining tube.

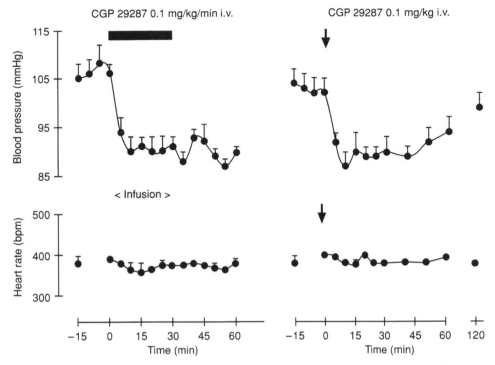

Figure 16.4 Mean arterial blood pressure and heart rate in five furosemide-treated restrained marmosets before, during and after either infusion of CGP 29287 0.1 mg/kg/min (left) or intravenous bolus injection of CGP 29287 0.1 mg/kg (right). Values represent mean ± SEM.

his family) once he was returned to his home cage. However, 85% of the implantations we have performed to date have been successful.

Most of the research activities in our marmoset facilities were dedicated to screening of cardiovascular drugs to lower blood pressure. For this reason, an arterial catheter was implanted in the femoral artery of the animal at the same time as the venous catheter described before. The next day, the arterial catheter was connected to a pressure transducer while the conscious marmoset was placed in a restraining tube for continuous recording of blood pressure and heart rate data.

In the early 1990s, a miniaturised radiotelemetry system initially designed for remote measurements of cardiovascular parameters in conscious, unrestrained rats was adapted in our laboratory for the use in marmosets (Schnell and Wood 1993). Once the telemetry transmitter was implanted in the peritoneal cavity and the sensing catheter positioned in the descending aorta, remote monitoring of blood pressure, heart rate and motor activity in these animals was possible for a period up to 1 year. After the transmitter batteries ran flat, a new transmitter was reimplanted, allowing an additional 12 months' recording. Reimplantation has been successfully performed in our marmoset colony up to four times in the same animal. This new method considerably revolutionised the collection of cardiovascular parameters in conscious marmosets by reducing the variance and increasing the sensitivity, allowing the measurement of normal physiological data and finally reducing the number of animals used in an experiment, as well as improving their care and welfare. At this stage, we were still dissatisfied that our animals had to undergo anaesthesia shortly before a pharmacodynamic experiment. Therefore we decided to develop an alternative system for chronic catheterisation in conscious marmosets, using an adapted commercially available vein-flow cannula.

16.3.2 *Implantation of indwelling catheters in conscious animals*

A sterile 24-gauge over-the-needle catheter (Insyte-w®, Vialon®, 0.7×19 mm) was inserted in the upper femoral vein by direct puncturing. The upper femoral vein of the marmoset is mainly superficial and easy to see (dark blue line) and enter. However, especially in elderly and heavy animals, visual location can be difficult. In this case, the femoral artery (easily located by palpation) can be used as a reference point for the location of the femoral vein which runs parallel and caudal to the artery. The femoral venous puncture is performed directly through the skin with an assistant restraining the marmoset in one gloved hand and firmly extending the leg with the other (Figure 16.5). Leather gloves are worn in most facilities to prevent bites. Strict asepsis precautions must be taken at all times to avoid risk of infection. After disinfection of the point of entrance in the skin, the vein-flow catheter can be inserted a distance of 1 cm into the femoral vein. Particular attention has to be taken to insert the needle with the bevel facing upward, at the most acute angle possible with the vein in order to minimise the danger of vessel perforation, which will cause haemorrhage. Moreover, it is highly recommended to start cannulation as distally (farthest from the heart) as possible and move proximally in case venepuncture is disrupted by the animal flinching. Typically, a small quantity of blood will 'flash back' into the hub of the needle of the vein-flow system, indicating proper placement within the vein. By pressing the vessels in the groin region with one finger, it is possible to stop blood flow for a short time. Under these conditions, the trocar from the introducer set is then withdrawn from the distal end, leaving the catheter in place. The female Luer lock distal part of the inserted vein-flow catheter is then cut away with scissors while the proximal part (2 cm) of the catheter remains in the femoral vein. This distal end is then connected to a 200 mm length of PE 50 tubing (ID 0.58 mm; OD 0.96 mm; Laubscher, Basle, Switzerland) and secured with an adhesive tape. In addition, a small drop of tissue glue is placed on the catheter at the site of exit from the lateral side of the leg. This method of taping and gluing has been shown to retain the venous catheter within the vessel in all animals for several hours.

The marmoset is then placed for the duration of the experiment in the same restraining tube as described previously. Once the experiment is finished, the catheter is removed delicately from the vessel and adequate pressure applied to minimise haematoma formation. Provision of a reward (grapes, honey or banana milk-shake) as a routine after handling, palpation, bleeding, etc. is recommended, which elicits rapid co-operation and greatly reduces stress. After a recovery period of 2 weeks, the same animal can be re-used for a new catheterisation. In our experience, there is no limitation to the number of times this procedure could be repeated for the same marmoset.

16.3.3 *Subcutaneous vascular access port (VAP)*

Ports were first developed around 1980 to deliver chemotherapy to humans. They became popular for animal use around 1983. The VAP is a subcutaneously implanted

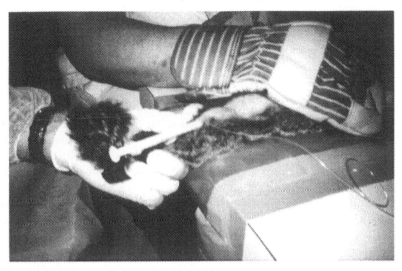

Figure 16.5 Intravenous injection through the femoral vein in a restrained conscious marmoset.

biomedical research device designed to be used where repeated access to the vascular system is desired. The port's distal catheter portion is identical to the externalised catheter's distal portion. The difference is that its proximal end terminates under the skin into a metal or plastic cylinder that has a rubber injection window that is accessed percutaneously with a hypodermic Huber needle. As ports are catheters that do not exit through the animal's skin, there is a little concern over the animal disturbing this implanted device. Because there is no chronic exit site wound, infection risks associated with ports are considerably lower than with external catheters. The use of ports in biomedical research has reduced animal use, allowed group housing and minimised stress. Recent studies indicate that subcutaneous vascular access ports have been successfully used in several species (dogs, swine, cats, rabbits, ferrets, woodchuck, hens and rats). In non-human primates, VAPs have been used for long-term access to the vascular system (14–18 months) in rhesus monkeys (Wojnicki *et al.* 1994) and squirrel monkeys (Kinsora *et al.* 1997). A commercially available VAP (V-A-P® Access Port Model SLA, Access Technologies, Skokie, IL, USA) has been used in individually housed marmosets for daily, slow intravenous dosing (Fitzgerald *et al.* 1996). The port consists of a dome-shaped reservoir body (0.8 cm high by 1.2 cm wide) on a flanged base (2.5 cm at the widest dimension). The base had suture holes for attachment to underlying tissue. The VAP was implanted in the femoral vein, with the port reservoir positioned subcutaneously between the scapulae. Catheter patency was retained for 3 weeks (1 week of post-operative recovery plus 2 weeks of daily intravenous dosing with test substance). However, problems related to the implants inducing skin abrasions over the port reservoir, bruising, attributed to the initial use of excess heparin in maintaining catheter patency, and catheter blockage, which was cleared by use of a thrombolytic agent such as urokinase, were reported. These findings suggested that a VAP with a much smaller reservoir (specifically, with decreased height resulting in a flatter profile) should be selected for use in marmosets to minimise potential cutaneous circulatory impairment from skin tension at the reservoir implantation site. There are presently no reports in the literature about long-term use of VAP in marmosets.

16.4 Infusion models

They are mainly three infusion models used in biomedical research at the present time: tethered, tetherless jacketed and implanted pump (Jacobson 1998).

16.4.1 Tethered infusion model

There are several reports in the literature showing the successful use of a tethered jacketed infusion model in large nonhuman primates, such as rhesus and cynomolgus monkeys (McNamee *et al.* 1984). Basically, the animals were fitted in a leather or denim jacket that was attached to a flexible stainless-steel cable. Through this conduit, an indwelling venous catheter was attached to a swivel unit located on the upper portion of the home cage, allowing complete freedom of movement for the monkey within the cage. However, this system is impracticable with group-housed monkeys like the marmosets. Moreover, the marmosets exhibit a very complex and rich behavioural repertoire that would not be compatible with any tethering device. For all these reasons, there is to date no published literature on the use of tethered infusion models in marmosets.

16.4.2 Tetherless jacketed infusion model

The tetherless jacketed infusion model (or 'ambulatory' model) consists of a catheter, a jacket, a pump contained in the jacket's pouch and a tubing joining the pump to the catheter. All components are housed in, or covered by, the jacket. A backpack system for long-term osmotic minipump infusions into unrestrained marmosets has been described (Ruiz De Elvira and Abbot 1986). In summary, the backpack was made from Velcro tape and essentially consisted of a closed compartment that was held secure by attachment to a vest with shoulder straps. The backpack was tailored to fit each individual monkey. A cannula was implanted subcutaneously in the back of the marmoset and the output end exteriorised between the scapulae. The cannula was secured by two sutures attaching the looped portion to the underlying muscle layer. The exteriorised distal end was connected to an Alzet osmotic minipump (model 2001) located in a compartment of the backpack. One advantage of the design of this lightweight backpack system (20 g) was that its content could be replaced without removal of the backpack from the animal. The marmosets used for the study were habituated to their backpacks for 2 weeks before insertion of the cannula and attachment of the pump system. Although they were not impeded in their mobility in any way (Figure 16.6), 2 of 10 monkeys removed their backpacks once during the first 2 weeks as a result of biting and pulling. After this time, interest in the backpacks faded and no further damage was observed. Each backpack remained functional for at least 3–4 months, and at this time no signs of major skin irritation or infections were observed. However, the chronic cutaneous exit site of the catheter will always be an potential source of infection. Thus, it can act as a wick for infection to track down to the deeper tissue. Moreover, another group (P. Hess 1997, personal communication) reported that several marmosets wearing a similar backpack system fitted with a telemetry device for long term study (> 6 months) in unrestrained, freely moving animals maintained in their home cage, developed severe skin irritation, leading to full-thickness necrosis.

Figure 16.6 An example of the backpack osmotic minipump system fitted to a female marmoset.

16.4.3 *Implanted pump infusion model*

The implantable infusion pump offers the advantage that the marmoset or its mate are not able to interfere with it. Therefore, group housing of animals during a study is possible, leading to an improvement in the well-being of the animals. Alzet® osmotic pumps have been used extensively over the last 20 years in pre-clinical research (see also Chapter 21). Basically, they are composed of three concentric layers, the drug reservoir, the osmotic sleeve containing a high concentration of sodium chloride and the rate-controlling, semipermeable membrane. The difference in osmotic pressure between the osmotic sleeve compartment and the implantation site drives the delivery of the test solution, and this is independent of the drug formulation used. These self-powered Alzet® osmotic miniature pumps are available with a variety of delivery rates between 0.25 and 10 μl/h and delivery durations of between 1 day and 4 weeks. When implanted subcutaneously or intraperitoneally, these pumps serve as a constant source for prolonged drug delivery. The usual site for subcutaneous implantation in the marmoset is on the back, just below the scapula, to one side of the spine. Therefore, the contents of the pump will be delivered into the local subcutaneous space resulting in systemic administration by local capillary absorption. Osmotic minipumps (model 2002, 0.48 µl/h) have been successfully implanted subcutaneously in marmosets for periods of 2–4 weeks without any adverse signs (P. C. Pearce 1999, personal communication). However, the large majority of pharmacological studies reported in the literature using osmotic minipumps in marmosets were performed using intraperitoneal implantation. Nearly all the models of Alzet osmotic pumps which are currently available have been successfully used for drug delivery in marmosets in our laboratories. We were interested in evaluating the hypotensive efficacy of compounds acting on the renin–angiotensin system (RAS) during chronic administration, namely angiotensin II antagonists, angiotensin-converting enzyme inhibitors, renin inhibitors and drugs that suppress the release of renin from the kidney. Interest in the marmoset for cardiovascular research began with the need for a primate model for evaluating the efficacy of high-potency renin inhibitors with marked primate specificity.

We conducted several multidose studies using the marmoset with several compounds acting on each level of the RAS. As an example, a converting-enzyme inhibitor (CEI), benazeprilat, was given continuously for 8 days to a group ($n = 6$) of sodium-depleted marmosets. Benazeprilat is the free acid, active form of the potent, orally active CEI benazepril, which has been used with success in the clinic for the treatment of essential hypertension. The marmosets had been maintained on a low-salt diet for a period of 8 days prior to the experiment to stimulate the RAS. Baseline values for mean arterial pressure (MAP) and heart rate (HR) were recorded continuously in freely moving animals maintained in their home cage via telemetry (Schnell and Wood 1993) for 8 days before benazeprilat was given and continued for 8 days thereafter. Delta values for MAP and HR were calculated using each marmoset as its own control (each post-administration diurnal profile was subtracted from the corresponding pre-administration diurnal profile). Benazeprilat was infused continuously by osmotic minipumps (2001, ALZET, Palo Alto, CA, USA) implanted in the peritoneal cavity. Implantation was performed under aseptic conditions and light anaesthesia (combination of 10 mg/kg ketamine (Ketalar; Parke-Davis, New York, USA), 0.15 mg/kg atropine, (Atropin; Sigfried, Zofingen, Switzerland) and 0.75 mg/kg diazepam (Valium; Roche, Basle, Switzerland) given intramuscularly) at 08.00 h on Day 0. The minipumps delivered at a constant rate of 1 ml/h for at least 7 days, and the average daily dose of benazeprilat was 10 mg/kg. A control group ($n = 5$) received saline only. After continuous intraperitoneal (i.p.) administration of the CEI benazeprilat, MAP decreased significantly (-29 ± 5 mmHg, $P<0.001$) and the hypotensive response was sustained during the 8 days of administration (Figure 16.7). Although baseline values were lowered, the day/night differences in MAP remained unaltered. There was a sustained increase in HR during the first day of treatment with a peak effect after 9–11 h ($+181 \pm 14$ beats/min, $P<0.01$), which corresponded to 21.00–23.00 h. This increase persisted during the next 48 h (Day 2 $+70 \pm 4$ beats/min, $P<0.01$; Day 3 $+45 \pm 1$ beats/min, $P<0.01$) but only during the night. MAP remained unchanged in the control group that received vehicle only. HR was slightly elevated in the control group during the first night after implantation ($+51 \pm 16$ beats/min) but returned to pre-treatment values thereafter. This kind of pivotal information about both the magnitude and duration of drug response obtained with remote monitoring under physiological conditions had not been possible with previously described methods for blood pressure measurements in conscious marmosets. Moreover, the significant night-time reflex tachycardia we observed after systemic delivery of the CEI benazeprilat during the first 3 days of infusion in the marmosets was never reported in the literature.

The animals tolerated the pumps very well and no adverse signs were noticed in our laboratory following intraperitoneal implantation (Wood *et al.* 1987; Schnell and Wood 1993). However, by choosing this route of drug administration, a majority of the dose may be absorbed via the hepatic portal circulation rather than by the capillaries. In the case of substances that are extensively metabolised by the liver (with high first-pass effect), the intraperitoneal route of administration may produce highly variable concentrations of compound in systemic blood and consequently highly variable pharmacological effects and should therefore be avoided. In addition to systemic administration, targeted delivery can be achieved by directing drug solutions

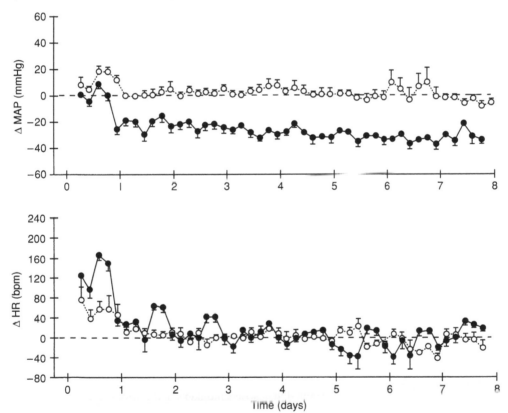

Figure 16.7 Effects of vehicle (open circles) and a converting enzyme inhibitor (CEI) (benazeprilat, filled circles) on mean arterial blood pressure (MAP) and heart rate (HR) in conscious marmosets freely moving in their home cages. Vehicle (saline, *n* = 6) and CEI (10 mg/kg/day, *n* = 6) were given by continuous intraperitoneal infusion for 8 days via osmotic minipumps, implanted at 08.00 h (Day 0). Values represent mean ± SEM of every fourth hour. Initial values for MAP were 101 ± 6 and 109 ± 8 mmHg for control and CEI 10 mg/kg/day respectively. Initial values for heart rate were 260 ± 23 and 276 ± 22 beats/min for control and CEI 10 mg/kg/day respectively.

to an area remote from the site of implantation (i.e. venous or arterial circulation, brain, or into any organ, lumen or solid tissue). This can be easily achieved by attaching a catheter to the pump. For example, Alzet® osmotic pumps have been shown to pump successfully against arterial pressure with no reduction in flow. However, osmotic minipumps have some limitations such as their inability to be refilled, the inability to change flow rates or infusates after starting, and the limited flow rates presently available which restricts their use to soluble compounds. Moreover, the osmotic attraction of water into the pump will continue after the pumping lifetime of the minipumps. Thereafter, the pump may swell and begin to leak a concentrated salt solution, resulting in local irritation of surrounding tissues. Therefore, osmotic minipumps should always be explanted after use, which will result in additional surgical stress to the animal.

16.5 Conclusion

In this chapter, an extensive overview was given of continuous infusion and chronic catheter access methodologies used in the past or presently in marmosets. A refined method used in our laboratories for implanting indwelling intravenous catheters in the marmoset was described in detail.

Undoubtedly, constant material and method refinement will lead to further improvements in the future. The instrumented laboratory marmoset of the future may be loaded with different miniature, lightweight, microchip-driven technologies, all working together and all controlled by an external computer allowing complete remote monitoring. The results of these product advances will be the collection of better and more physiological *in vivo* data as well as a reduction in animal usage, and enhancement of care and animal welfare.

References

Jacobson, A. (1998) Continuous infusion and chronic catheter access in laboratory animals. *Laboratory Animals*, 27 (7), 37–46.

Kinsora, J.J., Christoffersen, C.L., Swalec, J.M. and Juneau, P.L. (1997) The novel use of vascular access ports for intravenous self-administration and blood withdrawal studies in squirrel monkeys. *Journal of Neuroscience Methods*, 75, 59–61.

McNamee, G.A. Jr, Wannemacher, R.W. Jr, Dinterman, R.E., Rozmiarek, H. and Montrey, R.D. (1984) A surgical procedure and tethering system for chronic blood sampling, infusion, and temperature monitoring in caged nonhuman primates. *Laboratory Animal Science*, 34 (3), 303–307.

Ruiz de Elvira, M.-C. and Abbott D.H. (1986) A backpack system for long-term osmotic minipump infusions into unrestrained marmosets monkeys. *Laboratory Animals*, 20, 329–334.

Schnell, C.R. and Gerber, P. (1997) Training and remote monitoring of cardiovascular parameters in non-human primates. *Primate Report*, 49, 61–69.

Schnell, C.R. and Wood, J.M. (1993) Measurement of blood pressure and heart rate by telemetry in conscious, unrestrained marmosets. *American Journal of Physiology (Heart and Circulatory Physiology 33)*, 264, H1509–H1516.

Wojnicki, F.H.E., Bacher, J.D. and Glowa, J.R. (1994) Use of subcutaneous vascular access ports in rhesus monkeys. *Laboratory Animal Science*, 44 (5), 491–494.

Wood, J.M., Baum, H.P., Jobber, R.A. and Neisius, D. (1987) Sustained reduction in blood pressure during chronic administration of a renin inhibitor to normotensive marmosets. *Journal of Cardiovascular Pharmacology*, 10 (Suppl. 7), S96-S98.

Part 6

Continuous intravenous infusion in the minipig

The minipig is an emerging alternative non-rodent species to the dog for pre-clinical testing of pharmaceuticals. There are a limited number of laboratories with the expertise to conduct these studies, but there are many cases where this species is the most suitable model.

Minipigs are considered good models for humans because they share some important characteristics with man. The rationale for selection of the minipig is covered in Chapter 18. The species is accepted by regulatory authorities, but some justification for its use would be needed such as a close similarity in a principal target organ between the minipig and man, or that it provided a more suitable or appropriate metabolic or pharmacokinetic profile for the test compound.

The surgical preparation and conduct of general toxicology studies is described in Chapters 17 and 18. The minipig is also used as an alternative to the rabbit for reproductive studies by other routes, but not to date via continuous intravenous infusion. The recent advances that have been made that could enable teratology studies to be carried out by this route of administration are described in Chapter 19.

17 Surgical preparation of the minipig

P. Brinck

17.1 Introduction

In this chapter, various techniques for preparing minipigs for continuous intravenous infusion will be described. Both the techniques routinely employed by the author and a review of techniques that have been published are presented.

17.2 Surgical facilities

Adequate facilities include a pre-operative room, a room for the surgery and an area where recovery from anaesthesia can be monitored. The room to be used for surgery should comply with normal standards for operating rooms.

17.3 Staff

Generally, it is required to have three staff performing surgical implantation of an indwelling intravenous catheter in minipigs. The staff needed include as a minimum, an experienced surgeon, an experienced veterinary nurse and operating room assistant. All of them should have received a proper education in the specific procedure.

17.4 Standards of asepsis

Aseptic methods should be used throughout the implantation. All materials used should be sterile. The surgical area should be cleansed properly with soap and water, and swabbed with, for example, 70% ethanol and covered with sterile dressings. The surgeon and assistants should prepare themselves by performing a surgical handwash just prior to surgery and by wearing sterile gloves. Additionally, the use of surgical clothes, hair nets and facemasks should be considered within the operating room.

17.5 Role of prophylactic antibiotic treatment

The use of antibiotics during and after catheter implantation is controversial. The aim of antibiotic treatment is to reduce the incidence of infections associated with the surgery. Several laboratories as standard practice use antibiotics on the day of surgery and for the following 2 or 3 days. However, the following aspects should be considered: the risk that the antibiotic may induce alterations in the animal that may affect the value of the animal in ensuing experiments, the risk to personnel of developing an allergy against the particular antibiotic agent and the fact that every use of an antibiotic increases the risk that this antibiotic will become ineffective.

17.6 Preparation of animals before surgery

The minipigs should be acclimatised at the facility for at least a few days. During the last days before surgery, the minipigs should be observed daily for possible clinical signs of cardiovascular or pulmonary diseases and their food consumption should be checked and the pens should be inspected for the presence of diarrhoea. When minipigs are fed standard portions of standard minipig chows (for instance Altromin 9023, Altromin, Lage, Lippe, Germany) it takes more than 12 h for the stomach to empty. The minipigs should therefore be fasted for at least 12 h before surgery to reduce the possibility of aspiration of stomach contents into the lungs. The animals should have *ad libitum* access to drinking water until surgery. If hay or straw is used as bedding, this should probably also be removed from the pen during the fasting period, because the animals may ingest it.

17.7 Anaesthesia

In this section the technique currently used at Scantox will be described first. Thereafter, some other techniques will be reviewed.

The animal is pre-medicated with atropine sulphate, 0.05 mg/kg body weight by intramuscular injection (Svendsen 1997; Swindle 1998), in order to abolish vagal effects on the heart and respiratory airways. Sedation is efficiently achieved with azaperone, 10 mg/kg body weight (Sedaperone® Vet) given by intramuscular injection. Atropine and azaperone may be mixed and given as a single injection. This should be performed in the animal's usual environment, as this will facilitate anaesthesia. After 10–15 min, the animal should be given an intraperitoneal injection of metomidate, 10 mg/kg body weight (Hypnodil® Vet), which will lead to anaesthesia within 10 min. Alternatively, the animal may receive the metomidate by intravenous injection, most convenient through a marginal ear vein. At this time the animal may also be given a potent analgesic, for instance buprenorphine, 0.01 mg/kg body weight. Thereafter the animal is allowed some minutes before it is taken to the preparation room.

The depth of the anaesthesia is assessed by the reaction to stimuli and the respiratory pattern. The pedal reflex is particularly helpful, and is positive if the animal withdraws its leg upon pinching of the interdigital skin. Respiration should be both thoracic and abdominal. When the thoracic component disappears, the anaesthesia has become too deep. The depth of anaesthesia may also be assessed by monitoring the blood pressure and heart rate (Swindle 1998).

Other protocols for anaesthesia have been described in detail by Svendsen (1997). According to Svendsen, minor surgery may be performed with propofol (Rapinovet®, Diprivan®) anaesthesia. After pre-medication with atropine sulphate (0.05 mg/kg) and sedation with a mixture of azaperone (4 mg/kg) and midazolam (1 mg/kg) given intramuscularly, anaesthesia can be induced by intravenous injection of propofol 5 mg/kg. The anaesthesia may be maintained by injections of propofol (0.2 mg/kg) every 10 min.

Another protocol suggested by Svendsen (1997) and Svendsen and Rasmussen (1998) includes endotracheal intubation. After pre-medication with atropine sulphate

(0.05 mg/kg) and sedation with a mixture of azaperone (4 mg/kg) and midazolam (1 mg/kg) given intramuscularly, anaesthesia may be induced by rapid intravenous injection of propofol (2.5 mg/kg). Additional propofol is injected slowly until the mouth can be opened without resistance and muscular contractions of the tongue have disappeared.

17.7.1 Endotracheal intubation

The minipig is placed in left lateral recumbency and the neck is extended. A laryngoscope with a 20-cm straight blade is placed at the base of the tongue. In minipigs, the epiglottis is usually right behind the soft palate. It is therefore necessary to lift the soft palate and press the epiglottis forward onto the base of the tongue by use of the endotracheal tube stiffened with a stylet. The endotracheal tube is introduced into the trachea during an expiration. A slight rotation of the tube facilitates the introduction. Endotracheal intubation is usually followed by apnoea, so assisted ventilation is critical at this stage.

17.7.2 Inhalation anaesthesia

Anaesthesia can be maintained by the use of halothane 1% in an oxygen–nitrous oxide (1:1) mixture. Respiration should be assisted by artificial ventilation with a minute volume of 8–12 ml/kg and frequency of 15–20 times per minute.

Swindle (1998) has given descriptions of anaesthetic protocols, two of which use inhalation anaesthesia and one which does not use inhalation anaesthesia. One protocol is as follows: induction with ketamine (33 mg/kg), acepromazine (1.1 mg/kg) and atropine (0.05 mg/kg) intramuscularly. Maintenance with isoflurane 0.5–1.5% in an oxygen–nitrous oxide (1:2) mixture. With this protocol, it is not possible to perform intubation without administration of isoflurane 3–5% via a facemask during induction. The use of facemasks for inhalation anaesthetics should be given careful consideration, because this type of administration will expose the operating room personnel to the anaesthetic agent. Buprenorphine 0.05 mg/kg may be administered intraoperatively and may reduce the percentage of isoflurane required.

The second protocol described by Swindle (1998) is as follows: induction by administration of isoflurane 3–5% via a facemask. Maintenance with isoflurane 0.5–1.5% in an oxygen–nitrous oxide (1:2) mixture.

The third protocol described by Swindle (1998) does not use inhalation anaesthesia: induction by intravenous infusion with sufentanil (0.015 mg/kg/h). After infusion for 5 min, a bolus of 0.07 mg/kg sufentanil is administered. Maintenance is with sulfentanil (0.015–0.030 mg/kg/h). Atropine sulphate (0.02 mg/kg) may be given intravenously as pre-medication, to avoid bradycardia, which may be profound during the induction.

17.8 Post-surgical analgesia

Recovery from surgery is enhanced by appropriate treatment of pain. Post-surgical pain should be treated appropriately. Signs of discomfort include reduced food consumption, abnormal behaviour and abnormal posture. A recommended regimen may include the use of opiates during surgery and the first day or the first days thereafter, for instance of buprenorphine (0.05–0.1 mg/kg body weight/12 h). The following guide for swine may be used (Swindle 1998):

- fentanyl 0.02–0.05 µg/kg i.m. (30–100 µg g/kg/h i.v. drip)
- sufentanil 5–10 µg/kg/ i.m. (10–15 µg /kg/h i.v. drip)
- meperidine 10 mg/kg
- pentrazocine 1.5–3.0 mg/kg i.m.
- oxymorphone 0.15 mg/kg i.m.
- butorphanol 0.1–0.3 mg/kg i.m.
- buprenorphine 0.05–0.1 mg/kg body weight/12 h
- continuous i.v. drip with opioids
- aspirin 10–20 mg/kg p.o. q.i.d
- ketaprofen 1–3 mg/kg

17.9 Pre-operative preparation of the animal

In the preparation room, the surgical area is clipped free of hair, rinsed thoroughly with soap and water and shaved. Thereafter, the animal is taken to the operating room and placed on an insulated mattress. The surgical area is soaked with ethanol 70%, betadine or another antiseptic.

17.10 Surgical technique

In this section a description of a technique for insertion of a permanent catheter in the jugular vein with exteriorisation on the back of the animal is first given, followed by a description of implantation of a vascular access port (VAP). Thereafter, the choice of vein is discussed, and finally a description of a non-surgical method of catheterisation is given.

Catheterisation is based on the method described by Bailie *et al.* (1986). For a right-handed operator using this technique, it is more convenient to insert a catheter in the right jugular vein than in the left. For catheterisation of the right jugular vein, the animal is placed in left dorsolateral recumbency with the neck somewhat extended, the lower foreleg pulled forwards and the upper foreleg pulled backwards. An incision is made over the dorsal part of the right scapula. Another incision is made over the jugular vein in the right fossa jugularis on the ventral side of the neck. By blunt and sharp dissection the jugular vein is localised. The jugular vein of minipigs is located fairly deep in the neck compared with most other laboratory animals. Long forceps are introduced in the tissues close to the jugular vein, and advanced subcutaneously to the incision over the scapula. The tip of the catheter is grasped with the long forceps, and the catheter is gently pulled from the incision over the scapula to the tissues close to the jugular vein. The tunnelling of the catheter may alternatively be performed via a trocar introduced from the incision over the scapula, through the subcutaneous tissues to a position close to the jugular vein. The jugular vein is carefully dissected free of perivenous tissue for a length of 3–4 cm. Loose ligatures (silk 1–0) are placed around the proximal and distal ends of the isolated part of the jugular vein, and the vein is gently pulled to the surface of the incision. The vein is secured in this position by inserting a pair of forceps below the vein. A haemostat may be applied on the proximal end of the exposed part of the vein. The jugular vein is grasped with a small pair of forceps in the operator's left hand, and a small incision in the vein is made a few millimetres cranial to this. The catheter is introduced through this incision into the vein, and advanced about 7–10 cm. The introduction of the catheter may be facilitated by the use of a vessel dilator. Alternatively, the catheter may be introduced by means of a guidewire as follows: insertion of a needle into the vein, insertion of a guidewire through the needle, removal of the needle, insertion of the catheter over the guidewire, and removal of the guidewire. The tip of the catheter should be located within the cranial vena cava at a position very close to the heart. In the author's experience, this position of the tip of the catheter, or a position within the right atrium of the heart, has been associated with fewer complications than other positions. Once in place, the position of the catheter should be secured by tightening the proximal suture around the vein and the catheter. The distal suture on the jugular vein is then closed. If the catheter has a suture disc mesh attached, this should be sutured to the deep surface of the dermis in the incision at the catheter exteriorisation site above the scapula. The incisions in the subcutaneous tissue and in the skin are sutured with continuous or interrupted sutures using, for instance, Dexon® 2–0.

For the implantation of a VAP, the catheter is inserted and secured in the vein as described above. The distal end of the catheter is tunnelled to the incision over the scapula. Here, the catheter is not exteriorised but is connected to a subcutaneously located access port, which is sutured with a few stitches to the fascia overlying the muscles over the dorsal aspect of the scapula.

In minipigs, no non-surgical method for implantation of catheters for continuous intravenous infusion has been published. However, in young pigs a non-surgical method for catheterisation of the jugular vein has been described (Carrol *et al.* 1999). Piglets weighing 1.5–5.75 kg were anaesthetised and catheterised using a Cook single-lumen central venous catheter kit. In principle, the jugular vein was located transcutaneously by inserting an access needle in front of the manubrium sterni, inserting a guidewire through the needle, removing the needle, inserting the catheter over the guidewire, removing the guidewire, and fixing the catheter to the skin with stitches. The catheters were used for blood sampling for up to 24 h. Elements of this technique, perhaps in a modified version, could possibly be applied to continuous intravenous infusion.

17.11 Catheterisation of alternative veins

Traditionally the jugular vein has been selected for continuous intravenous infusion. This site is convenient from a surgical point of view and also for post-implantation maintenance. Experience has shown that catheterisation of the jugular vein gives superior performance (Webb 1997). In cases where blood samples for toxicokinetics are critical for the study, the use of an alternative route for intravenous infusion should be considered, because blood samples for toxicokinetics will normally be taken from the jugular vein very close to the site of infusion – which may lead to spurious high concentrations of the test material. Use of the femoral vein as the infusion site should be considered in such cases. Cannulation of the femoral vein may be performed according the method described by Svendsen and Rasmussen (1998), in which an incision is made perpendicular to the vessel approximately 2 cm from the groin. The gracilis and sartorius muscles are located and separated bluntly, and the femoral vein is located and dissected. Using a long trocar, the catheter is introduced from an appropriate site (the flank or the side of the thorax) to the incision. The vein is cannulated as described for the jugular vein.

Also, according to Svendsen and Rasmussen (1998), the cephalic vein may be catheterised. The animal is placed in dorsal recumbency, and a 3-cm-long incision made parallel to and 3 cm lateral to the manubrium sterni. The vein is located. Using a long trocar the catheter is introduced from the side of the thorax to the incision. The vein is cannulated as described for the jugular vein.

17.12 Recovery from implantation of a catheter

The period of recovery after implantation of a catheter into the jugular vein may be divided into two components, namely recovery from anaesthesia and recovery from the tissue damage caused by the surgery. During the recovery phase, the animal should be placed in a warm pen to avoid hypothermia. The recovery after anaesthesia includes the passage of the animal through the excitation phase of the anaesthesia. The animal should be observed during this period, and prevented from hurting itself. Vomiting may occur on rare occasions. The animal may suffocate from aspiration of vomit or from erroneous swallowing into the trachea, if the animal is allowed access to food during the first few hours after anaesthesia. Following anaesthesia with azaperone (10 mg/kg body weight, Sedaperone® Vet) given by intramuscular (i.m.) injection, metomidate (10 mg/kg body weight, Hypnodil® Vet) given intraperitoneally and buprenorphine (0.05 mg/kg body weight, i.m.), most of the animals will be walking around in the pen and eating within 3–4 h after the end of the surgical procedure.

The author is unaware of any published data on recovery time from the tissue damage caused by the surgical implantation of the indwelling catheter in minipigs. The non-surgical implantation of catheters in piglets caused increased concentration of cortisol, a standard indicator of stress and discomfort, for only 2 h after completing the implantation (Carrol *et al.* 1999). Skin sutures should not be removed earlier than 7–10 days after surgery, especially at sites where the skin is rather thick, for example over the dorsal aspect of the scapula. However, a recovery period of at least 5 days after surgery before the start of treatment in a toxicity study has been used successfully in the laboratory of the author.

17.13 Recommended specifications for catheters

Silicone is an excellent material for catheters in continuous intravenous infusion. It is flexible and very well tolerated by the tissues. For minipigs in sizes routinely used in toxicity testing, i.e. in a weight range from 6 to 25 kg, the following catheter can be recommended for implantation in the jugular vein: silicone, 7 French diameter, 45 cm length, a suture disc mesh glued at 12 cm from the distal end of the catheter, retention beads for securing the catheter in the vein and a rounded tip. Such catheters are commercially available from, for instance, Access Technologies, Skokie, IL, USA.

17.14 Recommended specifications for VAPs and catheters

In minipigs, the VAP should be fairly flat. Tall VAPs with steep sides may initially cause thinning of the skin above the port. This is because the skin in minipigs is particularly thick and stiff at the implantation site over the dorsal aspect of the scapula. The following specifications for catheters may be recommended for implantation in the jugular vein: silicone, 7 French diameter, retention beads for securing the catheter in the vein, and a rounded tip. Combinations of VAPs and catheters are commercially available from the supplier mentioned previously, where the catheter can be snapped onto the VAP. This allows for intra-operative trimming of the catheter to a suitable length.

17.15 Exteriorisation of catheter versus usage of VAP

The advantage of the VAP relative to the exteriorised catheter is that it is very unlikely that the VAP is accidentally pulled out of the vein. There are several disadvantages: the presence of a needle passing through the skin into the VAP may be associated with local skin problems, the line from drug formulation to vein is more easily broken using a VAP than an exteriorised catheter because the needle is fairly easily disconnected from the VAP, and an occluded VAP is more difficult to flush than an occluded exteriorised catheter. Exteriorised catheters are therefore considered to be more suitable for continuous intravenous infusion in minipigs than are VAPs.

17.16 Infusion equipment

Continuous intravenous infusion in minipigs may be performed with either the animal carrying the pump or by the use of a tethered system connected to a pump that is not carried by the animal. The advantages of a tethered system include that the pump may be placed outside the animal's cage allowing easy access to the pump, the pump may be large as it is not carried by the animal, and the amount of infusate may be large as it need not be carried by the animal. The major disadvantage is that the animal may not move completely freely around, which it may when the animal carries an ambulatory infusion pump. Using ambulatory infusion equipment, the major limitations are that the animal needs to be handled to access the pump, and the weight and size of the pump and infusate are limited to what the animal may carry.

17.17 Maintenance of patent catheters

After implantation, catheters are most easily kept patent until start of dosing by infusion of physiological saline at a low rate, for instance, 1 ml per hour. Before start of treatment, catheter patency may also be maintained by flushing the catheter twice weekly with physiological saline, upon which the catheter is filled with a heparin-lock: physiological saline containing 100 international units of heparin per ml.

17.18 Complications

Complications may be divided into three categories: those related to local effects of the infusion equipment/procedures, those related to remote effects of the infusion equipment/procedures and those related to the dysfunction of the equipment.

Local effects of the infusion equipment include inflammation and wounds in the skin caused by the weight or pressure of the equipment on the tissues. Damaged sites may be infected, which may affect the animal's use in a toxicity study. The catheter may cause thrombosis within the vein, which may mask possible local effects of the infusate. The tip of the catheter may theoretically be dislodged to the heart, where it may induce arrhythmia.

Remote effects of the infusion equipment/procedures include emboli from thrombosis at the site of the catheter. Such emboli will be carried with the bloodstream to the right ventricle of the heart, from which they are pumped to the lungs, where they are caught in the small vessels. Here they may become infected causing thromboembolic pneumonia. Catheter-related sepsis (CRS) is the chief and most significant complication (Webb 1997).

The third type of complication includes various dysfunctions of the materials. The pump may fail to deliver the intended amount of infusate for example because the catheter has kinked. The catheter may break, which is not easily observed if it occurs inside the animal.

17.19 Conclusion

Continuous intravenous infusion is a well-established technique in minipigs. The techniques resemble those employed in other laboratory animal species, and the results and complications also resemble those in other species.

References

Bailie, M.B., Wixson, S.K. and Landi, M.S. (1986) Vascular access port implantation for serial blood sampling in conscious swine. *Laboratory Animal Science*, 36, 431–433.

Carrol, J.A., Daniel, J.A., Keisler, D.H. and Matteri, R.L. (1999) Non-surgical catheterisation of the jugular vein in young pigs. *Laboratory Animals*, 33(2), 101–200.

Svendsen, P. (1997) Anaesthesia and basic experimental surgery of minipigs. *Pharmacology and Toxicology.* 80, Suppl. II, 23–26.

Svendsen, P. and Rasmussen, C. (1998) Anaesthesia of minipigs and basic surgical technique. *Scandinavian Journal of Laboratory Animal Science,* 25, Suppl. 1, 31–43.

Swindle, M.M. (1998) Anaesthesia and analgesia. In *Surgery, Anesthesia, and Experimental Techniques in Swine.* Ames: Iowa State University Press, pp. 33–63.

Webb, A.J. (1997) Venous catheter implantation in the minipig. Surgical approaches and refinements. *Pharmacology and Toxicology*, 80, Suppl II, 1997.

18 Multidose infusion toxicity studies in the minipig

P. Brinck

18.1 Introduction

In this chapter, a brief introduction to the minipig will be given, because in the past the minipig was not commonly used in toxicity testing. Thereafter, the use of minipigs in toxicology will be described, with special emphasis on the advantages and disadvantages of the species. Finally, the special requirements in continuous intravenous infusion toxicity studies will be described.

18.2 The minipig as a model in biomedical research

The pig has a long history in biomedical research. As long ago as the eighteenth century, John Hunter, a reputed British medical doctor and surgeon, recognised the pig as one of the best animal models in physiological research (Bustad 1966). However, the size and weight of the pig are impractical, and several breeds of miniature pigs have been developed for experimental use. Several breeds of minipigs exist, including Göttingen, Yucatan minipig, Yucatan Micropig®, Sinclair, Hanford, Pitman-Moore, Hormel, Ohmini, Kangaroo, Lee Sung, Miniature Siberian, and Troll. In Europe, the Göttingen minipig followed by the Yucatan Micropig® are the most commonly used breeds in toxicity testing. The Göttingen minipig weighs about 7–9 kg at an age of 4 months, which is an age when they may be included in general toxicity tests. Thereafter, they gain approximately 1–2 kg body weight per month (Figure 18.1). Minipigs are generally good models for humans because they share important characteristics with man.

18.2.1 Cardiovascular system

The cardiovascular system resembles that of man (Swindle 1998). The heart is anatomically similar to that of the human, with the exception of the presence of a left azygos vein (hemiazygos). There are no pre-existing collateral arteries in the heart. The swine is the species most used as a cardiovascular model, and this is because of these anatomical and physiological characteristics (Swindle 1998). Minipigs are not as susceptible to sympathomimetic drugs as are dogs. Lehmann (1998) tested a sympathomimetic drug in minipigs and dogs, and found myocardial necrosis in dogs as expected but no histopathological changes in minipigs.

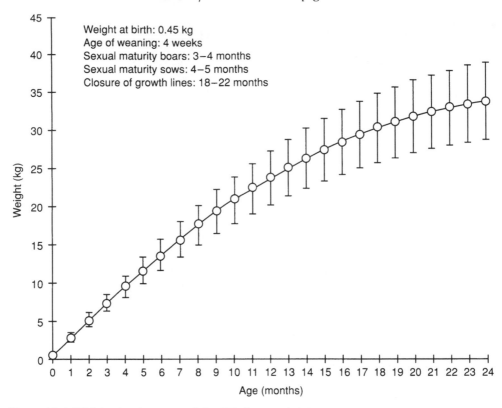

Figure 18.1 Weight development of the Göttingen minipig.

18.2.2 Gastrointestinal tract

The gastrointestinal tract of minipigs, particularly the caecum and colon, differs anatomically from that of man in several respects. However, the minipig is omnivorous, as is man, and the physiology and histology resemble that of man quite closely. Transit time through the small intestine is similar in the two species.

18.2.3 Kidneys

Anatomically and physiologically, the kidneys of pigs are more like those of humans than in most other species, even closer than in other primates (Swindle 1998).

18.2.4 Skin

In the following section, a description of the minipig in dermal toxicity is given. This is because the use of minipigs in toxicity testing originated mainly with studies employing the dermal route of administration, because the skin of pigs is more like that of man than other animal species (Vogel and Singer 1997). Mortensen *et al.* (1998) have given a thorough description of the minipig in dermal toxicology.

Macroscopically, porcine skin resembles human skin. The hair coat is sparse in both species. The surface of the skin is carved by fine intersecting lines in characteristic patterns. Both human and porcine skin may or may not be pigmented depending on race. The texture of the porcine skin varies with the region, the dorsal skin being rather coarse and the abdominal skin or the skin of the ears more finely structured.

Microscopically there are both similarities and differences (Fowler and Calhoun 1964; Montagna and Yun 1964; Smith and Calhoun 1964; Marcarian and Calhoun 1966; Meyer *et al.* 1978; Monteiro-Riviere 1986; Meyer 1996). The epidermis is rather thick in both species, being approximately 70–140 μm in pigs and 50–120 μm in humans, with large variations from site to site. This should be compared with an epidermal thickness of approximately 10–20 μm in rats, which have a thick protective hair coat (Meyer *et al.* 1978; Monteiro-Riviere *et al.* 1990).

A Finnish group (Männistö *et al.* 1984; 1986; Hanhijärvi *et al.* 1985) used minipigs for long-term skin irritation studies with a number of anthraline derivatives intended for topical treatment of psoriasis. They reported that the skin of the minipig was less sensitive to irritants than rabbit skin and offered a better prediction for the adverse effects in humans than studies in rabbits. The local reactions comprised erythema, acanthosis, parakeratosis, inflammation and superficial abscesses. It was possible to differentiate between the analogues with respect to irritative properties. No systemic toxicity was reported for the compounds tested.

18.2.5 Regulatory acceptance

According to international guidelines, the minipig is fully acceptable as an alternative to the commonly used non-rodent species in regulatory toxicity testing. The minipig is specifically mentioned as a possible non-rodent species in Canadian, Japanese and OECD guidelines. However, it is important to justify the choice of any less frequently used species to the regulatory bodies, although this is also now required for selection of all species. However, the comment has been made that the FDA has found the pig, particularly the minipig, to be a useful model in regulatory toxicity testing (Ikeda *et al.* 1998). Good reasons for using the minipig could be a suitable (human-like) metabolic or pharmacokinetic profile for the test compound or a close similarity between a main target organ in minipig and man.

It should be noted, though, that there is less practical experience with toxicity studies in minipigs than in rodents, dogs or primates, and that published background data on haematology, blood chemistry and pathology are limited. This is a potential drawback, particularly if equivocal results are obtained in a study. When using the minipig for regulatory toxicity studies, it is therefore of utmost importance for the interpretation of the findings to build up a historical database in the laboratory and to include sufficiently sized control groups in the studies.

Despite the scarcity of published historical data there are no serious reasons not to conduct exploratory or regulatory toxicity studies in minipigs. The selection of the non-rodent species in toxicity testing should be based on both sound scientific and practical considerations. The scientific considerations may include similarities with humans with respect to physiology and anatomy, metabolism, possible target organs, and the practical considerations may include availability, size and technical possibilities. These considerations often lead to the choice of the minipig as the model in toxicity testing.

18.3 Continuous intravenous infusion

In the following sections, recent developments in techniques used at Scantox will be presented. The procedures needed to conduct continuous infusion toxicity studies in minipigs do not differ much from those used in toxicity studies in minipigs using other routes of administration, or in toxicity studies using Beagle dogs.

18.3.1 Caging

In the past, minipigs were housed in metal cages with a metal grid floor. Below the grid floor, a tray for collection of faeces and urine was placed. Softwood sawdust was placed in the tray in order to absorb urine, thus reducing the ammonia concentration in the air. Hay was supplied daily as a welfare measurement, and in order to supply sufficient fibre to the diet. Currently, the minipigs are preferentially housed in floor pens, with concrete covering most of the floor. Part of the floor is heated, while another part is not, which allows the animal the choice between a warm and a less warm floor area to lie on. The remainder of the floor is furbished with wide grids separated by narrow spaces. Water is supplied over this part of the floor, so that any spillage of water will result in only a restricted area of the floor becoming wet. The pens are supplied with softwood sawdust bedding. The bedding is changed as needed, which may vary from daily to twice weekly. Hay is provided daily. In solid-floor pens, the hay will not be lost between the grids, as may occur in cages with grid floors. Food is provided twice daily in beakers, and water is available *ad libitum* either in beakers or bottles or through an automatic watering system. These are the standard routines for all toxicity tests in minipigs, including continuous intravenous infusion studies.

18.3.2 Pumps, jackets and other infusion equipment

The author's laboratory currently uses CADD pumps (Deltec Inc., MN, USA). The CADD-1 pumps can deliver between 1 and 299 ml per 24 h. When bigger volumes are needed, then CADD Plus pumps are used. Both pumps deliver the flow with sufficient accuracy, i.e. within ± 6% of the nominal flow. A wide variety of other ambulatory infusion pumps are commercially available. Reliability, accuracy, flexibility, size and weight are key parameters in the selection of pumps.

The pumps are fitted with a medication cassette of either 50 or 100 ml that has previously been filled with the fluid to be infused. The medication reservoir is connected to the exteriorised part of the implanted catheter, and carried in a pocket in a jacket worn by the animal. Jackets may be obtained from, for instance, Alice King Chatham, CA, USA. A counterweight may be placed in the opposite pocket, to obtain equilibrium with respect to the weight of the equipment. Minipigs have a rather stiff neck and back, and they are not able to turn their head round and bite or chew at the jacket. Thus, the jacket is well protected when carried by a minipig. The pumps may also be connected to a remote reservoir. Typically, the remote reservoir is carried in one pocket in the jacket, while the pump is carried in opposite pocket (Figure 18.2).

18.3.3 Design of standard 4-week continuous intravenous infusion study

At the author's laboratory, treatment periods of 4 weeks for intravenous infusion toxicity studies are routine, whereas studies of longer duration are only occasionally

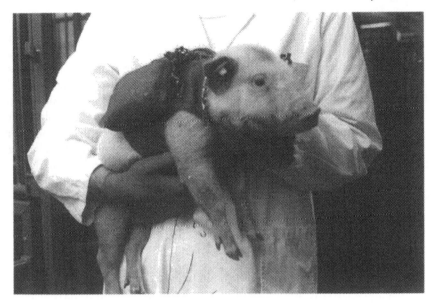

Figure 18.2 Minipig in an infusion pump jacket.

performed. The standard design of a continuous infusion toxicity study does not differ from the design of studies employing other routes of administration. A typical design of a 4-week continuous intravenous infusion toxicity study is as follows. The minipigs arrive at the facility and are acclimatised for a suitable period, after which the animals are prepared surgically (see Chapter 17). From implantation until the start of treatment, patency of the implanted catheter is maintained by infusion of physiological saline at a low flow rate, for instance at 1 ml/h. After a recovery period of about 1 week, pre-treatment examinations are performed: electrocardiography, ophthalmoscopy, blood sampling for haematology and clinical chemistry, and urine sampling for urinalysis. Thereafter, treatment is started and continued for 28 days. At the end of the treatment period, the same examinations are repeated. During the entire study period, clinical signs are recorded at least daily, food consumption daily and body weight at regular intervals. Blood samples for toxicokinetics may be taken during the study. After treatment is terminated, the animals are killed, subjected to a necropsy and examined microscopically. Clinical signs of toxicity, electrocardiography, ophthalmoscopy, laboratory analyses of blood and urine, as well as pathological and histopathological procedures follow the same routines as in other laboratory animals. Literature reports on background haematology and blood chemistry values in various strains of minipigs are available (Wrogemann and Holtz 1977; Brechbühler *et al.* 1984; Rispat *et al.* 1993; Ellegaard *et al.* 1995; Schedl *et al.* 1996; Petroianu *et al.* 1997; Jørgensen *et al.* 1998). Data on the microscopic pathology of normal (untreated) Göttingen minipigs have also been published (Svendsen *et al.* 1998).

All the above-mentioned examinations may be performed without interruption of infusion. Change of medication cassettes, inspection of catheter integrity and other procedures relating to the catheter and infusion pump are greatly facilitated by placing the animal in a sling during these manipulations. Standard procedures such as detailed clinical observations, food consumption recording, electrocardiography, ophthalmoscopy, blood sampling for haematology and clinical chemistry, and urine sampling in metabolism cages do not present problems with regard to continuous intravenous infusion. The body weight should be measured by weighing the animal wearing jacket, pump, medication cassettes, etc., and the true body weight should then be calculated by deducting the weight of this equipment from the total weight of animal and equipment. Simultaneous infusion into one jugular vein and blood sampling downstream from the infusion site – which is the normal site for sampling blood in minipigs – may cause high concentrations of the test material in the blood samples, without this being a true reflection of the systemic concentration of the test material.

Occasionally the implanted catheters may leak, become broken or damaged in other ways or become blocked. Such incidents require immediate and competent attention. Leaking, broken and otherwise damaged catheters may very often be repaired successfully using spare parts, glue and commercially available catheter repair sets.

18.3.4 Animal welfare – enrichment

Various types of enrichment in addition to hay have been attempted. Various types of toys have been given to the animals, but so far the animals seem to have benefited the most from access to adequate amounts of fresh, autoclaved hay. It may be possible to house more than one animal to a cage. However, there are several practical aspects that should be considered, of which the main one is the risk that one animal may damage the catheter or other parts of the infusion system of the other animal. With tethered animals, the tethers of the animals could become entangled.

18.3.5 Animal welfare – reduction of stress caused by continuous intravenous infusion procedures

The weight of the ambulatory infusion pump, jacket and other accessories carried by the animal should be reduced as much as practically possible, and should not restrain the animal's free movements in any way. The jacket, pumps, catheters, etc. may exert pressure on the skin, resulting after a while in ulceration. Therefore, the susceptible areas on the skin should be examined carefully on a daily basis, in order to correct and treat any such finding. The author has no experience in the use of tethered systems for intravenous infusion, but any restriction of the free movement of the animal caused by the use of a tethered system for continuous intravenous infusion should be considered carefully from an animal welfare perspective.

18.3.6 Animal welfare – conclusion

Generally, minipigs tolerate continuous intravenous infusion using ambulatory equipment very well. The welfare of animals in continuous intravenous infusion studies should be continuously monitored, and improved whenever possible. Establishment of an independent animal welfare group at each facility may be of benefit in this respect.

18.4 Conclusion

The conduct of toxicity studies using administration by continuous intravenous infusion is a well-established technique in minipigs. The techniques resemble those employed in other laboratory animal species, and the results and complications also resemble those in other species.

References

Bustad, L.K. (1966) Pigs in the laboratory. *Scientific American*, 214, 94–100.

Brechbühler, T., Kaeslin, M. and Wyler, F. (1984)Reference values of various blood constituents in young minipigs. *Journal of Clinical Chemistry and Clinical Biochemistry*, 22, 301–304.

Ellegaard, L., Jørgensen, K.D., Klastrup, S., Hansen, A.K. and Svendsen, O. (1995) Haematologic and clinical chemical values in 3 and 6 months old Göttingen minipigs. *Scandinavian Journal of Laboratory Animal Science*, 22, 239–248.

Fowler E.H. and Calhoun, M.L. (1964) The microscopic anatomy of developing fetal pig skin. *American Journal of Veterinary Research*, 25, 156–164.

Hanhijärvi, H., Nevalainen, T. and Männistö, P.T. (1985) A six month dermal irritation test with anthralins in the Göttingen miniature swine. *Archives of Toxicology*, 8 (Suppl.), 463–468.

Ikeda, G.J., Friedman, L. and Hattan, D.G. (1998) The Minipig as a model in regulatory toxicity testing. *Scandinavian Journal of Laboratory Animal Science*, 25 (Suppl. 1), 99–105.

Jørgensen, K.D., Ellegaard, L., Klastrup, S. and Svendsen, O. (1998) Haematologic and clinical chemical values in pregnant and juvenile Göttingen minipigs. *Scandinavian Journal of Laboratory Animal Science*, 25 (Suppl. 1), 181–190.

Lehmann, H. (1998) The minipig in general toxicology. *Scandinavian Journal of Laboratory Animal Science*, 25 (Suppl. 1), 59–62.

Marcarian, H.Q. and Calhoun, M.L. (1966) Microscopic anatomy of the integument of adult swine. *American Journal of Veterinary Research*, 27, 765–772.

Männisto, P.T., Havas, A., Haasio, K., Hanhijärvi, H. and Mustakallio, K. (1984) Skin irritation by dithranol (anthralin) and its 10-acetyl analogues in 3 animal models. *Contact Dermatitis*, 10, 140–145.

Männistö, P.T., Hanhijärvi, H., Kosma, V.M. and Collan, Y. (1986) A 6-month dermal toxicity test with dithranol and butantrone in miniature swine. *Contact Dermatitis*, 15, 1–9.

Meyer, W. (1996) Bemerkungen zur Eignung der Schweinhaut als biologisches Modell für die Haut des Menschen. *Hautarzt*, 47, 178–182.

Meyer, W., Schwarz, R. and Neurand, K. (1978) The skin of domestic mammals as a model for the human skin, with special reference to the domestic pig. *Current Problems in Dermatology*, 7, 39–52.

Montagna, W. and Yun, J.S. (1964) The skin of the domestic pig. *Journal of Investigative Dermatology*, 43, 11–21.

Monteiro-Riviere N.A. (1986) Ultrastructural evaluation of the porcine integument. In Tumbleson, M.E. (ed.) *Swine in Biomedical Research*. Plenum Press: New York, pp. 641–655.

Monteiro-Riviere, N.A., Bristol, D.G., Manning, T.O., Rogers, R.A. and Riviere, J.E. (1990) Interspecies and interregional analysis of comparative histologic thickness and laser Doppler blood flow measurements at five cutaneous sites in nine species. *Journal of Investigative Dermatology*, 95, 582–586.

Mortensen, J.T, Brinck, P. and Lichtenberg, J. (1998) The minipig in dermal toxicology. *Scandinavian Journal of Laboratory Animal Science*, 25 (Suppl. 1), 99–105.

Petroianu, G., Maleck, W., Altmannberger, S., Jatzko, A. and Rüfer, R. (1997) Blood coagulation, platelets and hematocrit in male, female, and pregnant Göttingen minipigs. *Scandinavian Journal of Laboratory Animal Science*, 24, 31–41.

Rispat, G., Slaoui, M., Weber, D., Salemink, P., Berthoux, C. and Shrivastava, R. (1993) Haematological and plasma biochemical values for healthy Yucatan Micropigs. *Laboratory Animals*, 27, 368–373.

Schedl, H.P., Conway, T., Horst, R.L., Miller, D.L. and Brown, C.K. (1996) Effects of dietary calcium and phosphorus on vitamin D metabolism and calcium absorption in hamster. *Proceedings of the Society for Experimental Biological Medicine*, 211, 281–286.

Smith, J.L. and Calhoun, M.L. (1964) The microscopic anatomy of the integument of newborn swine. *American Journal of Veterinary Research*, 25, 165–173.

Svendsen, O., Skydsgaard, M., Aarup, V. and Klastrup, S. (1998) Spontaneously occurring microscopic lesions in selected organs of the Göttingen minipig. *Scandinavian Journal of Laboratory Animal Science*, 25 (Suppl. 1), 231–234.

Swindle, M.M. (1998) *Surgery, Anesthesia, and Experimental Techniques in Swine.* Ames: Iowa State University Press.

Vogel, B. and Singer, T. (1997) The use of minipigs in dermal toxicity testing. *Pharmacology and Toxicology.* 80 (Suppl. II), 34.

Wrogemann, J. and Holtz, W. (1977) Blutuntersuchungen am Göttingen Miniaturschwein. *Zeitschrift für Versuchstierkunde* 19, 276–289.

19 Reproductive infusion toxicity studies in the minipig

P. A. McAnulty

19.1 Introduction

Minipigs are showing increasing popularity as a species for non-rodent toxicology testing. This was originally attributable to their suitability for performing dermal studies, but it has quickly become obvious that they can be used for all routes of administration, and in many cases are preferable to dogs or primates for metabolic or pharmacological reasons. Their use in general toxicology testing employing the continuous intravenous infusion route has been described in the previous two chapters, and the purpose of this chapter is to explore their potential for reproductive studies by this route.

In this introductory section, a general overview of the parameters relevant to reproductive toxicology testing will be considered. In the subsequent sections, a review of the techniques that have been published will be presented, and in a final section some information on the application of continuous infusion techniques in reproductive toxicology will be given.

The potential of minipigs in reproductive toxicology was a subject of a recent review (Jørgensen 1998a). Clearly, economic concerns dictate that, wherever possible, the smaller species should be used in reproductive toxicology, and therefore the vast majority of studies are performed in rats, mice and rabbits. However, there can be times when any or all of these traditional species are found to be unsuitable for conducting studies and an alternative needs to be considered. Primates have often been used as a result, but there may be problems with animal supply and conservation concerns, and the majority of species have only single offspring. Ferrets have also been suggested, but they are seasonal breeders and they may have an inappropriate metabolism for the product being tested. Minipigs have the advantages of being purpose-bred, relatively inexpensive compared with primates, having a number of metabolic and physiological similarities to humans and producing multiple offspring. Their disadvantage is the amount of test article required.

Minipigs also have the advantage of becoming sexually mature much earlier than primate species, at approximately 5 months of age. The oestrous cycle is 21–22 days, with oestrus lasting 3 days. There is only limited information in the literature on the duration of the spermatogenic cycle, and no staging information has been published for the seminiferous tubules.

Oestrus is best detected by the lordosis reaction, because the external signs of the cycle are minimal. Ovulation is spontaneous, and seasonality has been lost during domestication. The length of gestation is 112–114 days, and the majority of minipig strains have a litter size of five or six.

Minipigs, like conventional pigs, have a six-layered diffuse epitheliochorial placenta. The passage of various chemicals across this type of placenta resembles that in the human more than with some other types of placenta.

Embryonic and fetal development in minipigs is similar to that of the larger breeds. The following data in Table 19.1 have been published for the Göttingen minipig (Becze and Smidt 1968; Grote *et al.* 1977; Glodek and Oldigs 1981).

Göttingen minipig fetuses have been shown to be sensitive to several known teratogens. Single intraperitoneal doses of 6-aminonicotinamide on Day 10 or Day 16 of gestation have been found to cause multiple skeletal and visceral abnormalities, but no effects were obtained following injection on Day 23 (Sudeck and Grote 1974). Single doses of methylnitrosourea given in the first third of pregnancy cause brain malformations, and single doses of ethylnitrosourea cause extensive skeletal abnormalities. These effects are seen regardless of the route of administration used (Graw *et al.* 1975; Ivankovic 1979). Dietary administration of pyrimethamine, a folic acid antagonist, during the first part of the organogenesis period, results in a high incidence of cleft palate, club foot and micrognathia (Misawa *et al.* 1982; Yamamoto *et al.* 1984; Hayama and Kokue 1985; Ohnishi *et al.* 1989). Dietary administration of tretinoin from Day 11 up to Day 37 of gestation has been found to produce multiple malformations resembling retinoic acid teratogenicity in humans (Jørgensen 1998b). Göttingen minipigs have also been used to confirm absence of teratogenicity for compounds such as coumarin and troxerutin (Grote *et al.* 1977). Pitman–Moore miniature pigs have been shown to be sensitive to trypan blue; a single intravenous injection on Day 10 of gestation resulted in multiple cardiovascular and hindgut abnormalities (Rosenkrantz *et al.* 1970). In a study involving an unspecified strain of miniature pig, one case of anencephaly was reported after feeding a concentrate of blighted potato during the first half of gestation (Sharma *et al.* 1978).

Background data regarding fetal abnormalities in minipigs have been published. In Göttingen minipigs bred and maintained in Japan, there was a low incidence of palatal defects, agnathia, diaphragmatic hernia, oligo- and polydactyly, umbilical hernia and hypoplasia of the caudal lobe of the liver. The incidence of limb defects, at 1.6%, was

Table 19.1 Embryonic and fetal development in the minipig

Day of gestation	Developmental stage
1	Two-cell stage embryo
3	Morula
5–6	Blastocyst
7–9	Elongation of the blastocyst
8–12	Formation of primitive streak
14–16	Somite formation begins
16	Optic vesicles form
16–18	Mandibular and maxillary projections form; forelimb buds appear; neural tube closure begins
18–22	Hindlimb buds appear; organogenesis of brain, eyes, heart, liver, and kidneys
22–28	Palatal shelves form; hand and foot plates form, and digital rays appear; differentiation of sex organs
34–35	Closure of palate
36	Commencement of fetal development

higher than in other breeds of pig (Hayama and Kokue 1985). More recently published data for the same strain bred and maintained in Denmark have confirmed the relatively high incidence of limb defects (muscular–joint contractures 3.4%), with the other most frequent malformations being cryptorchidism (1.5%), pentadactyly (1.0%) and inguinal hernia (0.6%) (Jørgensen 1998a).

Sinclair miniature swine have been found to be very good models for studies of fetal alcohol syndrome because they will voluntarily consume ethanol in the drinking water at dose levels up to 4 g/kg/day. In serial pregnancy studies, it has been demonstrated that there is a progressive reduction in live litter size, attributable to an increase in abortions and perinatal deaths. Birth weight also became progressively lower (Dexter *et al.* 1980; 1983).

For investigations of female subfertility, the swine leucocyte antigen (SLA) inbred miniature swine has been found to be useful (Howard *et al.* 1982; 1983). This strain was developed for immunological studies of effects of tissue transplantation, and was found to have a lower ovulation rate and litter size than other breeds of miniature and conventional pigs. These reproductive parameters have been shown to improve following administration of agents such as altrenogest and exogenous gonadotrophins (Diehl *et al.* 1986).

19.2 Continuous infusion procedures

19.2.1 Pregnant sows

There is currently no published literature on continuous infusion procedures during the first two trimesters of pregnancy in minipigs. At Scantox we have performed some preliminary investigations, and have concluded that if the infusion is established before mating, then conception and early embryonic and fetal development should be unaffected by the procedure. The procedures would be the same as described in the previous two chapters.

Several studies have set up infusion during the third trimester of pregnancy. In one study, 2- to 3-year-old Göttingen minipigs were infused into the superior vena cava (Beller *et al.* 1985; Holzgreve *et al.* 1985). They initially received intramuscular injections of 0.5 mg atropine and 120–200 mg of azaperone. Anaesthesia was induced by intravenous administration of 2.5–10 mg/kg metomidat HCl via a cannula in one of the ear veins, and maintained by continuous administration of the same anaesthetic at 5–7.5 mg/kg/h in 0.9% saline. Following tracheotomy and intubation, the sows were ventilated on a pressure- and volume-controlled respirator. The catheter for infusion was guided into the superior vena cava, and in this particular investigation two other catheters were implanted. One was placed in the pulmonary artery via the jugular vein to monitor changes in pulmonary artery pressure and the second was implanted via the femoral artery to measure arterial pressure in the aortic arch. This second catheter had a thermistor sensor at the tip to measure cardiac output using the cold dilution method.

In another study, an infusion catheter was placed in the axillary vein of Yucatan minipigs on Day 76 of pregnancy (Laferrière *et al.* 1995). Premedication consisted of intramuscular injections of 15 mg/kg ketamine and 0.04 mg/kg atropine. Anaesthesia was induced and maintained by 1–5% inhaled isoflurane. After implantation of the catheter, it was tunnelled to exit between the shoulders, and attached via an extension to both a syringe pump and a programmable syringe pump. The catheter line was protected by a metal tether, fixed at the proximal end to a custom-made jacket that was fitted at the time of surgery, and at the distal end to the pumps. The pumps were programmed to deliver the test article in heparinised saline and various times each day during the remainder of pregnancy, and the pumps were also programmed to flush the catheter with heparinised saline before and after each test article administration. Topical and systemic antibiotics were administered during surgery and for 3 days afterwards.

Continuous ovarian infusion has been developed in non-pregnant Yucatan micropigs (Musah *et al.* 1994). The sows were 15 months old and were fasted for 24 h before surgery. Anaesthesia and laparoscopy were performed using a modification of a procedure developed in conventional gilts (Anderson *et al.* 1961). Following this, a 15-gauge stainless-steel hypodermic needle was inserted into the ovarian vein in the direction of the ovary about 1 cm from the bifurcation of the uterine and ovarian veins. Polyethylene tubing (ID 0.575 mm; OD 0.95 mm) was threaded through the needle into the vein and secured with silk thread knots, fastening the vein around the catheter. The hypodermic needle was removed, and the catheter attached to the broad ligaments with silk thread. The catheter was exteriorised through the mid-ventral incision and tunnelled under the skin of the flank region. The external part of the catheter was enclosed in a pouch attached to the dorsolateral flank.

A microdialysis system for individual corpora lutea has been developed in Göttingen minipigs (Jarry *et al.* 1990; Wuttke *et al.* 1998), and used to investigate various aspects of luteal function following administration of exogenous test articles such as tamoxifen and tumour necrosis factor. Sows were observed for behavioural oestrus, and 2–4 days later they were anaesthetised with halothane and a laparotomy performed. The ovaries were exposed, and easily accessible corpora lutea chosen for applying the microdialysis system (MDS). In the first of the reported studies, Vitafiber 3×50 dialysis tubing was used (OD 200 μm, molecular weight cut-off 50,000) and, in the second study, SPS 660 tubing was used instead (molecular weight cut-off 1,000,000); however, the fragility of the second type of tubing resulted in a 30% failure rate. The preparation of the MDS is shown in Figure 19.1. The dialysis tubing was cut into 2-cm-long pieces, and both ends were inserted into tightly fitting Silastic® tubing. The free surface of dialysis tubing was 4–5 mm long.

The metal part only of a 22-gauge hypodermic needle was connected to one of the Silastic® tubes on the MDS, and inserted through the two poles of a suitable corpus luteum. The MDS was pulled through until all of the free dialysis tubing surface was located within the corpus luteum. Tissue glue was applied at the two points where the tubing penetrated the corpus luteum. The two Silastic® tubes were glued to a Silastic® plate, and attached via the 22-gauge hypodermic needles to Teflon® tubing, which was also glued to the plate. The plate was folded to protect the connections, and fixed to the parametrium. All of the Teflon® tubes ran through a single piece of Latex tubing, and this was exteriorised through the skin of the back of the animal. After closure of the surgical incisions, the input tubing was connected to a peristaltic pump, and continuously flushed with Ringer solution at a flow rate of 2.4 ml/h. The infusion systems were maintained for up to 5 days, and at the end of this time the corpora lutea were found to show no signs of inflammation, and there was little or no connective tissue formation. It was also found that it was possible to implant two MDSs in a single corpus luteum.

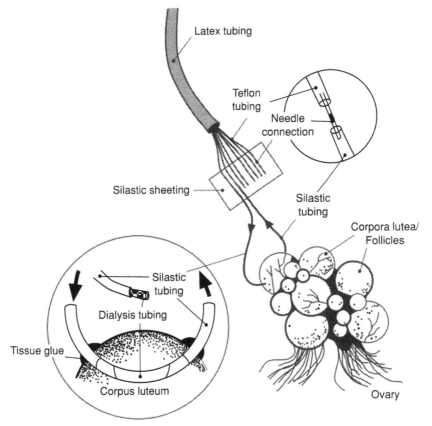

Figure 19.1 Schematic drawing of the microdialysis system (MDS; lower insert) and its localisation *in situ* within individual porcine corpora lutea. The inlet and outlet tubings of the MDS were exteriorised with Teflon® tubing, of which the ends are labelled with arrows. The connection of both types of tubing is detailed in the upper insert. (Reproduced with the permission of the Endocrine Society from Jarry *et al.* 1990.)

19.2.2 *Fetal piglets*

Procedures for setting up continuous infusion in fetal Yucatan miniature piglets in the third trimester of pregnancy have been described in great detail by Swindle *et al.* (1996). The Yucatan strain was chosen after initial studies in Hanford miniature swine had been less successful. Two days before the surgical procedures were performed, the sows were treated intramuscularly with 15–50 mg of medroxyprogesterone acetate if it had no potential effect upon the experimental protocol; this was repeated on the day of surgery and 2 days after surgery, and was intended to reduce the chances of abortion. The sows were prepared for surgery by intramuscular injections of ketamine (22 mg/kg), acetylpromazine (1.1 mg/kg) and atropine (0.05 mg/kg). Intravenous infusion of Ringer solution was established into an ear vein at a rate of 10 ml/kg/h. Surgical anaesthesia was induced by inhalation of isoflurane and, after intubation, anaesthesia was maintained with 0.5–2% isoflurane in a 2:1 oxygen–nitrous oxide mixture. The operating room was kept at a temperature over 29 °C to prevent hypothermia, and heated water blankets and a heat lamp were also used. Rectal temperature was monitored, and maintained in the range 37.8–39.5 °C. The sow was placed in right lateral recumbency, and the surgical site, a 25-cm incision dorsolateral to the mammary tissue, treated with betadine solution and rinsed with ethanol. For analgesia, 1 g of cefazolin sodium was administered intravenously before surgery, and 0.1 mg/kg pancuronium bromide was given intravenously every 30 min during surgery.

When the initial incision was made, care was taken to avoid the mammary tissue, which is highly vascularised in the third trimester of pregnancy. Following laparotomy, a hollow trocar was channelled from the abdominal cavity through the wall of the flank, just ventral to the wings of the lumbar vertebrae. A single fetus was selected, and the uterotomy site packed around with laparotomy sponges soaked with warm saline. The uterotomy was performed on the antimesometrial border along the avascular plane using a staple surgical incision device with absorbable staples. Care was taken to minimise loss of amniotic fluid and exteriorisation of the fetus, and in particular the fetal head was never allowed out of the amniotic fluid to prevent initiation of breathing. The left foreleg of the fetus was caudally retracted out of the incision, and a 2-cm left ventral paramedian neck incision was made along the jugular furrow. The internal carotid artery and external jugular vein were exposed by caudal displacement of the thymus and medial dissection dorsal to the body of the sternohyoideus muscle. The fetal catheters were custom-made Silastic® access ports: the carotid catheter was 4 French, 90 cm in length, with a 6-French cover which extended from the port to 5.5 cm from the tip, and Silastic® retention beads attached 3, 4 and 5 cm from the tip; the jugular catheter was identical, except that the internal catheter was 2 French. These catheters were passed into the abdomen via the trocar through the flank of the sow, after which the trocar was removed. Each catheter had a wire insertion aid at its tip, and the catheters were inserted beyond the first retention bead through a small incision in each blood vessel. The catheters were sutured in place, and the ends of the sutures were attached to the second retention bead.

The fetal incisions were closed, followed by closure of the uterotomy incision. To prevent potential mixture of the maternal and fetal circulation, as well as placental separation, sutures were passed through the myometrium into the lumen of the allantochorionic cavity. The allantochorion was penetrated on both sides of the incision, and the suture passed out through the uterine wall. The suture was then passed inwards again through the myometrium and endometrium, but not the allantochorion, and then back out through the endometrium and myometrium. When the sutures were tightened, the uterine walls were everted, and the allantochorion separated from the endometrium. The fetal catheters were sutured to the outer wall of the uterus with a Silastic® retention collar fastened to the 6-French catheter cover 10 cm from the tip. The vascular ports were sutured to the back of the sow using non-absorbable sutures, and these were covered by a custom-made canvas pouch, which was also sutured to the back of the sow.

Intramuscular buprenorphine was administered at 0.05–0.1 mg/kg before removal of the sow from gas anaesthesia, and administration of this analgesic was repeated every 8–12 h when indicated by clinical signs. Cephalexin was administered in the feed for 5 days after surgery at a dose level of 25–50 mg/kg. The fetal catheters were flushed every 48 h with 2 ml of saline followed by 1 ml of heparin (10 U/ml).

19.2.3 Neonatal piglets

A method for catheterising neonatal piglets of 1–2 weeks of age has been described by Rudinsky *et al.* (1994). After an initial intraperitoneal injection of 20 mg/kg ketamine, the trachea was intubated. This was followed by an intravenous injection of 20 mg/kg sodium pentobarbital, and 1 mg/kg D-tubocurarine for muscle relaxation. The level of anaesthesia was monitored from heart rate and blood pressure responses to noxious stimuli, and anaesthesia was maintained by intermittent bolus doses of sodium pentobarbital, delivering approximately 2 mg/kg/h.

Mechanical ventilation was used, and the initial settings were as follows:

Fractionated inspired oxygen concentration (F_{IO_2})	0.50
Positive end-expiatory pressure (PEEP)	4 cmH$_2$O
Tidal volume	15 ml/kg
Rate	10 breaths/min

Subsequently, the tidal volume was adjusted to maintain P_{CO_2} in the range of 35–40 mmHg. Warming blankets and overhead heating were used. In this study catheters were introduced into the right atrium of the piglets via the left external jugular vein, into the pulmonary artery and into the left atrium via left lateral sternotomy.

19.3 Applications

19.3.1 Teratology

There are no published studies that describe the use of continuous infusion in a teratology study performed in minipigs. However, as described above, as long as the initial surgery for implanting the catheters is performed before mating, such a procedure should not pose significant difficulties.

19.3.2 Behavioural teratology

For studies involving the development of the central nervous system, pigs are considered to be superior to other laboratory animal species, because the timing of neurological development with respect to the time of birth is very similar to that of humans (Dickerson and Dobbing 1967). Also, the development of the blood–brain barrier is similar to that of man, with both species having a relatively immature barrier compared with other species during the last trimester of pregnancy (Ikeda *et al.* 1985a, b).

The effect of prenatal cocaine on the behaviour of neonatal Yucatan minipigs has been investigated following intravenous administration to the sow during the last third of pregnancy, using the methodology described previously (Laferrière *et al.* 1995). The sows received 2 mg/kg cocaine as a 1% solution in heparinised saline over 10 min, four times a day from Day 76 of gestation until birth. The responses in an open-field situation suggested that there was a transient period of lower responses to spatial novelty.

19.3.3 Maternal–fetal pharmacokinetics

The fetal catheterisation method developed by Swindle *et al.* (1996) has been used to study maternal–fetal pharmacokinetics of class I-C antiarrhythmic agents such as moricizine in Yucatan minipigs (Swindle *et al.* 1993). This class of pharmaceutical is used for the treatment of fetal tachyarrhythmias, a cause of congestive heart failure, and the minipig model has been useful in predicting the kinetics of these agents in the human fetus.

19.3.4 Endotoxic shock

Several of the infusion procedures that have been developed have been used to investigate induction and treatment of septicaemia in pregnant sows and newborn piglets (Beller *et al.* 1985; Holzgreve *et al.* 1985; Rudinsky *et al.* 1994).

19.4 Conclusion

It is clear that the techniques for various types of continuous infusion have been developed and shown to be successful in pregnant minipigs, as well as both fetal and neonatal piglets. These techniques have already been used to investigate some aspects of behavioural teratology and endotoxic shock.

Minipigs have been used in a number of teratology studies, but in the published literature this is only with routes other than administration by continuous infusion. However, the availability of developed techniques in both areas suggests that it is possible to perform teratology studies by continuous infusion.

References

Anderson, L.L., Butcher, R.L. and Melampy, R.M. (1961) Subtotal hysterectomy and ovarian function in gilts. *Endocrinology*, 69, 571–580.

Becze, J. and Smidt, D. (1968) Die Entwicklung der äusseren Gestalt beim Göttinger Zwergschwein. *Dtsch. tierärtzl. Wschr.*, 75, 352–356.

Beller, F.K., Schmidt, E.H., Holzgreve, W. and Hauss, J. (1985) Septicemia during pregnancy: a study in different species of experimental animals. *American Journal of Obstetrics and Gynecology*, 151, 967–975.

Dexter, J.D., Tumbleson, M.E., Decker, J.D. and Middleton, C.C. (1980) Fetal alcohol syndrome in Sinclair (S-1) miniature swine. *Alcoholism*, 4, 146–151.

Dexter, J.D., Tumbleson, M.E., Decker, J.D. and Middleton, C.C. (1983) Comparison of the offspring of three serial pregnancies during voluntary alcohol consumption in Sinclair (S-1) miniature swine. *Neurobehavioural and Toxicological Teratology*, 5, 229–231.

Dickerson, J.W.T. and Dobbing, J. (1967) Prenatal and postnatal growth and development of the central nervous system of the pig. *Journal of Physiology (London)*, 166, 384–395.

Diehl, J.R., Stuart, L.D., Goodrowe, K.L. and Wildt, D.E. (1986) Effects of altrenogest and exogenous gonadotropins on ovarian function and embryo recovery in Swine Leukocyte Antigen inbred miniature swine as influenced by cystic endometrial hyperplasia. *Biology of Reproduction*, 35, 1261–1268.

Glodek, P. and Oldigs, B. (1981) *Das Göttinger Miniaturschwein*. Paul Parey: Berlin.

Graw, J., Ivankovic, S., Berg, H. and Schmähl, D. (1975) Teratogene Wirkung von Äthylnitrosoharnstoff an Göttinger Miniaturschweinen. *Arzneim.-Forsch.*, 25, 1606–1608.

Grote, W., Schulz, L.-C., Drommer, W., Überschär, S. and Schäfer, E.-A. (1977) Überprüfung einer Kombination der Wirkstoffe Cumarin und Troxerutin auf embryotoxische und teratogene Nebenwirkungen and Göttinger Miniaturschweinen. *Arzneim.-Forsch.*, 27, 613–617.

Hayama, T. and Kokue, E. (1985) Use of the Goettingen miniature pig for studying pyrimethamine teratogenesis. *CRC Critical Reviews in Toxicology,* 14, 403–421.

Holzgreve, W., Schmidt, E.H., Schlegel, W. and Beller, F.K. (1985) Prostaglandinveränderung während des Endotoxin-induzierten Schocks bei schwangeren Minischweinen. *Z. Geburtsh. u. Perinat.*, 189, 235–238.

Howard, P.K., Chakraborty, P.K., Camp, J.C., Stuart, L.D. and Wildt, D.E. (1982) Correlates of ovarian morphology, estrous behavior, and cyclicity in an inbred strain of miniature swine. *Anatomical Record*, 203, 55–65.

Howard, P.K., Chakraborty, P.K., Camp, J.C., Stuart, L.D. and Wildt, D.E. (1983) Pituitary–ovarian relationships during the estrous cycle and the influence of parity in an inbred strain of miniature swine. *Journal of Animal Science*, 57, 1517–1524.

Ikeda, G.J., Miller, E., Sapienza, P.P., Michel, T.C. and Sager, A.O. (1985a) Maternal–foetal distribution studies in late pregnancy. I. Distribution of [*N-methyl*-^{14}C] betaine in tissues of beagle dogs and miniature pigs. *Food and Chemical Toxicology*, 23, 609–614.

Ikeda, G.J., Miller, E., Sapienza, P.P., Michel, T.C., King, M.T. and Sager, A.O. (1985b) Maternal–foetal distribution studies in late pregnancy. II. Distribution of [1-^{14}C] acrylamide in tissues of beagle dogs and miniature pigs. *Food and Chemical Toxicology*, 23, 757–761.

Ivankovic, S. (1979) Teratogenic and carcinogenic effects of some chemicals during prenatal life in rats, Syrian golden hamsters, and minipigs. *National Cancer Institute Monograph No. 51. Perinatal Carcinogenesis*, 103–115.

Jarry, H., Einspanier, A., Kanngiesser, L., Dietrich, M., Pitzel, L., Holtz, W. and Wuttke, W. (1990) Release and effects of oxytocin on estradiol and progesterone secretion in porcine corpora lutea as measured by an *in vivo* microdialysis system. *Endocrinology*, 126, 2350–2358.

Jørgensen, K.D. (1998a) Minipig in reproduction toxicology. *Scandinavian Journal of Laboratory Animal Science*, 25 (Suppl. 1), 63–75.

Jørgensen, K.D. (1998b) Teratogenic activity of tretinoin in the Göttingen minipig. *Scandinavian Journal of Laboratory Animal Science*, 25 (Suppl. 1), 235–243.

Laferrière, A., Ertug, F. and Moss, I.R. (1995) Prenatal cocaine alters open-field behavior in young swine. *Neurotoxicology and Teratology* 17, 81–87.

Misawa, J., Kanda, S., Kokue, E., Hayama, T., Teramoto, S., Aoyama, H., Kaneda, M. and Iwasaki, T. (1982) Teratogenic activity of pyrimethamine in Göttingen minipig. *Toxicological Letters*, 10, 51–54.

Musah, A.I., Bloch, J.F., Baker, S.L. and Schrank, G.D. (1994) Human atrial natiuretic peptide infusion and ovarian venous progesterone secretion in Yucatan micropigs. *Journal of Reproduction and Fertility*, 101, 109–113.

Ohnishi, M., Kojima, N., Kokue, E. and Hayama, T. (1989) Experimental induction of splayleg in piglets by pyrimethamine. *Japanese Journal of Veterinary Science*, 51, 146–150.

Rosenkrantz, J.G., Lynch, F.P. and Frost, W.W. (1970) Congenital anomalies in the pig: teratogenic effects of trypan blue. *Journal of Pediatric Surgery*, 5, 232–237.

Rudinsky, B., Bell, A., Hipps, R. and Meadow, W. (1994) The effects of intravenous L-arginine supplementation on systemic and pulmonary hemodynamics and oxygen utilization during group B streptococcal sepsis in piglets. *Journal of Critical Care*, 9, 34–46.

Sharma, R.P., Willhite, C.C., Wu, M.T. and Salunkhe, D.K. (1978) Teratogenic potential of blighted potato concentrate in rabbits, hamsters, and miniature swine. *Teratology*, 18, 55–62.

Sudeck, M. and Grote, W. (1974) Phasenspezifisches Missbildungsmuster beim 'Göttinger Miniaturschwein' nach Verabreichung von 6-amino-nicotinsäureamid. *Z. Morph. Anthrop.*, 65, 265–275.

Swindle, M.M., Wiest, D.B., Smith, A.C., Garner, S.S. and Gillette, P.C. (1993) A maternal and fetal catheterization system for the study of antiarrhythmics in Yucatan miniature swine. *Journal of Investigative Surgery,* 6, 384.

Swindle, M.M., Wiest, D.B., Smith, A.C., Garner, S.S., Case, C.C., Thompson, R.P., Fyfe, D.A. and Gillette, P.C. (1996) Fetal surgical protocols in Yucatan miniature swine. *Laboratory Animal Science*, 46, 90–95.

Wuttke, W., Spiess, S., Knoke, I., Pitzel, L., Leonhardt, S. and Jarry, H. (1998) Synergistic effects of prostaglandin F2α and tumor necrosis factor to induce luteolysis in the pig. *Biology of Reproduction,* 58, 1310–1315.

Yamamoto, Y., Ohnishi, M., Kokue, E., Hayama, T., Aoyama, H., Kaneda, M. and Teramoto, S. (1984) Pyrimethamine teratogenesis by short term administration in Goettingen miniature pig. *Congenital Anomalies*, 24, 83–87.

Part 7

Continuous intravenous infusion
– general

The final section of this book covers more general themes that are relevant to many if not all of the species covered in Parts 1–6.

Firstly, there is a review of the common pathological findings that can be expected following catheterisation and infusion. There follows an overview of the use of mini-osmotic pumps, which can be implanted subcutaneously to dose via a suitable vein. This is a valid and often neglected alternative to using an external pump with its associated tubing and consequent limitations on the mobility of the animal. However, there is a limit on the volumes that can be administered from these devices.

The third chapter in this section examines suitable vehicles and recommended volumes and rates of administration. Problems with solubility can often limit the maximum dose of a drug, and many vehicles themselves exhibit some degree of associated toxicity. Finally, there is a comprehensive overview of the range of equipment that is currently available from commercial suppliers, including full contact details and website addresses.

20 Common pathological findings in continuous infusion studies

J. G. Evans and P. J. Kerry

20.1 Introduction

Venous catheters for use in humans were first introduced in 1945 (Meyers 1945), and the complications associated with the infusion of fluids through such catheters began to be reported a short time later (Neuhoff and Seley 1947; Hoshal 1972; Maki *et al.* 1973). The complications almost invariably involved sepsis, and much debate took place on the source of the infection – cutting down to the vessel under less than optimal surgical conditions was considered to be one but not the only source of bacterial contamination.

The experience of venous infusions in humans thus in large part preceded the experience of such techniques in veterinary medicine and toxicology. Not surprisingly, the problems in animals mirrored those in man, with infection and septicaemia being major issues.

Continuous intravenous infusion is a well-established procedure for the administration of test materials to laboratory animals. These methods are available for the rat, dog, mouse and rabbit and have also been applied to primates and other species. They have been used for routine toxicity studies and for reproductive studies. Detailed procedures are set out in other chapters. The major reasons for performing continuous infusion are as follows: continuous infusion is the intended clinical route of compound administration; compound bioavailability is low when administered by other routes; test compounds are irritant when given by other routes; and continuous infusion allows assessment of pharmacokinetic parameters under steady-state conditions. However, there is one major disadvantage to this route of administration in that it allows long-term access to the body, bypassing many of the normal body defence functions. This problem is exacerbated by the surgical intervention necessary to implant and secure the infusion cannulae; this procedure maintains a physical portal of entry into the body that may allow bacteria entry to the subcutaneous tissue from the skin. Therefore, there may be pathology that is associated with the procedure rather than with the administered compounds. This additional pathology may be attributed to:

1 the surgical procedure;
2 volume and rate of fluid administration;
3 physical and chemical properties of the infusate.

In this chapter, the pathological features that are associated with continuous intravenous infusion are summarised from over 60 studies conducted in these laboratories in a variety of species, together with a review of the available literature. Details of the surgical and husbandry procedures are set out in other chapters.

Many of the general pathology issues associated with intravenous catheterisation (short-term, long-term, interrupted or continuous infusion) are introduced in a useful review article (Bayly and Vale 1982), even though the article title targets the horse as the main species to be addressed. Bayly and Vale discuss under various headings the factors that may contribute to catheter failure. These include tissue trauma, microbial contamination, interaction between blood and materials foreign to the body and blood–vehicle interactions. These factors, singly or in concert, can contribute to various conditions such as phlebitis, thrombosis, embolisation and infection that in turn can result in vessel occlusion, organ infarction, thrombophlebitis and septicaemia. As an introduction, this paper is to be recommended.

20.2 Pathological assessment

20.2.1 Infusion methods

The most commonly used catheterisation sites for the infusion of vehicles and xenobiotics are the jugular, femoral or tail veins, usually via Silastic® or polyethylene cannulae. Procedural details are provided elsewhere in this book. Unless otherwise stated, the pathological response to cannulae of different composition appears to be similar. Silastic® cannulae are generally favoured on the basis that they tend not to adhere to the side of the vessel, they are less thrombogenic and the risk of phlebitis is reduced.

20.2.2 Macroscopic and microscopic sampling

Comprehensive tissue sampling of the infusion site from the point of entry of the cannula to the body (either the scapular region or tail) to the cannula tip is necessary to establish the extent of any problems. Such sampling should also include any macroscopic abnormalities along the subcutaneous route of the cannula. It is also advisable to sample the cannulated vein approximately 1 cm distal to the point of entry of the cannula tip into the vessel. Histological sections in this laboratory are prepared from all tissue samples submitted and sections routinely stained with haematoxylin and eosin (H&E). On occasion, the Martius, scarlet, blue (MSB) technique for fibrin and the periodic acid–Schiff (PAS) method are also useful for the evaluation of histology.

Irrespective of any study protocol tissue requirements, representative samples from a wide variety of tissues are needed in infusion studies (with particular emphasis on kidneys, heart, all major coronary vessels, lung and sections of all the major liver lobes) to fully evaluate infusion procedure-related pathology.

As a general rule, the method for tissue sampling and recording should be standardised as far as possible so that appropriate comparisons can be made between different experiments. However, when lesions are extensive the most appropriate estimate of *extent* may be made through macroscopic examination, histology confirming the cellular composition of the macroscopic change. When the lesion is extensive or severe, caution should be exercised when comparing histological and macroscopic assessment of severity unless multiple sections have been taken of the affected site. It should be emphasised that a good description of the macroscopic appearance of any lesion is a prerequisite for proper assessment, backed by photography if possible.

20.3 Pathogenesis of vascular injury

The structure of vessel walls is fairly simple and essentially consists of three layers. The *tunica intima* is the layer in direct contact with the blood and is composed of endothelium lying on a basal lamina. The *tunica media*, composed of longitudinal and circumferential muscle layers, is the next adjacent layer. In arteries, the intima and media are separated by a prominent elastic lamina. A limited *tunica adventitia* formed, in the case of veins, of a limited amount of connective tissue constitutes the outermost layer. This relatively simple structure belies the complex chemistry of vessels: the endothelium produces a variety of humoral factors in response to injury. These may be involved in the formation of thrombus, in proliferative activity in the vessels and in the control of haemodynamics within vessels. This notwithstanding, the response of vessels to injury is limited and generally easily recognised. This may range from slight swelling of endothelial cells to hypertrophy and hyperplasia. In severe vessel injury, necrosis of the endothelium may occur, resulting in ulceration of the vessel. Such changes are inevitably associated with clot formation, which itself results in the production of other chemokines. Severe and chronic injury may lead to secondary changes in the muscle and connective tissue layers: this may include hypertrophy and fibrosis of the muscle coat with infiltration of mononuclear inflammatory cells. In old thrombi and in areas of necrosis calcium salts may be deposited.

It is inevitable that the presence of a cannula within a vessel will result in a response by the vessel (foreign body reaction) while the delivery of material through the cannula may have a direct effect on the vessel wall due to haemodynamic factors. The chemistry of the delivered compound/vehicle may also induce a response in the vessel wall. Experience in humans indicates that a fibrin sleeve starting at the tip from the point of insertion of the cannula into the vein rapidly covers cannulae. In dogs this process also occurs and may be complete within 5–7 days (Hoshal *et al.* 1971). Fibrin deposition is the initial event in the development of thrombophlebitis, and minimising the deposition of fibrin may reduce the extent of inflammation. Furthermore, precipitation of the compound at the point of delivery may result in crystal formation at and around the cannula tip. If crystal formation is suspected, then additional specialist procedures may be required in order to investigate the nature of the crystalline material. This may involve alternative processing procedures including frozen sections and the utilisation of polarised-light microscopy.

20.4 Pathology associated with surgical procedure

No chronic surgical introduction of infusion material such as cannulae is ever entirely free of pathology as a consequence of tissues being exposed to foreign material. This includes the cannula itself and the suture material used to secure it in place. Furthermore, the suture material may act as a nidus for infection while the cannula may act as a conduit for the entry of bacteria into the body. This is controlled by good surgical procedure that includes minimal handling of tissues and by scrupulous aseptic technique. The use of intravenously administered antibiotics, as a preventive measure or to treat established infections, is questionable. Antibiotics have little effect on the frequency of positive cultures from catheter tips (Maki *et al.* 1973), and in toxicology studies there is also the potential for direct or indirect interaction between an antibiotic and the xenobiotic under test.

Local fibrosis, chronic inflammation and occasional local infection at needle or cannula entry points through skin are unavoidable owing to the continuous rubbing movement of the cannula and attachment equipment (e.g. skin buttons) at this site. Open sores and ulceration are quite common in the vicinity of indwelling needle and cannula entry points on the tails of rats and vary considerably in degree of severity. This is almost certainly related to the particular difficulty of keeping this site free from faecal and urine contamination (Figure 20.1).

Haematoma rarely produce a serious problem, although Stein and Gray (1995) have suggested that penile haematomas might be a potential complication when the corpus cavernosum is used for emergency vascular access in dogs (Stein and Gray 1995). These authors propose that this route of fluid administration might be of value in the human male when difficulties are experienced locating suitable peripheral veins, in hospital practice or on the battlefield.

20.5 Infusion pathology compounded by infection

Infection is potentially the most problematic issue in continuous infusion studies, and such problems, where they occur, can render entire studies valueless. As mentioned elsewhere in this volume, procedural asepsis is a requisite of infusion studies, but despite this local and systemic infections are not that uncommon.

Occasionally, acute inflammation and abscess formation are a problem at the cannula entry point through the skin and in association with sutures. This often foretells of more widespread problems along the cannula track and systemically. Where infection is a problem, this may be detected in-life and as macroscopic subcutaneous thickening along the track of the cannula. Microscopically, this may appear as chronic inflammation with extensive fibrosis and, on occasion, single or multiple abscess formation. Septicaemia and embolic spread of bacteria may result in abscess formation at distant sites. This may be particularly obvious in the lung, liver and kidney. These problems were exemplified in a 1-month continuous infusion study in 24 dogs. In this study, severe phlebitis with abscess formation was apparent at the infusion site. Seventeen of these animals had extensive haemorrhagic infarcts in the lung; in some

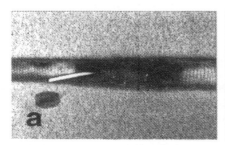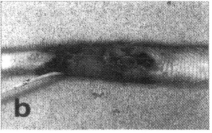

Figure 20.1 Indwelling catheters in the tails of rats and continuous intravenous infusion. (a) Evidence of epidermal necrosis (dark areas). (b) Full-thickness epidermal ulceration and necrosis in vicinity of cannula entry point through the skin.

animals there were extensive pleural adhesions and fibrosis. In addition, haemorrhagic glomerulitis, suppurative pancreatitis and suppurative meningitis were seen in a number of dogs. There was also a marked increase in inflammatory cell foci in the liver. Some of these changes are illustrated in Figure 20.2. It should be mentioned that these severe findings are not typical of those normally seen in infused control animals from this laboratory. They were almost certainly related to problems with a particular batch of cannulae.

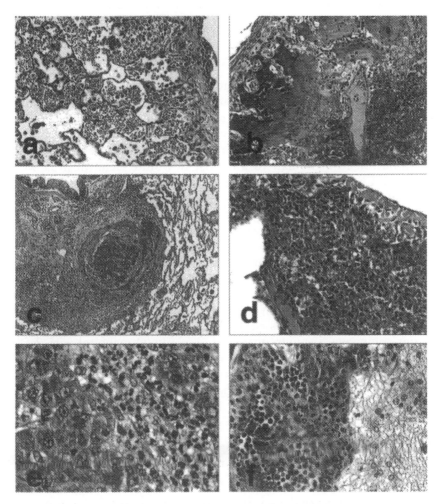

Figure 20.2 One-month continuous intravenous infusion in dogs. (a) Lung – alveolar haemorrhage and pleural fibrosis. (b) Lung – alveolar oedema and inflammation. (c) Lung – thrombosis and perivascular/alveolar inflammation. (d) Lung – pneumonia. (e) Exocrine pancreas – acute inflammation. (f) Brain – acute inflammation of the meninges. These changes are atypically severe; the majority of dog studies carried out in this laboratory have minimal findings associated with the infusion procedure.

20.6 Foreign body granulomas

Foreign body granulomas may occasionally be seen following continuous intravenous infusion, although they are far more common following infusion procedures involving minimal surgical intervention, e.g. venepuncture of tail vein in rats. They are more frequently found in studies involving multiple intravenous injections and repeated blood sampling than in studies utilising continuous intravenous infusion. Numerous papers describe the formation of hair granulomas in various organs. Lung, as might be expected, is a common site (Schneider and Pappritz 1976), but they can occur widely, e.g. in dog urinary bladder wall, following continuous intravenous infusion. Suture granulomas may occasionally be seen. In long-term studies in the rat there is an increased risk of sarcoma development associated with granulomas when there is extensive fibrosis.

20.7 The effects of vehicle infusion composition

Most of the available information on the adverse effects of intravenously administered solutions is to be found in the literature relating to human medicine. Very little published material has been found addressing this subject in animal toxicology studies. More often than not, vehicles are chosen purely on their ability to maximise the solubility of those notoriously insoluble compounds requiring toxicological evaluation. However, by extrapolation from human to animals, it would seem an area worthy of consideration. Chapter 22, by Hickling and Smith, addresses this topic more fully.

By way of example, unbuffered glucose solutions for infusion have a pH of 3–5; the acid pH is of value in preventing degradation of the solution during sterilisation. However, these low-pH solutions are thought to be linked with an increased incidence of post-infusion thrombophlebitis in human patients, a problem that can be overcome by raising the pH of infusion solutions to neutral with sodium bicarbonate (Fonkalsrud *et al.* 1968). There are few or no published data to support a similar relation between acid pH and thrombophlebitis in animal infusion studies. There is some information available showing that formulation of infusion vehicles in toxicology studies can have an effect on the pathology expressed. In this laboratory for example, work has been carried out comparing 5% (w/v) mannitol, 2.5% (w/v) mannitol with 2.5% (w/v) glucose and polyethylene glycol (PEG400) at 20% or 40% (w/v) as potential vehicles in a rat 1-month continuous infusion study. Mannitol and mannitol plus glucose produced little histopathology that could be related to the respective vehicles. However, PEG400 at both concentrations produced abscessation and necrosis at the infusion site (in the vicinity of the cannula tip).

20.8 Continuous infusion pathology in different species

Species, or rather the size of the animal and the relative size and 'toughness' of infusion blood vessels, can influence the severity of the response to infusion procedures and any problems arising from them. In our experience with rats and dogs, the latter have rarely shown any procedure-related pathology, other than slight chronic inflammation and fibrosis where the infusion cannula passes through the skin, and reaction (chronic inflammation and fibrosis) at various suture sites (see below). In rats surgically prepared and maintained for infusion studies via the femoral vein/lower abdominal vena cava,

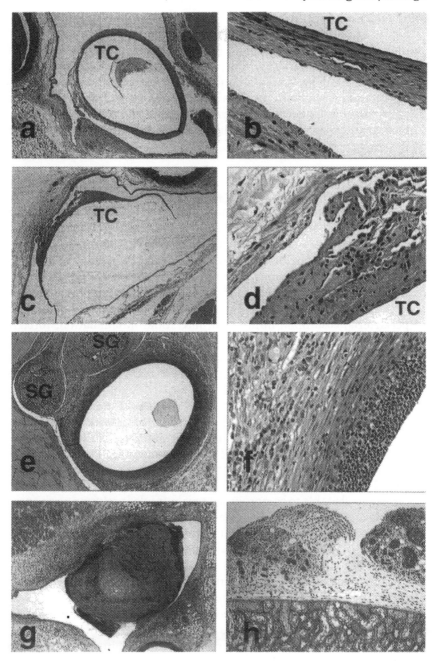

Figure 20.3 Continuous intravenous infusion in the rat: (a)–(d) The 'cleanest' pathology at the cannula tip of the infusion site achievable in our experience. TC, thrombus cuff (or sleeve), with some evidence of organisation. Note the minimal inflammation of the vessel wall. (e)–(f) A more severe inflammatory reaction at the infusion site, with neutrophils lining the vascular lumen, surrounded by fibrosis/ granulation tissue and mononuclear cells. SG, suture granuloma. (g) Thrombophlebitis at the infusion site. (h) Chronic inflammation of the perirenal connective tissue, tracking up from the infusion site (femoral vein/lower abdominal vena cava).

the 'least' changes reported were minimal compacted thrombi, forming a thin collar around the portion of the cannula sited in the vessel (cannula tip) (Figure 20.3). These are the minimal baseline changes seen in this system, in effect the 'cleanest' rat infusion model achievable in our experience. The picture is different when rat cannulation is via the jugular vein, with the cannula tip located 20 mm into the vein just above the heart; this technique may result, again in our experience, in a zone of oedema and necrosis in the vessel wall near the end of the cannula.

The published literature relating to continuous infusion pathology in different species is relatively sparse.

20.8.1 Rat

Morton *et al.* (1997) describe pulmonary histological effects associated with intravenous infusions of large volumes of isotonic saline solutions in rats for 30 days. Although large volumes were administered (up to 80 ml/kg at a rate of 0.25–1.0 ml/min), daily infusions were only for as long as it took to administer the required volume (i.e. these infusions were not continuous for the 30-day duration of the study). Nonetheless, this regimen did produce pulmonary pathology that for some findings increased in severity with increasing infusion volumes. The changes reported included periarterial eosinophil infiltration, interstitial inflammation, foreign body granulomas, alveolar histiocytosis and arterial changes (endothelial hypertrophy/hyperplasia, medial thickening, thrombosis). Inflammatory lesions at the injection site in the tail were the only other pathology reported and considered to be a sequel to catheter insertion.

Francis *et al.* (1992) reported the results of a 6-month continuous intravenous infusion study in Fischer 344 rats. Sterile saline was infused at approximately 0.8 ml/h for up to 6 months. According to the authors, the only pathology of importance was that at the injection site (acute haemorrhage, chronic inflammation and focal thrombosis). However, there were five decedent deaths in the study. All five had acute renal tubular necrosis, which Francis and his co-workers suggested had an ischaemic origin attributable to compromised renal vascular perfusion as a consequence of barbiturate anaesthesia (used periodically in the assessment of cannula patency). Other histopathology changes considered as minor included kidney mineralisation, inflammation of the liver and multifocal degeneration of the heart. Of the surviving 35 terminal animals, approximately 50% showed signs of septicaemia as gauged by the extent of interstitial inflammation in the lung. Surgical intervention was blamed for this surprisingly high incidence of systemic infection.

Acute renal tubular necrosis has also been recorded in a 28-day rat infusion study (S. Bjurström, personal communication). Although the cause of this lesion (Figure 20.4) is not clear, patency check procedures involving barbiturate anaesthesia were not implicated since these played no part in the experimental procedure.

The impact of stress on rodents subjected to continuous intravenous infusion has been studied, the stress factors being the restraint procedures or the plastic skin button used to secure the infusion cannula, rather than solution infusion *per se* (Birkhahn *et al.* 1976; 1979). These studies show that chronically infused rats exhibit stress as measured by reduced body weight gains. Histopathological markers of stress were not reported (from the papers it seems probable that they were not looked for), but

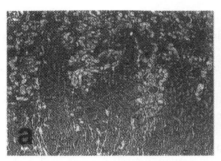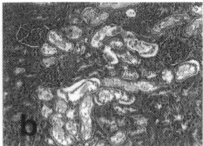

Figure 20.4 Continuous intravenous infusion in the rat: renal cortical necrosis of proximal tubules. Sections courtesy of Dr Sivert Bjurström, AstraZeneca R&D, Södertälje.

our own studies over many years have not revealed changes (e.g. gastric erosions and ulceration, thymic involution) that can be solely attributed to procedural stress.

20.8.2 Mouse

Published information on other rodents is limited. There is one report describing continuous long-term intravenous infusion in unrestrained mice (Lemmel and Good 1971). In this paper, mice were infused with unspecified vehicle at 0.2 ml/h for up to 35 days. These authors claim a success/survival rate greater than 95% (from a total of 200 mice), the only pathology reported being rare instances of otitis or cerebral thrombosis.

20.8.3 Dog

Thrombophlebitis, endocarditis and septicaemia have been reported in dogs with indwelling catheters (Cotton and Theran 1972; Wilkins 1973) and are consistent with the reports in man in which catheterisation has been used for therapeutic purposes. Burrows (1973) reported details of different routes of fluid administration together with some of the complications that may arise. Silicone-coated catheters are reported to reduce the amount of thrombus formation and fibrosis (Welch *et al.* 1974) and cause less endothelial damage in the dog (Mills and Simmons 1967).

20.8.4 Primates

Chronic indwelling venous catheters have been used in laboratory primates for toxicological investigations and, like the other species discussed, these animals have had their fair share of complications associated with the infusion procedure. DaRif and Rush (1983) reported problems of septicaemia in 20 rhesus monkeys (*Macaca mulatta*), 10 from each of two studies from two separate centres. The main pathogens isolated from blood cultures were *Klebsiella oxytoca* and *Staphylococcus aureus* and, although antibiotic therapy helped control the septicaemia, the infection was not completely eliminated unless the infusion catheter was removed. No histopathology was recorded in this study (DaRif and Rush 1983).

Baboons (*Papio cynocephalus*) are also used in studies employing long-term intravascular catheterisation, and two separate centres have described renal pathology associated with the infusion procedure. Heidel *et al.* (1981) detected renal disease at necropsy in 27 of 60 catheterised baboons, consisting of infarction, septic embolic nephritis and mesangioproliferative glomerulonephritis (the last associated with severe proteinaemia, hypoalbuminaemia and generalised oedema). A third of animals with glomerulonephritis developed uraemia and died of renal failure. Bacteria were isolated from blood cultures in a proportion of the baboons tested, the pathogens isolated (in decreasing order of frequency) being *Herellea, Streptococcus, Klebsiella, Staphylococcus* and *Providencia* spp. Heidel *et al.* (1981) associated the glomerulonephritis seen in some animals with deposition of immunoglobulins and complement (IgG, IgM, IgA, C3 and C4) in the kidneys. The amount of IgG found in glomeruli correlated with the severity of the glomerulonephritis; bacterial antigens were also present in glomeruli in a proportion of the animals examined. The authors suggested that the glomerulonephritis was immunologically mediated and bore some morphological resemblance to 'shunt nephritis' in humans (with infected indwelling shunts and catheters). Leary *et al.* (1981) further developed the idea of immune complex glomerulonephritis in baboons with long-term indwelling intravenous catheters infected with *Staphylococcus aureus*. Symptoms included proteinuria, hypoalbuminaemia, reversed albumin to globulin ratios and elevated serum IgM, IgG, rheumatoid factor and liver enzymes. Major histopathology findings included glomerulonephritis, chronic active hepatitis, arthritis and chronic sialoadenitis. In addition, immunofluorescence revealed IgG, IgM, B1c and C4 staining of glomerular capillaries and electron microscopy confirmed the presence of dense deposits in the renal glomeruli. Leary *et al.* (1981) concluded that sepsis was a major factor in the renal pathology seen in the baboon.

These two laboratories, in separate studies, describe renal pathology in the baboon linked to septicaemia resulting from infected indwelling intravascular catheters and they postulate an immune-mediated mechanism. Apart from the brief reference above to shunt nephritis in humans (Heidel *et al.* 1981), there is little evidence in the literature to suggest a similar mechanism in other species. This is almost certainly because the appropriate investigations have not been carried out, at least not in routine toxicology infusions studies.

Other pathological findings related to the continuous infusion of primates are to be found elsewhere in this volume (Chapter 15).

20.9 Conclusions

Chronic infusion by way of an indwelling venous catheter is a practical and effective way of administering test compounds. Factors affecting the pathology are choice of species, infusion rate, vehicle, surgical technique and scrupulous attention to aseptic technique throughout the course of the study. If surgical preparation is of high order and infection is avoided, then the procedure itself should result in limited pathology confined to the site of entry of the cannula into the skin and to the tip of the cannula in the blood vessel.

References

Bayly, W.M. and Vale, B.H. (1982) Intravenous catheterization and associated problems in the horse. *Compendium on Continuing Education for Practising Veterinarians*, 4, S227-S237.

Birkhahn, R.H., Bellinger, L.L., Bernardis, L. and Border, J.R. (1976) The stress response in the rat from harnessing for chronic intravenous infusion. *Journal of Surgical Research*, 21, 185–190.

Birkhahn, R.H., Long, C.L., Fitkin, D. and Blakemore, W.S. (1979) Stress induced by light weight back button used to prepare the rat for continuous intravenous infusion. *Journal of Parenteral and Enteral Nutrition*, 3, 421–423.

Burrows, C.F. (1973) Techniques and complications of intravenous and intraarterial catheterization in dogs and cats. *Journal of the American Veterinary Medical Association*, 163,1357–1363.

Cotton, R.C. and Theran, P. (1972) Intensive care. *Veterinary Clinics of North America*, 2, 419–432.

DaRif, C.A. and Rush, H.G. (1983) Management of septicemia in Rhesus monkeys with chronic indwelling venous catheters. *Laboratory Animal Science*, 33, 90–94.

Fonkalsrud, E.W., Pederson, B.M., Murphy, J. and Beckerman, J.H. (1968) Reduction of infusion thrombophlebitis with buffered glucose solutions. *Surgery*, 63, 280–284.

Francis, P.C., Hawkins, B.L., Houchins, J.O., Cross, P.A., Cochran, J.A., Russell, E.L., Johnson, W.D. and Vodicnik, M.J. (1992) Continuous intravenous infusion in Fischer 344 rats for six months: a feasibility study. *Toxicology Methods*, 2, 1–13.

Heidel, J.R., Giddens, W.E. and Boyce, J.T. (1981) Renal pathology of catheterized baboons (*Papio cynocephalus*). *Veterinary Pathology*, 18, 59–69.

Hoshal, V.L. (1972) Intravenous catheters and infection. *Surgical Clinics of North America*, 52, 1407–1417.

Hoshal, V.L., Ause, R.G. and Hoskins, P.A. (1971) Fibrin sleeve formation on indwelling subclavian central venous catheters. *Archives of Surgery*, 102, 253–258.

Leary, S.L., Sheffield, W.D. and Strandberg, J.D. (1981) Immune complex glomerulonephritis in baboons (*Papio cynocephalus*) with indwelling intravascular catheters. *Laboratory Animal Science*, 31, 416–420.

Lemmel, E.M. and Good, R.A. (1971) Continuous long-term intravenous infusion in unrestrained mice – method. *Journal of Laboratory and Clinical Medicine*, 77, 1011–1014.

Maki, D.G., Golman, D.A. and Rhame, F.S. (1973) Infection control in intravenous therapy. *Annals of Internal Medicine* 79, 867–887.

Meyers, L. (1945) Intravenous catheterisation. *American Journal of Nursing*, 45, 930–931.

Mills, L.J. and Simmons, D.H. (1967) An arterial catheter for chronic implantation in dogs. *Journal of Applied Physiology*, 23, 285–286.

Morton, D.M., Safron, J.A., Glosson, J., Rice, D.W., Wilson, D.M. and White, R.D. (1997) Histologic lesions associated with intravenous infusions of large volumes of isotonic saline solution in rats for 30 days. *Toxicologic Pathology*, 25, 390–394.

Neuhoff, H. and Seley, G.P. (1947) Acute suppurative phlebitis complicated by septicaemia. *Surgery*, 21, 831–842.

Schneider, P. and Pappritz, G. (1976) Hairs causing pulmonary emboli. *Veterinary Pathology*, 13, 394–400.

Stein, M. and Gray, R. (1995) Corpus cavernosum as an emergency vascular access in dogs. *Academic Radiology*, 2, 1073–1077.

Welch, G.W., McKeel, D.W., Silverstein, P. and Walker, H.L. (1974) The role of catheter composition in the development of thrombophlebitis. *Surgery, Gynecology and Obstetrics*, 138, 421–424.

Wilkins, R.J. (1973) *Serratia marcescens* septicaemia in the dog. *Journal of Small Animal Practice*, 14, 205–215.

Selected additional reading

Chang, S. and Silvis, S. E. (1974) Fatty liver produced by hyperalimentation of rats. *American Journal of Gastroenterology*, 62, 410–418.

Darbyshire, P.J., Weightman, N.C. and Speller, D.C.E. (1985) Problems associated with indwelling central venous catheters. *Archives of Disease in Childhood*, 60, 129–134.

Derrick, J. R. (1963) Effect of prolonged use of indwelling aortic catheters. *Surgery*, 54, 343–346.

Fuchs, P.C. (1971) Indwelling intravenous polyethylene catheters. Factors influencing the risk of microbial colonization and sepsis. *Journal of the American Medical Association*, 216, 1447–1450.

Hysell, D.K. and Abrams, G.D. (1967) Complications in the use of indwelling vascular catheters in laboratory animals. *Laboratory Animal Care*, 17, 273–280.

Kuwahara, T., Asanami, S., Tamura, T. and Kaneda, S. (1998) Effects of pH and osmolality on phlebitic potential of infusion solutions for peripheral parenteral nutrition. *Journal of Toxicological Sciences*, 23, 77–85.

Kuwahara, T., Asanami, S., Tamura, T. and Kubo, S. (1998) Dilution is effective in reducing infusion phlebitis in peripheral parenteral nutrition: an experimental study in rabbits. *Nutrition*, 14, 186–190.

Manenti, A., Botticelli, A., Buttazzi, A. and Gibertin, G. (1992) Acute pulmonary edema after overinfusion of crystalloids versus plasma: histological observations in the rat. *Patholologica*, 84, 331–334.

Nomura, G. and Yanagita, T. (1983) Assessment of intravenous toxicity as a function of infusion speed. *Developments in Toxicology and Environmental Science*, 11, 551–554.

Popp, M.B. and Brennan, M.F. (1981) Long-term vascular access in the rat: importance of asepsis. *American Journal of Physiology*, 241, H606-H612.

Steiger, E., Daly, J.M., Alien, T.R., Dudrick, S.J. and Vars, H.M. (1973) Postoperative intravenous nutrition: effects on body weight, protein regeneration, wound healing, and liver morphology. *Surgery*, 73, 686–691.

21 The use of mini-osmotic pumps in continuous infusion studies

L. Perkins, C. Peer and P. Murphy-Hackley

21.1 Introduction

When evaluating a compound for clinical development, it is essential to obtain predictive information on efficacy and toxicity. Although bolus injections can generate usable data for some drugs, continuous infusion is critical for others. After injection, plasma levels rise markedly, only to fall, more or less, very quickly. This allows the possibility that drug levels will become negligible in the intervals between injections. Fluctuating plasma levels, which depend on the half-life of the agent relative to the length of the intervals between injections, can suggest intermittent efficacy, and high peaks in concentration may trigger unwanted side-effects. Continuous drug exposure can obviate these problems. Additional key information on pharmacodynamics and pharmacokinetics can be obtained by comparing differences between the infusion and injection of a drug. Also, data from continuous infusion experiments can be used to hone drug selectivity, and may be helpful when choosing the mode of administration and dosing frequency in a clinical setting.

Compounds with short half-lives *in vivo* are notoriously difficult to evaluate, in that drug presence may not be maintained long enough to assess efficacy or safety. Differences in half-life between species can also create discrepancies between data generated from animals and humans. Small animals frequently metabolise and clear drugs very quickly, as their metabolism is accelerated relative to humans. If a drug with likely clinical value has a short half-life in the chosen animal model, and it is administered by injection, it may produce disappointing results. Continued drug presence in an animal model can help to diminish these complications by allowing the generation of key information regarding the full pharmacodynamic profile of a drug. Altering the route of delivery or schedule of administration of a drug can widen its therapeutic index, painting a more definitive picture of clinical efficacy.

Osmotic pumps have been used to deliver agents continuously, as cited in over 6000 published research papers (references from which can be found in the bibliography section of www.alzet.com). These references include articles on many candidates for clinical development, in addition to citations on proteins and peptides such as growth factors, cytokines and hormones.

The pumps are powered by the osmotic difference between the pump and the body fluid of an animal, and thus require no external power source. The ability to implant these pumps under the skin with or without a catheter connection can minimise the chance of animal interference and infection, and allows unrestrained movement. The stress that an animal undergoes when subjected to repeated injections or connected to an external infusion pump can be avoided when osmotic pumps are used.

ALZET pumps can be implanted subcutaneously or intraperitoneally, or used with a catheter to infuse a vein, artery or other target tissue such as the brain. The ability to target flow from the osmotic pump has allowed researchers to localise drug delivery, avoiding potentially high levels of systemic toxicity or untoward side-effects.

In the case of many compounds, pharmacokinetics are very similar for agents administered continuously via intravenous or subcutaneous infusion. Subcutaneous pump implantation is quicker and less stressful to the animal than intravenous catheterisation, and obviates potential problems with preservation of catheter and vein patency. Additionally, osmotic pumps provide long-term infusion for up to 28 days (or longer if implanted serially), which can contribute to a more complete pharmacokinetic profile than data obtained from a short infusion period.

This chapter reviews many toxicology and pre-clinical experiments. After summaries of the principle of operation and range of products available, techniques for optimising the drug administration regimen are discussed. This section is followed by surgical and animal care techniques related to pump implantation. Applications follow, which are related principally to intravenous infusion studies for toxicological and pharmacological purposes, but also briefly address the subcutaneous route.

21.2 Principle of mini-osmotic pump operation

ALZET osmotic pumps consist of three concentric cylinders. The outermost layer is a semipermeable membrane constructed of cellulose materials. Within this cylinder is a supersaturated salt solution, which surrounds the innermost impermeable drug reservoir (Figure 21.1). The pump is filled by introducing a 25- or 27-gauge filling tube connected to a drug-filled syringe into this reservoir, and the drug is delivered from the same port.

When implanted, interstitial fluid enters the pump via the semipermeable membrane because of the osmotic difference between this fluid and the salt solution in the pump. This fluid causes expansion of the salt layer, which compresses the flexible drug reservoir and forces solution out of the delivery portal. As the outer membrane is rigid and cannot expand, the rate at which fluid enters the pump is the same as the rate at which the solution is delivered from the pump, which provides constant and predictable delivery. The drug solution used to fill the pump does not itself need to pass through any membrane, which allows the delivery of compounds of any molecular weight, including proteins and peptides.

21.3 Range of products available

ALZET pumps have been designed in a range of sizes to account for differences in agent solubility, potency and desired experimental duration. Ten models of pumps are available currently. These pumps provide continuous delivery for 1–28 days at infusion rates of 0.25–10 µl/h. The model of pump chosen depends on the size of the research animal, the solubility of the agent, the desired experimental duration and the route of administration. Lower flow rates may be desired for intracerebral or solid tissue microperfusion, whereas higher flow rates may be advantageous when infusing agents intra-arterially. For longer-duration studies, pumps can be implanted serially. Table 21.1 lists the recommended animal size for each pump model. Figure 21.2 shows the range of pumps available.

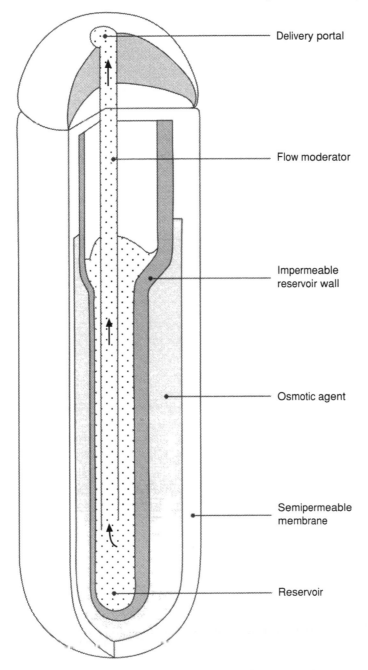

Figure 21.1 Cross-section of an ALZET® osmotic pump showing its design, components and mechanism of operation. (© DURECT Corporation).

DURECT also offers two brain infusion kits that can be used to infuse agents intracerebrally via the ventricles or to a specific site in the brain. These kits consist of 28-gauge stainless-steel cannulae, which can penetrate 3–5 mm below the surface of the skull, and polyvinyl tubing to connect the cannula to the osmotic pump. These

Table 21.1 Estimated minimum animal size for implantation of ALZET pumps

Model	Mice		Rats	
	Subcutaneous	*Intraperitoneal*	*Subcutaneous*	*Intraperitoneal*
1003D 1007D 1002	10 g	20 g	10 g	20 g
2001D 2001 2002 2004	20 g	Not recommended	20 g	150 g
2ML1 2ML2 2ML4	Not applicable	Not applicable	150 g	300 g

Note
The minimum animal size estimates are based on experience with male Sprague–Dawley rats and Swiss Webster mice. When using the pumps with other types or genders of rats and mice, or with animals other than rats and mice, these guidelines should be modified accordingly.

kits enable the researcher to keep all infusion instrumentation under the skin, minimising the chance for infection or animal interference.

21.4 Surgical implantation techniques

ALZET pumps can be implanted subcutaneously or intraperitoneally, and can easily be used with a catheter. Further information on implanting these pumps can be attained from ALZETTechnical Services (alzet@durect.com). A free videotape is available which shows how to implant the pumps for subcutaneous, intravenous, intraperitoneal, intragastric and cerebral infusion.

This section lists a variety of considerations to weigh when using osmotic pumps and contains information on the selection of the appropriately sized pump, choice of implantation site and/or route of delivery, anaesthetic, choice of surgical instruments and equipment, procedural recommendations for external jugular vessel cannulation and post-operative care.

21.4.1 Selection of ALZET pump model (ALZA 1997)

The smallest size pump allowable by agent solubility and experimental duration should be used and should be appropriate for the desired delivery rate and duration. Pumps can be implanted serially if longer delivery is required. Fill and prime the pump maintaining aseptic conditions, and filter sterilise the solution as needed. A sterile, particulate-free solution can obviate microbial contamination and resultant compound deterioration. The stability of the solution at 37 °C for the duration of the experiment should be evaluated. The functionality of the pump can be evaluated by weighing the pump before and after filling, and aspirating the residual volume at the end of the experiment. If an assay is available, serial plasma levels are the best technique for assessing constant drug delivery.

To deliver compounds that are incompatible with the pump reservoir (i.e. agents

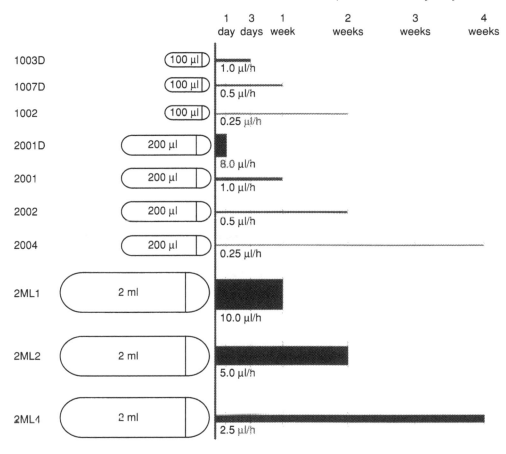

Figure 21.2 Rates and durations of ALZET® osmotic pumps. (© DURECT Corporation).

soluble only in oil or > 50% DMSO), one can utilise catheter tubing. This involves filling the appropriate length of polyethylene or other tubing with the compound to be delivered, and then filling the pump with saline such that the pump displaces the compound out of the reservoir tubing. When working with catheter tubing, the stability and compatibility of the solution to the catheter material must also be considered. Additionally, tubing may contain plasticisers, which may leach into the drug solution.

21.4.2 *Implantation/route of delivery considerations (ALZA 1997)*

Intravenous (i.v.) delivery requires a catheter tubing attachment (such as polyethylene or other tubing) to a SQ-implanted pump. The external jugular vessel is usually chosen for cannulation. Technical skill is required for this type of surgery, but it can be done with the correct instruments, training and practice. SQ-implanted pumps (with or without cannula attachments) can be replaced with newly primed pumps, preferably in a slightly different location (such as a new pocket on the other side of the midline) to extend the duration of infusion. If replacing a pump that is attached to a catheter, snip a small length of tubing off the existing catheter and dispose of both the old pump and its flow moderator.

21.4.3 Anaesthetic considerations (ALZA 1997; ILAR 1996)

Volatile inhalation anaesthetics are best for most indications in most species as induction and recovery times are shorter and the surgical plane can be maintained for a short or long duration. Injectable anaesthetics are an option in some instances.

Anaesthetic and analgesic protocols for different species and preparation of the animals prior to surgery are covered in detail elsewhere in this book (Chapters 1, 2, 6, 9, 10, 11, 14, 16 and 17).

21.4.4 Instrument and equipment considerations (Stepkowski et al. 1994; Tu et al. 1995; ILAR 1996; D. Brammer 1999, personal communication)

Although the pumps are manufactured and packaged sterilely, if they are not filled, primed, handled and implanted with aseptic technique, they can act as a nidus for infection. Additionally, reservoir contamination can lead to the microbial contamination of the agent used to fill the pump. Anaesthesia delivery equipment and surgical equipment suitable for different species are covered in earlier chapters of this book.

21.4.5 Jugular vessel cannulation procedures (Popesko et al. 1992; Stepkowski et al. 1994; Tu et al. 1995; Alza 1997; D. Brammer 1999, personal communication)

Position the animal in dorsal recumbency and secure its head and the anaesthetic delivery apparatus in place. In rats, it may be helpful to place a piece of rolled gauze beneath the dorsal surface of the neck to support the neck and aid in extending it to expose the jugular veins. Place traction on the upper incisors with a soft band or smooth bar to immobilise the head and neck region without inhibiting respiration or interfering with the surgical site and incorporate a support bar for the upper incisors into the nose cone scavenging device (D. Brammer 1999, personal communication).

The skin incision should be made from the ramus of one side of the jaw to the tip of the sternum just lateral to the trachea/midline (Popesko *et al.* 1992). Use a small, sharp scalpel blade in a single cut to facilitate closure and healing, as scissors generally do not make straight incisions and can crush tissues at wound edges, delaying healing. Gently dissect down through the salivary and lymphoid glands, adipose tissue and fascia to the external jugular vein, which is superficial to most of the neck musculature. Take care not to stretch or traumatise the vessel as this may cause vasoconstriction (Popesko *et al.* 1992). Electrocautery can be used if performed carefully. Delicate hand-held units can be used on small branches off the vessel (D. Brammer 1999, personal communication). Identify where the external jugular splits into two branches at the level of the mandibular and parotid salivary glands (maxillary and lingofacial branches). Place and tie off one branch with a ligature of fine (4–0 or 5–0) silk, and leave tails 4–5 inches long to use as handles to place gentle traction on the vessel while dissecting and inserting the cannula (Popesko *et al.* 1992; D. Brammer 1999, personal communication). Place two more loose ligatures, one at the rostral end and one at the proximal end of the exposed external jugular vein.

Apply a few drops of lidocaine or other vasodilatory substance (at body temperature), and allow time for vasodilatation. When small vessels appear in the field, the lidocaine has had time to act, and will inhibit, but not eliminate, vasoconstriction as the vessel is punctured and the cannula is passed (D. Brammer 1999, personal communication). Use a fine-gauge needle (25–23 gauge for mice and 22–20 gauge for rats) bent at an approximate 90° angle to pierce the vessel. Alternatively, a small ellipsoidal piece can be cut from the ventral aspect of the vessel with very fine iris scissors. Do not cut so much tissue as to weaken the vessel such that it breaks when traction is applied via the rostral ligature ends while passing the cannula. (ALZA 1997; D. Brammer 1999, personal communication). Once the vessel has been pierced, control haemorrhage with gentle traction on the rostral ligature ends (ALZA 1997; D. Brammer 1999, personal communication). The cannula can then be passed directly or with the aid of a plastic catheter introducer or fine vessel dilator. Pass the cannula far enough into the vessel to ensure stability and patency but not so far as to enter the thorax or interfere with cardiac function. It may be helpful to do preliminary non-survival practice surgery to confirm cannula placement in the rodent model that is used (ALZA 1997; D. Brammer 1999, personal communication).

If it is acceptable to delay the initial compound delivery, the cannula can be attached to a 1-cc syringe filled with physiological saline via a blunt needle adapter. Successful cannulation can then be confirmed, and any microclots can be dislodged by flushing a small amount of saline after the cannula is placed and before all ligatures are secured (D. Brammer 1999, personal communication). After successful insertion of the cannula, secure the proximal and distal ligatures around the vessel and cannula. Trim the ends close to the knot so as not to interfere with closure or to delay healing, but not so short as to risk the integrity of the knot (ALZA 1997; D. Brammer 1999, personal communication). Reattach the cannula to the pump and use fine blunt haemostats to make a subcutaneous tunnel from the neck incision to the dorsal scapular region, making sure to create a sufficient pocket for pump positioning (1 cm of space around each side of the pump is recommended). Progressive insertion of closed haemostats, followed by the opening of the haemostats, and their withdrawal while still open, results in blunt dissection through the fascial planes of least resistance (ALZA 1997; D. Brammer 1999, personal communication).

The caudal end of the pump is then passed through this tunnel into the pocket. For jugular cannulation in mice, one may use a dab of sterile silicone glue to splice the polyethylene PE-50 or 60 size tubing that attaches the pump into the smaller PE-10 tubing used to cannulate the vessel. This small bleb of silicon can also be used as an anchor by tying one ligature on the pump side to the fascia or muscle, thus inhibiting traction on the cannulated portion as the animal heals and moves about. Take care not to tighten ligatures around the tubing so that they cause strictures and inhibit flow (D. Brammer 1999, personal communication). Use a two-layer closure, with one layer of suture of the underlying fascial tissues and one of the skin. The deep layer should be closed with 4–0 or 5–0 absorbable material in a simple continuous or interrupted stitch, but silk is acceptable for short-term survival studies of 2–4 weeks. The skin can be closed with the same material, non-absorbable suture, or stainless-steel wound clips. In mice, sutures are recommended for comfort. Wound clips or ligatures in the skin should be removed in 1–2 weeks if the animals are to survive longer than 2–4 weeks (ALZA 1997; D. Brammer 1999, personal communication).

21.4.6 Post-operative considerations (ALZA 1997; ILAR 1996)

Anaesthetised animals should be placed on a clean, smooth, and absorbent surface such as a paper towel or other disposable material. Conventional rodent bedding can obstruct airways and adhere to moist incisions. Reapply ophthalmic lubricant during recovery. Keep surgical patients warm, and observe them until they are fully ambulatory and exhibiting near-normal exploratory or grooming behaviours (ILAR 1996). After normal behaviour is observed, return the animals to standard housing or as appropriate for the research protocol. Add a few moistened blocks of rodent chow on the bottom of the cage to facilitate feeding during the immediate post-operative period. Animals should be observed by qualified individuals who are familiar with normal rodent behaviour at least once a day for at least 3–7 days following any surgical procedure. Additional post-operative analgesics are not usually required for SQ implantations and peripheral vein cannulations, but, if animals appear dehydrated, guarded, aggressive or have lost considerable weight, additional analgesics or other veterinary intervention is indicated (ILAR 1996). It should be noted that animal stress from osmotic pump implantation is rare. For more information, please consult the bibliography section of www.alzet.com.

21.4.7 Summary of surgical techniques for intravenous infusion

As with any technique, osmotic pump implantation and vascular cannulations should be practised initially on non- or short-term survival animals designated for training, under the direction of an individual experienced with the specific technique in the given species or strain indicated on an approved protocol. Depending on the level of previous experience and the manual dexterity of the personnel, the number of practice sessions required to establish proficiency will vary. Techniques related to anaesthesia, aseptic protocols, pump, instrument or animal preparation can lead to the success or failure of the technique in a given situation.

21.5 Optimising a drug delivery regimen

Compounds whose efficacy depends upon their administration schedule demonstrate a shift of the dose–response curve to the right or left according to the time pattern of drug administration. This means that different effects can be achieved with the same dose of drug by altering the administration pattern. For some compounds, administration by daily injections produces the desired biological effect. However, this regimen can produce widely fluctuating plasma levels which are inappropriate for many other compounds, leading either to a lack of apparent efficacy or to side-effects. Lack of efficacy due to serum peaks and troughs may occur simply because the drug is not maintained long enough at a therapeutic concentration, or because a critical event during the treatment period occurs when the serum level of the experimental compound is outside its effective range.

ALZET pumps provide continuous infusion, a regimen that can be conceptualised as injections administered at the smallest possible intervals. Injections can be administered with intervals of various lengths based upon the experimental design (Huber *et al.* 1993; Figures 21.3a and b). Injections typically produce a high plasma level which rapidly declines, as can be seen in Figure 21.3c. The fluctuation in serum

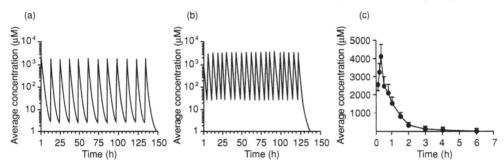

Figure 21.3 (a and b) The pharmacokinetic profile of 5-fluorocytosine administered to nude mice was determined following a single i.p. injection of 500 mg/kg to nude mice. These data were used to simulate plasma concentrations that might result from injections of 5-fluorocytosine given twice daily (b.i.d.) or four times daily (q.i.d.) to nude mice. (c) Plasma levels following a single i.p. injection of 500 mg/kg 5-fluorocytosine to nude mice. (Reproduced with permission from Huber *et al.* 1993.)

levels that occurs during an injection regimen depends upon the half-life of the agent relative to the length of the intervals between injections. Figure 21.4 shows a simulation of serum peaks and troughs following injections at regular intervals. In this example, the peaks greatly exceed the therapeutic range, which is represented by the indicated horizontal band. In addition, as the half-life of the compound is much shorter than the injection interval, the serum level falls to insignificant levels between injections (Horton *et al.* 1989). One approach is to assume that, if the interval between injections exceeds four times the half-life of the drug, plasma concentrations will fall rapidly to zero between doses (Fara and Urquhart 1984). This suggests that compounds with short half-lives may not achieve sufficient concentrations unless administered by continuous infusion.

Slate *et al.* (1993) compared the effects of injection versus infusion when studying the chemotherapeutic combination of doxorubicin and verapamil. ALZET pumps were used 'in order to mimic prolonged i.v. co-administration in the clinic, and to compare the efficacy and toxicity of this delivery scheme with more traditional bolus IP regimens'. These researchers suggested that 'continuous delivery may produce a greater cell kill for doxorubicin and verapamil by eliminating variants that are resistant to repeated dosing' (Slate *et al.* 1993). Prior work on the regimen dependence of verapamil had demonstrated that, although ALZET pump infusion produced steady-state plasma levels, intraperitoneal injection achieved a similar level for less than 1 h, and was limited by lethal side-effects to half of the daily infusion dose (Horton *et al.* 1989).

The ideal administration regimen maintains the drug at a therapeutic concentration for the time period required to elicit the desired effect, which can be challenging for proteins and peptides with short half-lives *in vivo*. In addition, the ideal regimen maximises the drug's therapeutic index, which is the ratio of therapeutic to adverse effects. However, in a pre-clinical situation, the precise half-life and optimal therapeutic concentrations may not be well understood. A useful approach compares the dose–response relationship when the same total dose of drug is administered over several days via once- or twice-daily injections versus multiday, constant-rate infusion. As

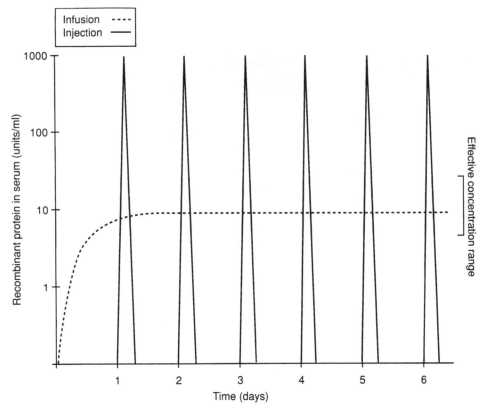

Figure 21.4 Injections of short half-life proteins can result in great variations in protein concentration in serum and tissues. Immediately after injection, serum concentrations commonly exceed effective levels. Rapid elimination of the injected protein results in long periods between injections when the protein is absent from serum and tissues, hence is unable to exert an effect. (© DURECT Corporation).

Urquhart (1985) noted in his review of pharmacology for drug development and pre-clinical research, 'this injection–infusion comparison (IIC) protocol may well be the logical starting point in the systematic pharmacodynamic study of many classes of drug, and a step towards basing dosage form and regimen design on pharmacodynamic data'.

The method of drug delivery must be considered when predicting a drug's effects in humans based upon results obtained from an animal model. Many drugs are metabolised and excreted in small animals more quickly than in humans, as seen in Figure 21.5 (Nau 1985a; Prevo 1993). By way of estimation, the ratio of the half-lives of a compound in two different species is generally proportional to the one-fourth power of the ratio of the average weights of the two species (Fara and Urquhart 1984; Fara and Mitchell 1986). Using this rule, the plasma half-life of a drug in a 10-g mouse would be 1/10 the half-life in a 100-kg human.

A pioneering study of schedule dependence in chemotherapeutics was conducted by Sikic *et al.* (1978). This group compared the effectiveness of the injection versus the infusion of bleomycin, which has a half-life of less than 2 h in humans and mice. Tumour-bearing mice received one of three bleomycin regimens for 5 days: (1) injection

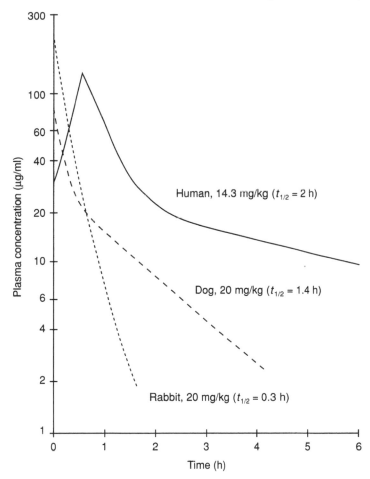

Figure 21.5 Interspecies differences in half-lives of a cephalosporin. (Reproduced with approval from Bristol-Myer-Squibb.)

every 12 h, (2) injections on Days 1 and 4 or (3) continuous subcutaneous infusion via ALZET pumps. The same total dose was used in each group. Lung toxicity was measured using lung hydroxyproline content as a marker for fibrosis. The injection groups had improved survival rates but significant pneumotoxicity. Continuous infusion significantly reduced tumour growth compared with the injection groups ($P < 0.05$; see Figure 21.6), and did not result in pulmonary fibrosis. Continuous infusion improved the therapeutic index, as the dose–response curve for bleomycin toxicity shifted to the right, while the dose–response curve for efficacy shifted left. Sikic *et al.* (1978) concluded that 'the decreased pulmonary toxicity associated with continuous infusion clearly indicates that the therapeutic effects of bleomycin may be selectively improved by dosage schedule'.

Many other studies show improved efficacy for compounds when administered by continuous infusion rather than injection. These include IGF-1 (Tomas *et al.* 1996), heparin (Edelman and Karnovsky 1994), growth hormone and IGF-I (Gargosky *et al.* 1994), interferon γ (Flynn *et al.* 1993), and morphine (Yoshimura *et al.* 1995).

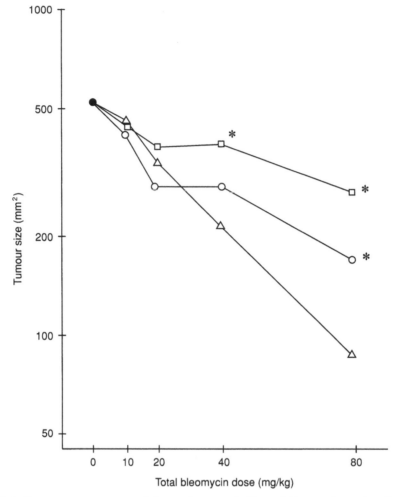

Figure 21.6 Dose–response curve of the anti-tumoral effect of bleomycin against Lewis lung carcinoma. Three schedules of administration were compared. Measurements were made on Day 15 after osmotic pump implantation, but are representative of differences which existed throughout the course of tumour growth. Open squares represent subcutaneous (s.c.) injections given twice weekly; open circles represent s.c. injections 10 times weekly; open triangles represent continuous s.c. infusion. *Significantly different from continuous infusion, $P < 0.05$. (Reproduced with permission from Sikic *et al.* 1978.)

In some instances, manipulation of the administration schedule elicits effects that cannot be demonstrated at all by injections. For example, Albiston *et al.* (1995) showed that injections of growth hormone had no effect on hepatic expression of the enzyme 11βHSD2, whereas infusion via osmotic mini-pumps significantly decreased both enzyme mRNA levels and enzyme activity compared with untreated, hypophys-ectomised controls. An administration regimen comprising injections only might have led these authors to conclude that growth hormone had no effect on 11βHSD2 enzyme expression in the liver. Instead, these results emphasise that the schedule of administration can be integral to demonstrating the efficacy of a compound.

Characterising the rate dependence of drug action is one means by which its therapeutic index can be optimised. For many compounds, continuous infusion should be considered integral to the pre-clinical development effort.

21.6 Preclinical pharmacology and toxicity applications: intravenous infusion

Continuous infusion can be invaluable in toxicology experiments when the effects of the compound are area-under-the-curve dependent. The use of intravenous infusion studies can provide useful data for those who wish to mimic the clinical dosing of drugs. The lack of data from continuous infusion studies has hampered the clinical testing of some drugs, as the relevant toxicity information that elucidated side-effects was not generated during standard toxicology testing.

Milacemide, or 2-*n*-pentylamineacetamide HCl, was originally used as an antiepileptic treatment. It was also found to deliver glycine to the brain, which interacted with N-methyl-D-aspartate receptors to cause anti-anoxic and memory-enhancing effects. When tested in patients suffering from severe depression and Alzheimer's disease, it caused an unexpected elevation in plasma transaminases, an indication of drug-induced liver toxicity. Rogiers and her group designed an experiment that combined *in vivo* and *in vitro* approaches to detect early alterations in key markers of metabolic and liver function (Rogiers *et al.* 1997).

Before clinical testing on milacemide started, 2 years of pre-clinical research had been conducted. In rats and cynomolgus monkeys, milacemide was given via gavage at multiples of the expected therapeutic dose by the route intended for man. There were no signs of liver toxicity in these experiments. Rogiers *et al.* conducted their experiments by infusing 250 and 500 mg/kg/day milacemide to rats via i.v. jugular infusion by osmotic pump for 7 days. When 500 mg/kg/day milacemide was infused, hepatocyte triglyceride levels increased 3.1-fold. Electron and light microscopy on both total liver and isolated hepatocytes showed a concentration-dependent accumulation of lipid droplets, numerous vacuoles in the cytoplasm and other structural abnormalities. These findings demonstrated the potential hepatotoxic properties of milacemide. The authors of this study deduced that the short half-life of this drug of 0.5–1 h in rats, combined with the gavage used in the preliminary toxicological assessment, maintained systemic plasma levels at the appropriate level for only a few hours per day. They also stated that 'the continuous intravenous infusion of 2-*n*-pentylamineacetamide HCl used here and the associated exposure levels reflect better the therapeutic situation in man and could therefore be a determining factor to explain the difference of effects recorded in regular toxicity studies' (Rogiers *et al.* 1997).

Rodriguez-Martin *et al.* (1993) assessed the unexpectedly toxic effects of anilide by using continuous intravenous delivery. This group predicted that total absorption would be attained if anilide were infused at a constant and predictable rate, and that possible toxic and cumulative effects would not be affected by luminal hydrolysis or intestinal absorption. In 1981, a toxic oil syndrome was epidemic in Spain, and was traced to rapeseed oil contaminated with aniline. The rapeseed oil had been treated with 2% aniline for industrial purposes, was reprocessed to remove the aniline and was then sold as cheap 'olive oil'. With refinery processing, large amounts of unsaturated fatty acid anilides formed, the amount of which correlated with the risk of illness. A host of symptoms were seen with this illness, including weight loss (Bell *et al.* 1992).

Most of the studies conducted on anilide toxicity in rodents prior to this time relied upon acute injection or chronic administration in the diet, neither of which produced side-effects in animals. Rodriguez-Martin *et al.* infused ^{14}C-labelled oleylanilide into the vena cava of rats by introducing catheters attached to ALZET pumps. The group of animals that received anilide rather than vehicle consumed more food and lost more weight than control animals, and their energy balance showed high inefficiency. These findings suggested that anilides could produce an emaciating effect in rats which was similar to that seen in humans afflicted with toxic oil syndrome (Rodriguez-Martin *et al.* 1993).

Continuous i.v. delivery was also used by Houghton *et al.* (1988) when this group wished to maintain tightly controlled plasma concentrations of gentamicin. These workers investigated whether a constant infusion of gentamicin for up to 6 months would cause chronic tubulointerstitial nephropathy. Patients who suffer from cystic fibrosis or osteomyelitis often require long-term treatment with gentamicin or other aminoglycosides, though the toxicity of long-term gentamicin treatment has limited its usage in other than extreme cases of infection. When clinicians prescribe this class of drug, it is given within very narrow therapeutic levels, and serum creatinine and creatinine clearance levels are monitored carefully to avoid nephrotoxicity (Houghton *et al.* 1988).

To evaluate acceptable treatment ranges in rats, Houghton and co-workers administered 20 mg/kg/day gentamicin via an i.v. jugular catheter connected to an ALZET pump. Pumps were changed every 28 days. The chosen dose was low enough to avoid acute tubular necrosis and renal failure. During treatment, serum creatinine and creatinine clearance overestimated glomerular filtration rate, and no differences were seen between treated animals and controls. However, in the month after the 6-month treatment regimen ended, tubular microcystic changes and active tubulointerstitial nephritis developed, with a continued fall in inulin clearance. This group concluded that careful maintenance of gentamicin serum levels did not preclude nephrotoxicity. By using continuous infusion, this group was able to answer key questions about clinical drug treatment modality and side-effects (Houghton *et al.* 1988).

Continuous i.v. infusion with ALZET pumps was also used to assess the effects of a gonadotrophin-releasing hormone (GnRH) antagonist. Previous compounds of this class have had clinical potential which has been limited by allergic side-effects. Gordon *et al.* (1994a) assessed the pre-clinical potential of A-75998, a fourth-generation GnRH antagonist. This agent was promising because of its enhanced water solubility, high affinity for the GnRH receptor and resistance to endopeptidase action (compared with other such antagonists). It was also thought to have less than one-tenth of the histamine stimulation seen with other such antagonists. This group tested the effects of A-75998 when given via injection and infusion to female cynomolgus monkeys. When given for 7 days via ALZET pumps connected to a catheter in the jugular vein, 0.025 and 0.05 mg/kg/day A-75998 was fully effective in suppressing serum oestradiol levels. The suppression of serum oestradiol levels to below 30 pg/ml indicated that A-75998 had clinical potential as a safe and potent GnRH antagonist (Gordon *et al.* 1994a).

Gordon and his group conducted additional studies on intact male cynomolgus monkeys. When this antagonist was administered via ALZET pumps attached to a jugular catheter, concentrations of 0.1 and 0.2 mg/kg/day A-75998 suppressed serum testosterone levels fully. In addition, no overt toxicity was seen in any monkey treated with A-75998, indicating that it merited clinical evaluation (Gordon *et al.* 1994b). Finally, Leal *et al.* (1994) tested A-75998 and a series of GnRH antagonists to assess their effects on testosterone suppression in male dogs. This group examined AUC compared with peak concentrations, and also regimen-dependent toxicity, and concluded that a lower dose of A-75998 was needed when this compound was infused subcutaneously (Leal *et al.* 1994).

Intravenous delivery using ALZET pumps allowed these researchers to see effects that were not apparent when the selected compounds were administered in a different fashion pre-clinically. In the case of milacemide, highly detrimental clinical effects were examined using long-term infusion studies that provided more information than standard long-term toxicity tests. Tightly controlled plasma levels of anilides and gentamicin yielded critical information about clinical exposure and side-effects. A series of studies using GnRH antagonist A-75998 provided information about its clinical potential.

21.7 Preclinical pharmacology and toxicity applications: alternative infusion routes

21.7.1 *Subcutaneous infusion*

Subcutaneously implanted osmotic pumps have been used to evaluate the clinical potential of many compounds, as the resultant kinetics are usually closely similar to those of intravenous infusion. Subcutaneous infusion studies have also been conducted to evaluate teratological effects, as resultant steady-state plasma levels can provide pertinent information on controlled agent exposure during select periods of organogenesis.

Plowman *et al.* (1987) evaluated the pharmacokinetics and anti-tumoral effects of deoxyspergualin. As this drug is rapidly cleared from plasma, optimal anti-tumoral activity was seen when plasma levels were maintained. There are many other examples of pre-clinical toxicity studies in which schedule dependence data from infusion studies proved valuable in altering the anti-tumoral activity or toxicity of anti-cancer agents (Collins *et al.* 1987).

Nau and his colleagues used the 'injection–infusion comparison' protocol (Fara and Urquhart 1984) to create a better embryotoxicity model for a number of suspected teratogens (Nau *et al.* 1985). Valproic acid (VPA) is rapidly cleared from plasma after injection, with a half-life in humans that is 10-fold higher than that in mice. Nau *et al.* (1981) used Alzet pumps to maintain plasma levels similar to human therapeutic levels and compared those with daily injections in pregnant mice. This group concluded that the doses or AUC values did not correlate with the teratogenic response of the different regimens of administration. Steady-state concentrations of the drug produced embryolethality and fetal weight retardation primarily, while intermittent injections produced a high incidence of exencephaly (Nau 1985b). In later experiments, this group saw that the effects of cyclophosphamide were related to maternal AUC values, and not to peak drug levels. Very low and sustained steady-state concentrations of cyclophosphamide were as teratogenic as more than 100 times higher peak concentrations given by injection (Reiners *et al.* 1987).

Nau *et al.* (1985) concluded 'we believe that the injection–infusion comparison protocol is a valuable technique to study if the peak drug levels (such as the case of VPA) or the AUC values (such as the case with cyclophosphamide) will predominantly determine the toxicity of a drug'. This group intuited that, if peak levels were of primary importance, toxicity could be related to absorption and distribution. If AUC values were the determining factor, toxicity could be related to drug clearance.

Drugs of abuse have also been studied in teratology models. Slotkin and his colleagues at Duke University (Slotkin 1998) chose continuous infusion over injection because steady-state plasma levels more akin to human exposure are attainable with ALZET pumps. The placenta provides fetal protection by metabolising a portion of drugs such as nicotine, and by introducing a phase delay between the maternal and fetal circulation. As a result, intermittent drug delivery produces less penetration into the fetal compartment than does continuous infusion. The protective role of the placenta is further compromised because steady-state maternal plasma drug levels cause equilibration of all fluid compartments to the same final concentration. Slotkin concluded that 'nicotine infusion paradigms... produce drug exposure without the confounds of other components of tobacco'.

As small animals eliminate chemicals much more rapidly than humans, Clarke (1993) and others have argued that constant infusion compensates for this difference more effectively than bolus delivery. The effect of low, steady-state plasma levels on toxicity during development, and the importance of total AUC exposure, may be examined and compared with the high plasma levels attained by bolus dosing (Clarke 1993; O'Flaherty *et al.* 1994).

21.7.2 *Cell proliferation studies*

Pre-clinical drug development requires consideration of cancer risk, especially for agents likely to be dosed chronically. The timing of this evaluation varies based on laboratory protocol and when toxicity is identified. For example, hepatotoxicity identified early in development might trigger earlier carcinogenicity screening. Cell proliferation is a key factor in both the initiation and promotion of clonal growth of cancerous cells. Measurement of induced cell proliferation should be a component of a coordinated carcinogenicity evaluation (Butterworth *et al.* 1992).

Cell proliferation can be assessed by administering a labelling agent, such as [³H]thymidine or bromodeoxyuridine (BRDU), during or after exposure to an experimental agent. Cells in the replicative DNA synthesis phase, or S-phase, incorporate one of these base analogues into their chromosomes. Label incorporation is demonstrated using autoradiographic or immunochemical methods. Traditionally, labelling agents were injected, which detects only cells in S-phase at that moment. Poorly timed label administration could cause critical cell proliferation to be overlooked. (Wilson and Eldridge 1989). Through extensive experimentation, researchers at the Chemical Industry Institute of Toxicology (CIIT) optimised and standardised cell labelling methods using continuous infusion (Wilson and Eldridge 1989; Eldridge *et al.* 1990; Goldsworthy *et al.* 1991; Weghorst *et al.* 1991; Butterworth *et al.* 1992).

The use of ALZET pumps to measure cell proliferation proved critical to the commercialisation of omeprazole (Prilosec® or Losec®), an inhibitor of gastric acid secretion, which went on to become the largest selling pharmaceutical in the history of the industry (*Med Ad News* 1997, 1998, 1999). Omeprazole development was suspended in the 1980s, when high-dose toxicology studies identified apparent gastric mucosal tumours, an effect not seen with Glaxo's competitive drug, ranitidine (Zantac®). Further investigation revealed that when ALZET pumps were used for continuous administration of ranitidine, which has a short half-life, equivalent acid suppression was achieved compared with omeprazole. Using ALZET pumps to help measure cell turnover, researchers then observed gastric hyperplasia after administration of both compounds. The hyperplasia was determined to be a physiological effect of hypergastrinaemia, rather than a unique toxic effect of omeprazole, and the drug was commercialised (SCRIP 1988).

21.8 Conclusion

Continuous infusion has proven helpful in scores of pre-clinical and toxicology experiments. Cell proliferation labelling using ALZET pumps filled with BRDU or [^{3}H]thymidine has provided toxicologists and cancer researchers with a new tool which improves results over the injection of the labelling agent. In teratology studies, pronounced teratogenic effects were investigated by infusing pregnant animals during select periods of organogenesis. ALZET pumps are helpful in obviating differences in species clearance, and in generating data on the specific pharmacokinetics of compounds. The importance of teratology research can never be overemphasised.

With the long time frames necessary in the drug development process, tools that enable pharmaceutical and biotechnology companies to direct their development efforts more effectively are invaluable. When conducting pharmacokinetic and pharmacodynamic experiments by using bolus injections, data may be confounded by the lack of consistency in plasma levels throughout the dosing period. Data about likely toxicity or hyperplasia, in addition to the best mode of clinical administration, are helpful early in the discovery process. Protocols such as the injection–infusion comparison allow researchers to amass data on the pharmacokinetics of a series of compounds. Adjusting the route of administration can provide additional data, elucidating effects that would not be apparent with systemic administration. Research on dozens of peptides and proteins would not have been possible without a continuous infusion tool that circumvented problems with assessing the efficacy of agents with short half-lives.

When conducting pre-clinical and toxicology research, it is helpful to evaluate all of the tools which will direct the research most successfully. Continuous infusion is a tool that can help create a specific pharmacokinetic profile of a drug of interest.

Acknowledgements

The authors would like to thank Dr John Urquhart, Marie Barry and Mary Prevo for their invaluable assistance with this chapter.

References

Albiston, A.L., Smith, R.E. and Krozowski, Z.S. (1995) Sex- and tissue-specific regulation of 11β-hydroxysteroid dehydrogenase mRNA. *Molecular and Cellular Endocrinology*, 109, 183–188.

ALZA Corporation (1997) Technical information manual, ALZET osmotic pumps. Palo Alto.

Bell, S.A., Hobbs, M.V. and Rubin, R.L. (1992) Isotype-restricted hyperimmunity in a murine model of the toxic oil syndrome. *Journal of Immunology*, 148(11), 3369–3376.

Butterworth, B.E., Popp, J.A., Conolly, R.B. and Goldsworthy, T.L. (1992) Chemically induced cell proliferation in carcinogenesis. In Vainio, H., Magee, P.N., McGregor, D.B. and McMichael, A.J. (eds.) *Mechanisms of Carcinogenesis in Risk Identification*. International Agency for Research on Cancer: Lyon, pp. 279–305.

Clarke, D.O. (1993) Pharmocokinetic studies in developmental toxicology: practical considerations and approaches. *Toxicology Methods*, 3(4), 223–251.

Collins, J.M., Leyland-Jones, B. and Grieshaber, C.K. (1987) Role of preclinical pharmacology in phase I clinical trials: considerations of schedule-dependence. In Muggia, F.M. (ed.) *Concepts, Clinical Developments, and Therapeutic Advances in Cancer Chemotherapy*. Martinus Nijhoff Publishers: Boston, pp. 129–140.

Edelman, E.R. and Karnovsky, M.J. (1994) Contrasting effects of the intermittent and continuous administration of heparin in experimental restenosis. *Circulation*, 89, 770–776.

Eldridge, S.R., Tilbury, L.F., Goldsworthy, T.L. and Butterworth, B.E. (1990) Measurement of chemically induced cell proliferation in rodent liver and kidney: a comparison of 5-bromo-2'-deoxyuridine and [^{3}H]thymidine administered by injection or osmotic pump. *Carcinogenesis*, 11(12), 2245–2251.

Fara, J. and Mitchell, C. (1986) Osmotic systems for rate-controlled drug delivery in preclinical and clinical research: therapeutic implications. In Struyker-Boudier, H.A.J. (ed.) *Rate-controlled Drug Administration and Action*. CRC Press: Boca Raton, FL, pp. 115–142

Fara, J. and Urquhart, J. (1984) The value of infusion and injection regimens in assessing efficacy and toxicity of drugs. *Trends in Pharmacological Science*, 55(1), 21–25.

Flynn, J.L., Chan, J., Triebold, K.J., Dalton, D.K., Stewart, T.A. and Bloom, B.R. (1993) An essential role for interferon γ in resistance to *mycobacterium tuberculosis* infection. *Journal of Experimental Medicine*, 178, 2249–2254.

Gargosky, S.E., Tapanainen, P. and Rosenfeld, R.G. (1994) Administration of growth hormone (GH), but not insulin-like growth factor-I (IGF-I), by continuous infusion can induce the formation of the 150-kilodalton IGF-binding protein-3 complex in GH-deficient rats. *Endocrinology*, 134(5), 2267–2276.

Goldsworthy, T.L., Morgan, K.T., Popp, J.A. and Butterworth, B.E. (1991) Guidelines for measuring chemically-induced cell proliferation in specific rodent target organs. *Progress in Clinical and Biological Research*, 369, 253–284.

Gordon, K., Williams, R.F., Greer, J., Bush, E.N., Haviv, F., Herrin, M. and Hodgen, G.D. (1994a) A-75998: a fourth generation GnRH antagonist: II preclinical studies in female primates. *Endocrine*, 2, 1141–1144.

Gordon, K., Williams, R.F., Greer, J., Bush, E.N., Haviv, F., Herrin, M. and Hodgen, G.D. (1994b) A-75998: a fourth generation GnRH antagonist: I preclinical studies in male primates. *Endocrine*, 2, 1133–1139.

Horton, J.K., Thimmaiah, K.N., Houghton, J.A., Horowitz, M.E. and Houghton, P.J. (1989) Modulation by verapamil of vincristine pharmacokinetics and toxicity in mice bearing human tumor xenografts. *Biochemical Pharmacology*, 38(11), 1727–1736.

Houghton, D.C., English, J. and Bennett, W.M. (1988) Chronic tubulointerstitial nephritis and renal insufficiency associated with long-term 'subtherapeutic' gentamicin. *Journal of Laboratory and Clinical Medicine*, 112, 694–703.

Huber, B.E., Austin, E.A., Good, S.S., Knick, V.C., Tibbels, S. and Richards, C.A. (1993) In vivo antitumor activity of 5-fluorocytosine on human colorectal carcinoma cells genetically modified to express cytosine deaminase. *Cancer Research*, 53, 4619–4626.

ILAR (Institute of Laboratory Animal Resources) (1996) Commission on Life Sciences, National Research Council. *Guide for the Care and Use of Laboratory Animals.* National Academy Press: Washington, DC.

Leal, J.A., Bush, E.N., Holst, M.R., Cybulski, V.A., Nguyen, A.T., Rhutasel, N.S., Diaz, G.J., Haviv, F., Fitzpatrick, T.D., Nichols, C.J., Swenson, R.E., Mort, N., Carlson, R.P., Dodge, P.W., Knittle, J. and Greer, J. (1994) A-75998 and other GnRH antagonists suppress testosterone in male beagle dogs. A comparison of single injection, multiple injections and infusion administration. *Endocrine*, 2, 921–927.

Med Ad News, May, 1997, p. 16.

Med Ad News, May, 1998, p. 18.

Med Ad News, May, 1999, p. 10.

Nau, H. (1985a) Improvement of testing for teratogenicity by pharmacokinetics. *Concepts in Toxicology*, 3,130–137.

Nau, H. (1985b) Teratogenic valproic acid concentrations: infusion by implanted minipumps vs. conventional injection regimen in the mouse. *Toxicology and Applied Pharmacology*, 80, 243–250.

Nau, H., Zierer, R., Spielmann, H., Neubert, D. and Gansau, C. (1981) A new model for embryotoxicity testing: teratogenicity and pharmacokinetics of valproic acid following constant-rate administration in the mouse using human therapeutic drug and metabolite concentrations. *Life Sciences*, 29, 2803–2814.

Nau, H., Trotz, M. and Wegner, C. (1985) Controlled-rate drug administration in testing for toxicity, in particular teratogenicity: toward interspecies bioequivalence. In Breimer, D.D and Speiser, P. (eds.) *Topics in Pharmaceutical Sciences*. Elsevier: Amsterdam, pp. 143–157.

O'Flaherty, E.J., Clarke, D.O. (1994) Pharmacokinetic/pharmacodynamic approaches for developmental toxicity. In Kimmel, C.A. and Buelke-Sam, J. (eds.) *Developmental Toxicology*, 2nd edn. Raven Press: New York.

Plowman, J., Harrison, Jr., S.D., Trader, M.W., Griswold, Jr., D.P., Chadwick, M., McComish, M.F., Silveira, D.M. and Zahaiko, D. (1987) Preclinical antitumor activity and pharmacological properties of deoxyspergualin. *Cancer Research*, 47, 685–689.

Popesko, P., Rajtova, V. and Horak, J. (1992) *A Color Atlas of Small Laboratory Animals*, Vol. 2, *Rat, Mouse and Hamster*. Wolfe Publishing: London.

Prevo, M.E. (1993) An evaluation of devices for continuous administration. In Niemi, S.M. and Willson, J.E. (eds.) *Refinement and Reduction in Animal Testing*. Scientists Center for Animal Welfare: Bethesda, pp. 41–55.

Reiners, J., Wittfoht, W., Nau, H. Vogel, R., Tenschert, B. and Spielmann, H. (1987) Teratogenesis and pharmacokinetics of cyclophosphamide after drug infusion as compared to injection in the mouse during Day 10 of gestation. In Nau, H. and Scott, W.J. (eds.) *Pharmacokinetics in Teratogenesis*, Vol. II, *Experimental Aspects in Vivo and in Vitro*. CRC Press: Boca Raton, FL, pp. 41–48.

Rodriguez-Martin, A., Remesar, X. and Alemany, M. (1993) Rates of utilization of intravenous oleylanilide administered chronically to the rat. *Food and Chemistry Toxicology*, 31(1), 37–40.

Rogiers, V., Vandenberghe, Y., Vanhaecke, T., Geerts, A., Callaerts, A., Carleer, J., Roba, J. and Vercruysse, A. (1997) Observation of hepatotoxic effects of 2-n-pentylaminoacetamide (Milacemide) in rat liver by a combined in vivo/in vitro approach. *Archives of Toxicology*, 71, 271–282.

Ruers, T.J.M., Buurman, W.A., Smits, J.F.M., Van der Linden, C.J., Van Dongen, J.J., Struyker-Boudier, H.A.J., Kootstra, G. (1986) Local treatment of renal allografts, a promising way to reduce the dosage of immunosuppressive drugs. *Transplantation*, 41(2), 156–161.

SCRIP, 1145, October 13, 1988, pp. 28–29.

Sikic, B.I., Collins, J.M., Mimnaugh, E.G. and Gram, T.E. (1978) Improved therapeutic index of bleomycin when administered by continuous infusion in mice. *Cancer Treatment Report,* 62(12), 2011–2017.

Slate, D.L., Fraser-Smith, E.B., Rosete, J.D., Freitas, V.R., Kim, Y.N. and Casey, S.M. (1993) Modulation of doxorubicin efficacy in P388 leukemia following co-administration of verapamil in mini-osmotic pumps. *In Vivo,* 7, 519–524.

Slotkin, T.A. (1998) Fetal nicotine or cocaine exposure: which one is worse? *Journal of Pharmacology and Experimental Therapeutics,* 285, 931–945.

Stepkowski, S.M., Tu, Y., Condon, T.P. and Bennett, C.F. (1994) Blocking of heart allograft rejection by intercellular adhesion molecule-1 antisense oligonucleotides alone or in combination with other immunosuppressive modalities. *Journal of Immunology,* 153, 5336–5346.

Tomas, F.M., Lemmey, A.B., Read, L.C. and Ballard, F.J. (1996) Superior potency of infused IGF-I analogues which bind poorly to IGF-binding proteins is maintained when administered by injection. *Journal of Endocrinology,* 150, 77–84.

Tu, Y., Stepkowski, S.M., Chou, T.-C. and Kahan, B.D. (1995) The synergistic effects of cyclosporine, sirolimus, and brequinar on heart allograft survival in mice. *Transplantation,* 59(2), 177–183.

Urquhart, J. (1985) Drug development and preclinical research applications of steady-state pharmacology. In Prescott, L.F. and Nimmo, W.S. (eds.) *Rate Control in Drug Therapy.* Churchill Livingstone: London, pp. 19–29.

Weghorst, C.M., Henneman, J.R. and Ward, J.M. (1991) Dose response of hepatic and renal DNA synthetic rates to continuous exposure of bromodeoxyuridine (BrdU) via slow-release pellets or osmotic minipumps in male B6C3F1 mice. *Journal of Histochemistry and Cytochemistry,* 39, 177–184.

Wilson, D. and Eldridge, S. (1989) Pulse and pump labeling methods. In Goldsworthy, T., Morgan, K., Popp, J. and Butterworth, B. (eds.) *Measurement of Chemically-induced Cell Proliferation in Specific Rodent Target Organs (Technical Notes for the CIIT Workshop on Cell Replication).* Chemical Industry Institute of Toxicology: Research Triangle Park, pp. 4–13.

Yoshimura, N., Lee, C.J., Shiho, O., Kita, M., Imanishi, J. and Oka, T. (1995) Local immunosuppressive therapy with monoclonal anti-T-cell antibody on renal allograft survival in the rat. II. Phenotypic and functional assessment of spleen cells. *Transplantation Proceedings,* 27(1), 390–391.

22 The contribution of vehicles, rates of administration and volumes to infusion studies

K. Hickling and D. Smith

22.1 Introduction

The ability to recover from toxic insult determines the ultimate outcome of toxicity studies (Soni *et al.* 1999). Therefore, derivation of meaningful and consistent toxicity data in pre-clinical evaluation requires the use of healthy animals that are housed in optimal conditions for the species. But are animals that are prepared and subjected to continuous intravenous infusion during toxicity studies capable of resisting toxic insult to the same degree as their non-infused counterparts? Surgical procedures, aseptic techniques, analgesia, choice of catheter and husbandry procedures are carefully selected and refined by experience to minimise any burden that the infusion technique presents to the animal. However, the subsequent contribution of the infusion vehicle, its rate of administration and therefore dose of its constituent excipients to this burden is often underestimated and sometimes totally ignored.

The purpose of this chapter will be to illustrate that the manifestation of toxicity of a given test article is a function of its administration conditions. We will propose a scheme for selection of vehicles and recommend 'best practice' infusion volumes and rates of administration that should not be exceeded without prior assessment of the consequences, both to the animal and the evaluation of the safety of the test article. We will not recommend a list of 'inert' vehicles and administration rates because, in the authors' opinion, such things simply do not exist. Every vehicle and infusion rate has a cost, the scale of which ranges from impairment of ability to recover and resist insult to frank, dose-limiting toxicity.

The choice of vehicle and its rate of administration therefore require careful cost–benefit analysis on a case-by-case basis. The cost is borne by the animal; it may be species, age and sex specific, but will ultimately be reflected in the derived maximum tolerated dose of the article under test. The case-by-case acceptability of a given vehicle is also a function of the purpose and duration of a study.

22.2 The burden of continuous infusion

Catheter implantation into a major vein under general anaesthesia, followed by continuous infusion of fluids, can hardly be described as ideal preparation for a toxicity study. Fortunately, control data from our in-house studies demonstrate that, following initial weight loss or reduced weight gain in the week after surgery, an animal that is continuously infused with isotonic saline at 2 ml/kg/h will be the same weight as an aged-matched control animal within 4 weeks (Figure 22.1). However, if an animal

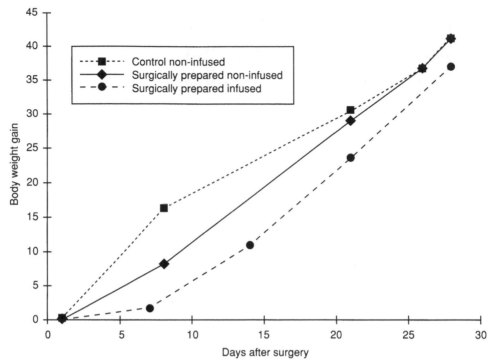

Figure 22.1 Comparison of weight gain in animals surgically prepared for continuous infusion studies. The weight gain of singly housed female animals that had been surgically implanted with catheters and were subsequently subjected to continuous infusion of saline at 2 ml/kg/h was compared with that of animals which underwent surgery but were not subsequently infused and with similarly housed control animals which did not undergo surgery. The data indicate that body weight recovers at a faster rate in non-infused animals than in their infused counterparts.

undergoes surgery but is not subsequently infused it will recover the lost weight within a shorter period of time.

There are a number of possible explanations for this delay in recovery. The most likely is the stress associated with permanent attachment to the infusion apparatus. Another possible contributory factor could be the sodium and chloride load that infusion of isotonic saline presents to the animal. The standard laboratory rat diet provides a sodium intake of around 250 mg/kg/day. A rat infused with 0.9% saline at 2 ml/kg/h receives an additional 430 mg/kg/day. Although there is no evidence that these levels are toxic, the animals still have to expend energy removing the excess. It is this energy that may slow recovery.

In general terms, these data imply a small but measurable consequence of continuous infusion, of even the most innocuous vehicles, on the animals under test. It does not indicate toxicity *per se* but suggests a burden on the capacity of an animal to recover from the insult of surgery. It is an observation that can presumably be extended to other insults, including those inflicted by the article under test.

22.3 The contribution of infusion volumes and rates to vehicle tolerability

For continuous infusion, the infused volume and administration rate are often synonymous. The infusion rate will, however, determine the steady-state plasma concentration of the test article although the relationship is non-linear (Claassen 1994). This same relation applies to vehicle excipients as they are typically present in fixed concentration. A general principle should be to infuse at the lowest possible rate that would allow acceptable exposure to the test article. However, in reality, for poorly soluble test articles, the dose rate is the limiting factor in determination of exposure.

Isotonic (0.9%) sodium chloride is widely used as the default vehicle for continuous infusion. The tolerability of animals to large intravenous volumes of saline administered at high rate is demonstrated in numerous publications (Zaucha *et al.* 1990; Cave *et al.* 1995; Morton *et al.* 1997a). These studies testify to the relative safety of isotonic saline as an infusing vehicle.

For continuous infusion, infused volume can be considered in terms of the total blood volume that an animal possesses. The blood volume of common laboratory species ranges from 56 to 85 ml/kg (Technical Subgroup of EFPIA/ECVAM 2000). Total replacement of this blood volume over 24 h would therefore be achieved using infusion rates of between 2 and 3 ml/kg/h. Several authors and ourselves regard the replacement of 1–1.5× blood volume each day as a 'best practice guide' for continuous infusion of 0.9% saline across most species (see Chapters 5 and 9).

Within our laboratory, a 'best practice' maximum continuous infusion rate to rats and dogs of 2 ml/kg/h is applied for any vehicle. At this infusion rate we have successfully completed studies of up to 3 months' duration, and studies over periods in excess of 6 months are reported elsewhere (Pecheur *et al.* 1995; Barrow and Guyot 1996). Some authors set a higher 'best practice' rate for isotonic solutions (see Chapter 9). However, evidence that may support an even lower infusion rate limit comes from infusion of isotonic saline to pregnant rats – an increase in embryofetal toxicity was observed at infusion rates above approximately 1 ml/kg/day (Barrow and Heritier 1995; Leconte *et al.* 1997). This re-emphasises the burden of continuous infusion on the animal.

Infusion of saline at 2 ml/kg/h will cause no haemodilution (O.P. Green, H.H. Edgar, L.M. Cawley and R.J. Harling, unpublished) and little background pathology. Our in-house data show that urine output over time is generally equivalent to that of non-infused animals. Rats compensate for the increased volume load of infusion by reducing water intake (Table 22.1).

Reduced water intake as a compensatory adjustment to increased fluid load has also been reported in rats and mice in other laboratories (Chapter 7) (O.P. Green, H.H. Edgar, L.M. Cawley and R.J. Harling, unpublished) and illustrates a primary long-term homeostatic mechanism that is initiated in preference to diuresis, which appears to be a shorter-term homeostatic process.

Table 22.1 Water intake and urine output of infused (2 ml/kg/h) versus comparable non-infused female rats over a 5-day period

	Urine output (ml/kg/day)	*Water intake (ml/kg/day)*
Non-infused (*n* = 5)	Average 38	Average 127
Infused (*n* = 30)	Average 41	Average 106

Unfortunately, the capacity of this long-term compensatory mechanism appears limited as water intake also accompanies food intake. This was demonstrated in a recent study in our laboratory, in which animals had access to water *ad libitum* but water was additionally administered by oral gavage to stimulate urine output. In spite of gavage dosing a typical daily water intake over 24 h, the animals still drank an almost normal quantity of water (data not shown), presumably to aid ingestion of dry laboratory diet.

Maintenance of a constant urine output therefore appears to be the default physiological choice that is normally regulated by oral water intake. In our laboratory, an increase in the infusion rate from 2 to 3 ml/kg/h results in no further reduction of oral water intake but a compensatory increase in urine output, i.e. a shift away from the default physiological process (Figure 22.2).

Within published literature and in previous chapters, infusion rates higher than 2 ml/kg/h have been used for studies in rats and other species without significant effect other than diuresis and haemodilution (Mann and Kinter 1993; Morton *et al.* 1997a, b). Although short-term episodes of diuresis represent normal physiological adjustment and are said to be of 'no toxicological significance', the consequences of long-term diuresis are less well understood and at the very least imply a cost of the procedure on the animal. A further additional cost of high infusion rates in rats, in our experience, was an unacceptable increase in failure of the infusion system. We would therefore recommend that increases in rate, above what is considered normal

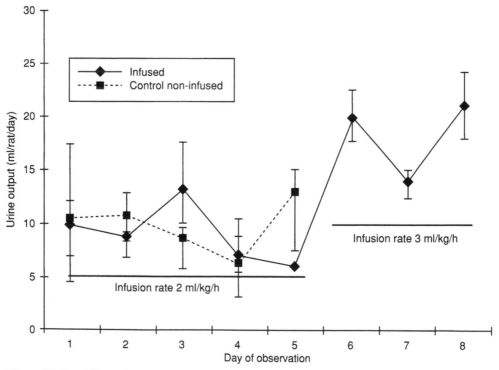

Figure 22.2   Effect of infusion rate on urine output.

Table 22.2 Recommended repeated intravenous dose volumes/rates (Technical Subgroup of EFPIA/ECVAM 2000)

Daily infusion period (h)	Mouse	Rat	Rabbit	Dog	Large primate	Minipig
	Total daily volume (ml/kg)					
4	–	20	–	20	–	–
24	96	60	24	24	60	24
	Rate (ml/kg/h)					
4	–	5	–	5	–	–
24	4	2.5	1	1	2.5	1

in the particular laboratory, should be gradually introduced and tested to ensure the rigour of the catheter implantation technique and infusion equipment.

Table 22.2 lists values for best practice infusion rates and volumes that have recently been recommended by the Technical Subgroup of EFPIA/ECVAM (2000). Reported maximum infusion rates that have been used in each species are also included. It should be borne in mind that these maximum figures are usually derived from studies using saline as a vehicle and, as illustrated by examples below, should not be extrapolated to other vehicles.

Table 22.3 lists some of the many factors that determine vehicle tolerability. The contribution of the test article to these parameters as well as its direct effects on the test animal will also influence vehicle tolerability.

22.4 The contribution of excipients

In pre-clinical studies, excipients are added to vehicles to alter their physicochemical properties, primarily to enhance the solubility and stability of the test article and to maintain a formulation that is as close as possible to physiological. These requirements are not always compatible. No vehicle is truly physiological, and it is the degree of incompatibility that determines the burden to the animal.

Excipients are commonly classed into groups that describe their function. The most commonly encountered in continuous infusion studies are tonicity modifiers such as the previously described sodium chloride, but also glucose and other sugars and sugar

Table 22.3 Factors that contribute to vehicle tolerability

Procedure/equipment	Animal	Vehicle	Excipient
Duration of infusion	Species	pH	Distribution
Continuous infusion	Age	Osmolality	Metabolism
Intermittent Infusion	Sex	Buffering power	Excretion
Rate of infusion	Weight	Temperature	Interactions
Cannula material	Plane of nutrition	Presence of particulates	Toxic properties
Cannula diameter	Health status		Solubility
	Calibre of infused vessel		

alcohols. Other classes include buffer systems, co-solvents, complexing agents and surfactants. Comprehensive reviews are published elsewhere (Wang and Kowal 1980; Sweetana and Akers 1996). On infusion, excipients, like any other xenobiotic, can induce both local and systemic effects. They can also act synergistically with other excipients or the test compound. The toxic effect of the vehicle may, conversely, mask drug effects.

22.5 Local effects

In pre-clinical evaluation, the most frequent dose-limiting effect of vehicles in continuous infusion studies is local intolerance at the site of administration.

Some of the known local effects of common excipients are pain on injection (P.K. Gupta *et al.* 1994), irritation and thrombosis (Sweetana and Akers 1996). Direct injury to capillary endothelial cells and an associated inflammatory reaction will inevitably result from the trauma of the implanted cannula. A suitable period of recovery from surgery will allow repair from the original lesion, but local abrasion and a chronic response to the very presence of the cannula will induce a continued minimal inflammation within the cannulated vessel. Local intolerance to continuously infused vehicles results from an exacerbation of this cannula-associated trauma. This is a mechanism peculiar to continuous infusion studies, and excipients that are not obviously irritant on bolus injection, even at relatively high concentration, may become unexpectedly irritant when continuously infused.

Local effects may become dose limiting only after many days of infusion and are often species, sex and age specific. An important factor is the weight of the animal as this determines the actual infusion rate and therefore the local concentration at the site of administration. The severity of local effects can be reduced by infusion into large vessels with greater flow rates. This may explain the species-specific nature of the phenomenon and why many vehicles that are approved for human use are poorly tolerated by smaller animals. The injury manifests as phlebitis and consequent redness, swelling, pain and localised temperature increases at the infusion site that can arise over a period of hours or many days. Mildly irritant excipients can also potentially act synergistically with test compounds and other excipients to induce a pronounced and unpredicted local response.

Local precipitation, usually of the test article, is a major factor in induction of local effects. Test material precipitation can also result in erratic or reduced bioavailability and a compromised study. The potential for local precipitation rises with decreasing solubility of the test article and the consequent inclusion of solubility enhancing excipients or non-physiological pH adjustment. Although precipitation typically occurs at the cannula tip, it may also occur within the formulation itself prior to or during administration. As particulates up to 25 µm in diameter are not visible (Claassen 1994), the use of in-line filters in infusion systems is to be encouraged. Recent infusion pumps also contain pressure alarms, which may provide early warning of impaired flow.

Test compound precipitation can be predicted on the basis of water solubility, protein binding and pK_a value. Simple screening tests, in which drug solubility is plotted as a function of dilution, and also *in vitro* dilution experiments, have been used to evaluate the ability of a pH-solubilised drug to remain in solution (Surakitbanharn *et al.* 1994; Myrdal *et al.* 1995). An *in vitro* dynamic flow model that incorporates spectrophotometric detection of precipitates was developed by Yalkowsky *et al.* (1983). Unfortunately, these tests are only partially predictive of local intolerance *in vivo* as precipitation represents only one possible induction mechanism. Increasing the concentration of mildly irritant excipients may paradoxically protect the test article from precipitation on infusion and more significant irritation (Yalkowsky *et al.* 1983). Increasing the infusion rate can also reduce drug precipitation *in vivo*. However, increasing the infusion rate may result in an increase in effects due to irritant vehicle components. Therefore, depending on the mechanism of irritation, either a reduction or an increase in infusion rate can exacerbate a local response. This illustrates the empirical nature of formulation development for continuous infusion.

The pH of vehicles can be adjusted to aid solubility of test materials. Examples of FDA-marketed parenteral products for human use range in pH from 2 to 11 (Nema *et al.* 1997). However, for biocompatibility reasons, formulation of bolus injectables within the pH range 4–9 are most common. For continuous infusion, tolerability to non-physiological pH depends upon local effects and systemic tolerability to the strength of buffer and the buffering system. The power of the buffer system that is tolerable to the animals decreases as the pH moves away from physiological. A review of human parenteral vehicles did not list any suitable alkaline buffer systems (Wang and Kowal 1980). Table 22.4 lists pH values and some buffer systems that have been used in continuous infusion studies.

The integrity of blood cells and endothelium depends on the tonicity of the surrounding fluid. Body fluids have an osmotic pressure that corresponds to that of 0.9% saline, and blood cells retain their normal size and shape within this fluid. However, iso-osmotic concentrations of other salts can induce haemolysis, and many excipients induce haemolysis even though they are present in iso-osmotic solutions. Reed and Yalkowsky (1985) have refined the *in vitro* haemolysis methodology to characterise the haemolytic nature of various co-solvents. Clinical publications caution against the use of solutions with high osmolality (Wright 1996). Haemolysis will usually occur local to the cannula but may subsequently induce systemic events such as alteration in heart rate as a result of release of adenylic compounds and increases in bilirubin excretion (Claassen 1994). Practical limits on the range of tolerable osmotic concentration are difficult to define as the true osmolality of vehicle containing various concentrations of test articles is rarely recorded. Formulations at up to three times osmolar concentration have been administered by other laboratories (Chapter 5), and hypotonic (0.5%) saline was tolerated at 2 ml/kg/h in the authors' laboratory for 3 days. Large primates are reported to tolerate solutions with osmolalities between 100 and 600 mosmol (Chapter 15).

Table 22.4 Tolerated pH-adjusted or buffered systems used in continuous infusion studies

pH value	Buffer	Species	Duration	Rate (ml/kg/h)	Reference
4	Unbuffered saline	Rat	14 days	2	In house
4	Unbuffered 5% dextrose	Rabbits	No limit	Not specified	Chapter 9
4	Citrate-buffered saline	Rabbits	No limit	Intermittent use	Chapter 9
5	0.05 mol/l acetate	Dog	28 days	0.42	In house
5.45	27 nmol/l citrate	Rabbits	No limit	Intermittent use	Chapter 9
3–9	Unbuffered saline	Primates	No limit	Up to 4	Chapter 15

Although not usually considered, any temperature difference between the animal and the vehicle will induce local effects if the insult is prolonged. Temperature equalisation occurs more easily at slow infusion rates, particularly if the route from exterior to implanted vessel is tortuous. At high rates and via shorter routes the temperature difference at the site of infusion may be significant. Solutions stored at low temperature prior to infusion should at the very least be allowed to reach ambient temperature before use.

22.6 Systemic effects

Within published literature there are many examples of vehicle-induced systemic toxicity (Levy *et al.* 1995; Morshed and Nagpaul 1995; Sweetana and Akers 1996; American Academy of Pediatrics Committee on Drugs 1997). Table 22.5 lists the systemic toxic properties of some commonly used excipients. Systemic effects can be described in terms of maximum tolerable dose or a maximum rate of administration of a given concentration. The most unpredictable of systemic events that are induced by excipients is alteration of pharmacokinetic and pharmacodynamic properties of the test article such that thresholds for toxic effect are exceeded in target organs (Thiel *et al.* 1986). Excipients, such as mannitol, can increase exposure of the test article and/or metabolites to the kidney and lower urinary tract (Brunner *et al.* 1986). Excipients may also influence toxicity of test articles by inhibition or induction of metabolic pathways (Zysset *et al.* 1980).

The maximum tolerable dose and/or rate of administration of an individual excipient depend upon their distribution, metabolism and excretion within the animal. For example, sodium chloride, infused at isotonic concentration, is distributed rapidly into extracellular fluid and, because it is not metabolised, it is excreted via urine along with an osmotic equivalent, i.e. its infusion volume, of water. Thus, an animal

Table 22.5 Some systemic toxic properties of some excipients

Excipient	Systemic effect
Mannitol	Diuresis, reduction in kidney blood flow (Brunner *et al.* 1986; Liss *et al.* 1996)
Glycerol	Diuresis (Wade and Walker 1994)
Sorbitol	Acidosis, hypoglycaemia (human only) (Wang and van Eys 1981; Keller 1989)
Citric acid	Hypocalcaemia (Mollison *et al.* 1987)
Polyethylene glycol 400	Diuresis, kidney and liver toxicity (Rowe and Wolf 1982)
Propylene glycol	Acidosis, CNS depression (American Academy of Pediatrics Committee on Drugs 1997; Rowe and Wolf 1982)
Solutol®	Decreased serum triglyceride levels (Woodburn *et al.* 1995) Anaphylaxis (dogs)
Polysorbate (Tween) 80	Anaphylaxis (dogs) (Masini *et al.* 1985) Degenerative changes in heart, liver and kidney on chronic administration to rats (Nityanand and Kapoor 1979)
Dimethylacetamide	Liver and reproductive effects (Kim 1988; BIBRA 1989), Multiple haemorrhage (dogs)
Dimethylsulphoxide	Antioxidation, diuresis, haemolysis, ocular toxicity (Noel *et al.* 1975; Willhite and Katz 1984; Brayton 1986)

expends little energy in homeostatic adjustment and can tolerate staggering fluid loads of this vehicle (Zeoli *et al.* 1998). More complex, non-metabolised, osmolytes of higher molecular weight will distribute slowly with extracellular fluid and therefore cause prolonged haemodilution until they are cleared unchanged. They can be expected to have much lower tolerable doses and place a burden of energy expenditure on the animal to recover physiological conditions following administration. Metabolised osmolytes, such as glucose, also pose problems when infused into animals. This is because metabolism removes the osmolyte, leaving the equivalent free water in plasma. Removal of this water, although achievable in the dog following bolus doses as high as 18 ml/kg (Mann and Kinter 1993), is at considerable cost to the animal in terms of energy expenditure.

22.7 A strategy for vehicle selection

Within the early phase of pre-clinical assessment many disciplines aim to perform short-term evaluations of test articles prior to selection of that article for safety evaluation. During this time, it is not uncommon for a toxicologist to be asked to make judgements on the safety of untried vehicles or excipients for short-term (minutes), intermittent (hours) or continuous infusion (days). Strategies for selection of vehicles for short-term infusion are beyond the scope of this chapter. However, the same principles apply in continuous infusion, although the number of acceptable vehicles decreases with the duration of the study.

The strategy for vehicle selection should begin at the initial stages of development. Departments involved in molecular design of candidate compounds should have, as a design criterion, a minimal solubility in a neutral aqueous environment. If this is not possible then, as candidate compounds are selected, water-soluble forms of these articles should be considered for synthesis and evaluation. This approach is rarely considered as a plausible alternative to the inclusion of excipients in formulations as it has been shown that salt forms often behave differently *in vivo* because of the physical, chemical and thermodynamic properties they impart to the parent compound (Berge *et al.* 1977). Excipients are therefore included in formulations to achieve the required concentration of test articles. However, many clinical and pre-clinical examples illustrate that these excipients can alter test article behaviour within the animal in a similar manner to salt forms (Kobayashi *et al.* 1977; Lo *et al.* 1994; Farooqui *et al.* 1995; Koporec *et al.* 1995; Raymond and Plaa 1997).

Formulation scientists and toxicologists should generate a list of approved tolerable vehicles for the species under test. Unfortunately, formulation scientists still have to rely on an empirical approach for screening vehicles for solubilising water-insoluble test articles as no single theory adequately explains solubility. However, there are published approaches to developing formulations for parenteral administration (Sweetana and Akers 1996). These selected vehicles should be ranked and sequentially tested for suitability with the candidate compound prior to departing from the agreed list and considering vehicles of unknown effect. Departure from this list would prompt a literature review of the toxicity of proposed excipients and an estimate of a 'safe' dose. For continuous infusion, blood compatibility and microprecipitation studies similar to those described (Yalkowsky *et al.* 1983; S.L. Gupta *et al.* 1994) should also be performed before consideration for use.

This system has been implemented within our organisation and has significantly

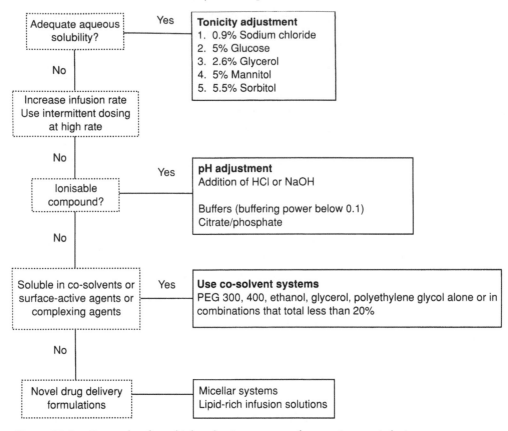

Figure 22.3 Example of a vehicle selection strategy for continuous infusion.

reduced the incidence of vehicle toxicity. However, it cannot predict interaction with test articles and vehicle components. An example of a flow chart for vehicle selection is shown in Figure 22.3.

For novel excipients there are guiding principles for their theoretical toxicological evaluation prior to the selection of likely maximum concentration that can be included in a proposed vehicle. However, a non-toxic dose of an excipient does not mean that the excipient does not present a significant burden to the animals under test. At a rate of 2 ml/kg/h it is worth reminding researchers that inclusion of any excipient above approximately 4% concentration will exceed the accepted UK daily limit dose of testing of test articles (i.e. 2 g/kg/day).

The consequence of failure of a study as a result of vehicle effects cannot be overemphasised, and novel vehicles should be subjected to preliminary studies to check for intolerance. These should be performed for the intended duration of the main study, as local changes may not become dose limiting for many days or even weeks into the study. If, as is often the case, it is not possible to perform such studies then, as a minimum, vehicle control animals should be included in early studies on the test article.

Estimates of safe dose should be conservative until in-house experience with the excipient justifies a dose increase. This should be based on all available information as the estimated 'safe' excipient dose usually determines the highest concentration and therefore dose of the test article. Conservative estimates of non-toxic doses can usually be derived from toxicity data following short-term or bolus administration of the excipient. A no-effect dose in mg/kg terms can be extrapolated to be administered over the infusion time or 24 h. Non-clinical experience of excipients that have been subsequently approved for human use can also be found across pharmaceutical companies. The authors hope that, for welfare reasons alone, communication of this commercially non-sensitive information becomes commonplace. Available data should hopefully have been derived in the species under test. Many known effects of excipients can be extrapolated across species. The common exceptions are surfactants such as Tween 80 (Masini *et al.* 1985), Cremophor (Lorenz *et al.* 1982) and others that induce histamine release in dogs and should not be considered for administration into this species. The organic solvent dimethylacetamide has, in our hands, also induced unpredictable effects in dogs such that its use at any concentration is prohibited. For combinations of excipients, a conservative approach would be to sum the total dose and regard it as a dose of the most toxic component. This approach incorporates a safety factor for possible synergistic effects. Substances that have not been administered intravenously before should not be considered for continuous infusion until evaluation following bolus or short-term infusion has been performed.

For long-term studies, selected formulations must be sterile and stable at room temperature for periods of a minimum of 30 h. This will allow normal husbandry and other study requirements to be performed before daily syringe changing. Unless photostability data have been generated, it is good practice to protect formulations contained within syringes from light during infusion. As well as checking animals and infusion equipment, technical staff should frequently assess the condition of the formulation being infused. Compatibility testing of the giving sets with the formulation should be performed, including assessment of compound adherence within the system and also possible leaching of substances from the equipment into the formulation.

Unfortunately, as the study length increases, the choice of tolerable vehicles decreases. Vessel size and flow rate appear to be major factors in determining the severity of effect and therefore species sensitivity. For the larger experimental animals, useful guides on suitable excipients can be found in clinical literature. For smaller species the choice is limited and dependent upon the duration of the study. In our hands, only tonicity modifiers such as 0.9% saline and 2.5% mannitol–glucose have proved suitable for studies over 28 days' duration in rats when infused at 2 ml/kg/h. For shorter-term studies, the inclusion of most types of excipient is possible at low concentration. The conditions of administration also appear to affect vehicle tolerability. For example, infusion of 5% glucose at 10 ml/kg/h for 1 h thrice daily could not be tolerated for more than 34 days (Salauze and Cave 1995) whereas, when infused at 1.67 ml/kg/h, it is the vehicle of choice for long-term reproductive studies in other laboratories (Barrow and Heritier 1995).

22.8 Conclusion

Continuous intravenous infusion studies of up to 6 months' duration can be performed in all common laboratory species, with minimal findings in animals exposed to vehicle alone. However, the contribution of the infusion vehicle, its rate and duration of administration to the derived maximum tolerated dose of the article under test should not be underestimated. We believe that every vehicle and infusion rate imposes a burden on the infused animal. The magnitude of the burden can range from impairment of ability to recover and resist insult to frank dose-limiting toxicity. Reference to best practice guidelines on infused volumes and rates, more open communication of data on excipients and molecular design of more soluble test articles should help reduce the incidence and severity of vehicle-induced effects in continuous infusion studies.

References

American Academy of Pediatrics Committee on Drugs (1997) 'Inactive' ingredients in pharmaceutical products: update (subject review). *Pediatrics*, 99, 268–278.

Barrow, P.C. and Guyot, J.Y. (1996) Continuous deep intravenous infusion in rabbit embryotoxicity studies. *Human and Experimental Toxicology*, 15, 214–218.

Barrow, P.C. and Heritier, B. (1995) Continuous deep intravenous infusion in rat embryotoxicity studies: the effects of infusion volume and two different infusion fluids on pregnancy. *Toxicology Methods*, 5, 61–67.

Berge, S.M., Bighley, L.D. and Monkhouse, D.C. (1977) Pharmaceutical salts. *Journal of Pharmaceutical Sciences*, 66, 1–19.

BIBRA (1989) *BIBRA Toxicity Profiles – Dimethylacetamide*. British Industrial Biological Research Association, BIBRA International: Carshalton.

Brayton, C.F. (1986) Dimethyl sulfoxide (DMSO): a review. *Cornell Veterinarian*, 76, 61–90.

Brunner, F.P., Hermle, M., Mihatsch, M.J. and Thiel, G. (1986) Mannitol potentiates cyclosporine nephrotoxicity. *Clinical Nephrology*, 25 (Suppl. 1), S130–S136.

Cave, D.A., Schoenmakers, A.C., van Wijk, H.J., Enninga, I.C. and van der Hoeven, J.C. (1995) Continuous intravenous infusion in the unrestrained rat – procedures and results. *Human Experimental Toxicology*, 14, 192–200.

Claassen, V. (1994) Intravenous drug administration. In Huston, J.P. (ed.) *Techniques in the Behavioural and Neural Scineces, Vol. 12: Neglected Factors in Pharmacology and Neuroscience Research*. Elsevier Science: Amsterdam, pp. 5–22.

Farooqui, M.Y., Ybarra, B., Piper, J. and Tamez, A. (1995) Effect of dosing vehicle on the toxicity and metabolism of unsaturated aliphatic nitriles. *Journal of Applied Toxicology*, 15, 411–420.

Gupta, P.K., Patel, J.P. and Hahn, K.R. (1994) Evaluation of pain and irritation following local administration of parenteral formulations using the rat paw lick model. *Journal of Pharmaceutical Science* and *Technology*, 48, 159–166.

Gupta, S.L., Patel, J.P., Jones, D.L. and Partipilo, R.W. (1994) Parenteral formulation development of renin inhibitor Abbott-72517. *Journal of Pharmaceutical Science* and *Technology*, 48, 86–91.

Keller, U. (1989) The sugar substitutes fructose and sorbite: an unnecessary risk in parenteral nutrition. *Schweizerische Medizinische Wochenschrift*, 119, 101–106.

Kim, S.N. (1988) Preclinical toxicology and pharmacology of dimethylacetamide, with clinical notes. *Drug Metabolism Reviews*, 19, 345–368.

Kobayashi, H., Peng, T.C., Kawamura, R., Muranishi, S. and Sezaki, H. (1977) Mechanism of the inhibitory effect of polysorbate 80 on intramuscular absorption of drugs. *Chemical and Pharmaceutical Bulletin*, 25, 569–574.

Koporec, K.P., Kim, H.J., MacKenzie, W.F. and Bruckner, J.V. (1995) Effect of oral dosing vehicles on the subchronic hepatotoxicity of carbon tetrachloride in the rat. *Journal of Toxicology* and *Environmental Health*, 44, 13–27.

Leconte, I., Barrow, P.C. and Descotes, J. (1997) Background control data for rat embryotoxicity studies using continuous and intermittent IV infusion. *Toxicologist*, 36, No. 1, Pt 2, 260.

Levy, M.L., Aranda, M., Zelman, V. and Giannotta, S.L. (1995) Propylene glycol toxicity following continuous etomidate infusion for the control of refractory cerebral edema. *Neurosurgery*, 37, 363–369.

Liss, P., Nygren, A., Olsson, U., Ulfendahl, H.R. and Erikson, U. (1996) Effects of contrast media and mannitol on renal medullary blood flow and red cell aggregation in the rat kidney. *Kidney International*, 49, 1268–1275.

Lo, H.H., Valentovic, M. A., Brown, P.I. and Rankin, G.O. (1994) Effect of chemical form, route of administration and vehicle on 3,5-dichloroaniline-induced nephrotoxicity in the Fischer 344 rat. *Journal of Applied Toxicology*, 14, 417–422.

Lorenz, W., Schmal, A., Schult, H., Lang, S., Ohmann, C., Weber, D, Kapp, B., Luben, L. and Doenicke, A. (1982) Histamine release and hypotensive reactions in dogs by solubilizing agents and fatty acids: analysis of various components in cremophor El and development of a compound with reduced toxicity. *Agents and Actions*, 12, 64–80.

Mann, W.A. and Kinter, L.B. (1993) Characterization of maximal intravenous dose volumes in the dog (*Canis familiaris*). *General Pharmacology*, 24, 357–366.

Masini, E., Planchenault, J., Pezziardi, F., Gautier, P. and Gagnol, J.P. (1985) Histamine-releasing properties of Polysorbate 80 in vitro and in vivo: correlation with its hypotensive action in the dog. *Agents and Actions*, 16, 470–477.

Mollison, P.L., Englefreit, C.P. and Contreras, M. (1987) *Blood Transfusion in Clinical Medicine*, 8th edn. Blackwell Scientific Publications: Oxford.

Morshed, K.M. and Nagpaul, J.P. (1995) Propylene glycol-induced changes in plasma and mucosal functions in rats: evidence of subchronic toxicity. *Toxic Substance Mechanisms*, 14, 13–25.

Morton, D., Safron, J.A., Rice, D.W., Wilson, D.M. and White, R.D. (1997a) Effects of infusion rates in rats receiving repeated large volumes of saline solution intravenously. *Laboratory Animal Science*, 47, 656–659.

Morton, D., Safron, J.A., Glosson, J., Rice, D.W., Wilson, D.M. and White, R.D. (1997b) Histologic lesions associated with intravenous infusions of large volumes of isotonic saline solution in rats for 30 days. *Toxicological Pathology*, 25, 390–394.

Myrdal, P.B., Simamora, P., Surakitbanharn, Y. and Yalkowsky, S.H. (1995) Studies in phlebitis. VII: In vitro and in vivo evaluation of pH-solubilized levemopamil. *Journal of Pharmaceutical Sciences*, 84, 849–852.

Nema, S., Washkuhn, R.J. and Brendel, R.J. (1997) Excipients and their use in injectable products. *PDA Journal of Pharmaceutical Science* and *Technology*, 51(4), 166–171.

Nityanand, S. and Kapoor, N.K. (1979) Effect of chronic oral administration of Tween-80 in Charles Foster rats. *Indian Journal of Medical Research*, 69, 664–670.

Noel, P.R.B., Barnett, K.C., Davies, R.E., Jolly, D.W, Leahy, J.S., Mawdesley-Thomas, L.E., Shillam, K.W.G., Squires, P.F., Street, A.E., Tucker, W.C. and Worden, A.N. (1975) The toxicity of dimethyl sulphoxide (DMSO) for the dog, pig, rat and rabbit. *Toxicology*, 3(2), 143–169.

Pecheur, C., Heriteur, B., Regnier, B. and Descotes, J. (1995) Eurotox '95. 34th European Congress of Toxicology. Prague, Czech Republic, 27–30 August 1995. Abstracts. *Toxicology Letters*, 78 (Suppl. 1), 1–96.

Raymond, P. and Plaa, G.L. (1997) Effect of dosing vehicle on the hepatotoxicity of CCl_4 and nephrotoxicity of CHCl3 in rats. *Journal of Toxicology* and *Environmental Health*, 51, 463–476.

Reed, K.W. and Yalkowsky, S.H. (1985) Lysis of human red blood cells in the presence of various cosolvents. *Journal of Parenteral Science* and *Technology*, 39 (2), 64–69.

Rowe, V. K. and Wolf, M. A. (1982) Glycols. In Clayton, G.D. and Clayton F.E. (eds.) *Patty's Industrial Hygiene and Toxicology*, 3rd edn. Interscience: New York, pp. 3844–3852.

Salauze, D. and Cave, D. (1995) Choice of vehicle for three-month continuous intravenous toxicology studies in the rat: 0.9% saline versus 5% glucose. *Laboratory Animals*, 29, 432–437.

Soni, M.G., Ramaiah, S.K., Mumtaz, M.M., Clewell, H. and Mehendale, H.M. (1999) Toxicant-inflicted injury and stimulated tissue repair are opposing toxicodynamic forces in predictive toxicology. *Regulatory Toxicology and Pharmacology*, 29, No. 2, Pt 1, 165–174.

Surakitbanharn, Y., Simamora, P., Ward, G.H. and Yalkowsky, S.H. (1994) Precipitation of pH solubilized phenytoin. *International Journal of Pharmaceutics*, 109, 27–33.

Sweetana, S. and Akers, M. J. (1996) Solubility principles and practices for parenteral drug dosage form development. *PDA Journal of Pharmaceutical Science* and *Technology*, 50, 330–342.

Technical Subgroup of EFPIA/ECVAM (2000) *A Good Practice Guide to the Administration of Substances and Removal of Blood, Including Routes and Volumes.*

Thiel, G., Hermle, M. and Brunner, F.P. (1986) Acutely impaired renal function during intravenous administration of cyclosporine A: a cremophore side-effect. *Clinical Nephrology*, 25 (Suppl. 1), S40–S42.

Wade, A. and Walker, P.J. (1994) *Handbook of Pharmaceutical Excipients*, 2nd edn. The Pharmaceutical Press: London.

Wang, Y.C. and Kowal, R.R. (1980) Review of excipients and pH's for parenteral products used in the United States. *Journal of the Parenteral Drug Association*, 34, 452–462.

Willhite, C.C. and Katz, P.I. (1984) Toxicology updates. Dimethyl sulfoxide. *Journal of Applied Toxicology*, 4, 155–160.

Wang, Y.M. and van Eys, J. (1981) Nutritional significance of fructose and sugar alcohols. *Annual Review of Nutrition*, 1, 437–475.

Woodburn, K., Sykes, E. and Kessel, D. (1995) Interactions of Solutol HS 15 and Cremophor EL with plasma lipoproteins. *International Journal of Biochemistry* and *Cell Biology*, 27, 693–699.

Wright A (1996) Reducing infusion failure: a pharmacological approach. A review. *Journal of Intravenous Nursing*, 19 (2), 89–97.

Yalkowsky, S.H., Valvani, S.C. and Johnson, B.W. (1983) In vitro method for detecting precipitation of parenteral formulations after injection. *Journal of Pharmaceutical Sciences*, 72, 1014–1017.

Zaucha, G.M., Frost, D.F., Omaye, S.T., Clifford, C.B. and Korte, D.W. (1990) Fourteen-day subacute intravenous toxicity study of hypertonic. *NTIS Government Reports Announcements and Index*. No. 7, 596.

Zeoli, A., Donkin, H., Crewell, C., Fetrow, N., Johnson, D.K., and Kinter, L.B. (1998) A limit rapid intravenous injection volume in dogs. *Toxicological Sciences*, 42 (Suppl. 1–5), 58.

Zysset, T., Preisig, R. and Bircher, J. (1980) Increased systemic availability of drugs during acute ethanol intoxication: studies with mephenytoin in the dog. *Journal of Pharmacology and Experimental Therapeutics*, 213, 173–178.

23 Equipment for continuous intravenous infusion

A. Jacobson

23.1 Introduction and history

While the demand for continuous intravenous infusion in laboratory animals has grown continuously over the past few decades, the pace of technology change has been relatively slow since the time when many of today's materials and methods were first introduced. The tethered infusion systems of today look remarkably similar to those of 1970. Alice King Chatham, considered to be the pioneer in laboratory animal jackets, developed a non-human primate restraint jacket while working on the NASA spaceshot programmes at Douglas Aircraft. In 1967, she left Douglas and started Alice King Chatham Medical Arts. In the late 1960s, Michael Loughnane at Temple University developed a single-channel rat swivel and tether based on Dr Jim Weeks' early swivel concept. In 1971, Loughnane started Instech Laboratories to commercialise the swivel. Harvard Apparatus began supplying its first syringe pump in the 1950s. In the late 1960s, Dr Robert Hickman published on his use of the tunnelled subcutaneous silicone catheter in humans. The marriage of the syringe pump, swivel, jacket, and tunnelled catheter began the proliferation of the laboratory animal tethered infusion model that has been used by virtually every animal infusion laboratory in the world.

The most significant product changes in the field are (1) use of subcutaneous access ports instead of externalised catheters, (2) the growth of the tetherless jacketed infusion model in large animals and (3) the enhanced programmability of infusion pumps. The proliferation of ports results from the vision of Michael Dalton of Norfolk Medical, who introduced the first laboratory animal access port in 1981 while awaiting FDA approval of the first human access port. Growth of tetherless jacketed infusions can be traced to the targeting of animal infusion laboratories by Pharmacia-Deltec (now Sims-Deltec), maker of the CADD infusion pump, which is designed principally for human use. Finally, the increased programmability of pumps is a natural outgrowth of increased microprocessor use in medical and laboratory devices.

Many researchers avail themselves of the 'high-tech' finished goods provided by manufacturers of laboratory animal supplies, whereas other researchers utilise 'hand-made' supplies. Pre-clinical safety testing utilises both the high-tech and hand-made supplies in reliable and repeatable models. Generally, however, although there is a variety of useful supplies, technology is lagging in this field. Implanted pumps are scarce, networking and high automation of pumps is limited, and technology miniaturisation is not keeping pace with the striking increased utilisation of transgenic mice. Clearly, the requisite technology and imagination exist to forge many useful new devices for the field. But the field represents a relatively small industry that does not attract enough development dollars from suppliers to fully utilise available technology in the manner that the human use infusion industry does.

23.2 Catheters

The component common to all continuous intravenous infusion models is the catheter (or 'cannula'). There is no standard laboratory animal infusion catheter, and there is a plethora of catheter configurations. Catheters fall into two categories: (1) finished catheters from catheter manufacturers and (2) 'hand-made' catheters fabricated by researchers from bulk rolls of tubing. As a rule, a finished catheter is more costly to acquire than a home-made catheter ($10–$40 for a sterile, finished catheter compared with $1 for hand-made catheter materials). However, when the variable of the laboratory's labour cost for hand-made catheter assembly is considered, the price gap is narrowed or erased. Catheter characteristics and sizes are shown in Tables 23.1 and 23.2.

23.2.1 Catheter properties

23.2.1.1 Biomaterials

The issue of catheter materials is straightforward as there are four principal materials utilised: silicone rubber (Silastic®), polyurethane, polyethylene (PE) and polyvinylchloride (PVC or Tygon®). Silicone and polyurethane are the predominant catheter materials employed, though the stiff PE is still common for small vessel access in rodents. Both silicone and polyurethane have notable drawbacks, however. Silicone's extreme softness is both its greatest benefit and its greatest limitation – its softness enhances haemocompatibility while also contributing to its low tear strength. Accordingly, polyurethane has rapidly accelerated its acceptance over the past decade because of its good haemocompatibility and resistance to tears. Additionally,

Table 23.1 Catheter material characteristics

	Silicone	Polyurethane	Poly-ethylene (PE)	Poly-vinylchloride (PVC)
Haemocompatibility	Excellent	Excellent	Fair	Fair
Compound compatiblity	Inert	Possible reactivity	Inert	Possible reactivity
Stiffness	Soft	Available soft or stiff	Stiff	Available soft or stiff
Ease of insertion	Difficult in small diameters	Moderate or easy	Easy	Easy
Sizes available	Excellent	Excellent	Excellent	Fair
Ease of bonding	Excellent	Fair	Poor	Good
Memory	Excellent	Poor	Poor	Poor
Tear strength	Poor	Excellent	Excellent	Excellent
Sterilisation	Ethylene oxide or steam	Ethylene oxide	Ethylene oxide or steam	Ethylene oxide or limited steam

Table 23.2 Guide to commonly available sizes of catheter for cannulation of the jugular vein

	Silicone	Polyurethane	Polyethylene (PE)	Polyvinylchloride (PVC)
Mice	–	0.2 mm ID × 0.4 mm OD 0.008″ ID × 0.016″ OD 1.2 French, 27 gauge OD	–	–
Mice, rats	0.3 mm ID × 0.6 mm OD 0.012″ ID × 0.025″ OD 1.9 French, 23 gauge OD	0.3 mm ID × 0.7 mm OD 0.013″ ID × 0.026″ OD 2.0 French, 23 gauge OD	0.3 mm ID × 0.6 mm OD 0.011″ ID × 0.024″OD PE-10, 1.8 French, 23 gauge	0.3 mm ID × 0.8 mm OD 0.010″ ID × 0.030″ OD 2.4 French, 22 gauge OD
Rats	0.5 mm ID × 0.9 mm OD 0.020″ ID × 0.037″ OD 2.8 French, 20 gauge OD	0.6 mm ID × 1.0 mm OD 0.024″ ID × 0.038″ OD 3.0 French, 20 gauge OD	0.6 mm × 1.0 mm OD 0.023″ ID × 0.038″ OD PE-50, 3.0 French, 20 gauge OD	–
Rats, rabbits, non-human primates	0.6 mm ID × 1.2 mm OD 0.025″ ID × 0.047″ OD 3.6 French, 18 gauge OD	0.7 mm ID × 1.2 mm OD 0.027″ ID × 0.047″ OD 3.6 French, 18 gauge OD	0.8 mm ID × 1.2mm OD 0.030″ ID × 0.048″ OD PE-60, 3.6 French, 18 gauge OD	–
Rabbits, non-human primates	0.7 mm ID × 1.7 mm OD 0.030″ ID ×0.065″ OD 5.1 French, 16 gauge OD	1.0 mm ID × 1.7 mm OD 0.040″ ID ×0.066″ OD 5.1 French, 16 gauge OD	1.2 mm ID × 1.7 mm OD 0.047″ ID × 0.067″ OD PE-190, 5.1 French, 15 gauge OD	1.0 mm ID × 1.8 mm OD 0.040″ ID × 0.070″ ID 5.4 French, 15 gauge OK
Dogs, pigs	1.0 mm ID × 2.2 mm OD 0.040″ ID × 0.085″ OD 6.6 French, 14 gauge OD	1.3 mm ID × 2.4 mm OD 0.052″ ID × 0.096″ OD OD 7.2 French, 13 gauge OD	–	1.3 mm ID × 2.3 mm OD 0.050″ ID × 0.090″ OD 6.9 French, 13 gauge OD
Dogs, pigs	1.6 mm ID × 3.2 mm OD 0.062″ ID × 0.125″ OD 9.6 French, 11 gauge OD	–	–	–

Note
3 French = 1 mm.

polyurethane catheters can be coated with advanced heparin coating and hydrogel coating, which can enhance catheter biocompatibility and haemocompatibility

23.2.1.2 Haemocompatibility

The issue of catheter haemocompatibility is multivariate. An intravascular device's blood compatibility relates to surgical techniques, infusate properties and device characteristics. Device characteristics are not limited to the device's biomaterial(s) of fabrication, but also include the diameter, tip geometry, surface finish, wall thickness and stiffness. Generally, however, the most important aspects of a long-term intravenous catheter's haemocompatibility are (1) material, (2) stiffness and (3) tip geometry. These catheter features become more critical as the duration of the infusion increases.

23.2.1.3 Compound compatibility

Researchers must assess a catheter's compound and vehicle compatibility for each study. Polyurethane and PVC have many leachable components, particularly plasticisers, which regulate the softness of a catheter. Inadvertent infusion of these catheter components might materially affect the animal's physiology. Further, compounds may react chemically with the catheter in such a way that the compound adsorbs onto the catheter wall, thereby reducing the compound dose infused into the animal. Finally, catheter materials have varying gas permeabilities which might permit some compounds to evaporate through the catheter wall, particularly if the infusion is at slow rates or through externalised catheters.

23.2.1.4 Stiffness and ease of insertion

Catheter stiffness presents an interesting trade-off – a stiffer catheter is more thrombogenic than a softer catheter yet a softer catheter is more difficult to thread into a vein, particularly in a rodent. However, any softer silicone catheter larger than 1.2 mm OD (outer diameter) is generally stiff enough to insert into the majority of laboratory animal veins. The stiffer catheter is more thrombogenic than a softer catheter as a result of trauma to the vein's intimal wall.

23.2.1.5 Size availability

Silicone, polyurethane and PE catheters are readily available in diameters ranging from 0.6 mm OD (suitable for mice and rats) up to 3.2 mm OD (suitable for large dogs and pigs). PVC tubing in the smaller sizes is less readily available. Polyurethane catheters are also available in sizes as small as 0.4 mm OD for use in mice and other small vein access.

23.3.2.6 Ease of bonding

For 'hand-made' catheters, researchers often need to bond catheter components together. As such, it is paramount to note that typical cyanoacrylate (CA) adhesives ('superglues') do not adhere well to silicone, polyurethane and polyethylene. CA adhesives usually establish a short-term mechanical bond to these materials, but not a strong, long-term chemical bond. Medical-grade silicone adhesives create a permanent bond to silicone catheters but not to the other materials. Polyurethane requires specialised UV-cured adhesives. Polyethylene is virtually impervious to effective adhesive or solvent bonding.

23.2.1.7 Memory

Memory of a catheter material refers to the extent to which it adopts a new shape when it is stretched or reshaped. Polyurethane, PE and PVC are thermoplastics which have bad memory. For example, if a catheter is inserted into a blunt hypodermic needle (Luer stub adapter) with an OD larger than a catheter's ID, the thermoplastic catheter will take the shape of that needle after a few days. That is, the compression that initially exists between the catheter and needle will be lost as the catheter relaxes into the shape of the needle. Leaking may result. Silicone, on the other hand, is a thermoset rubber with excellent memory; it will continue to maintain a good seal around a needle after extended periods.

23.2.1.8 Sterilisation

Most catheter materials are readily autoclaved, but polyurethane has a low melting point, so it must be gas sterilised by ethylene oxide (EtO).

23.2.1.9 Tear strength

Tear strength is critical, particularly for applications in rodents, as the smaller catheters have thinner walls that are more prone to tear. Tear strength describes how well a catheter resists bursting due to high infusion pressures and tearing due to contact with other parts of the infusion apparatus. Smaller-diameter silicone catheters are notorious for tearing and bursting, while the larger-silicone diameters (1.7 mm OD and larger) have ample wall thickness and tear strength. The other catheter materials have excellent tear strength properties for laboratory animal infusion applications. Polyurethane offers relative softness combined with high tear strength, a combination of properties that makes it increasingly popular among researchers in rodent infusion.

23.2.2 Catheter configurations

23.2.2.1 Catheter tips

The intravascular end of the catheter is typically referred to as the 'tip' or 'distal tip'. These tips are fabricated as either radiused (rounded), straight (blunt) or bevelled at a 30–60° angle. Straight and bevel cut catheters are used extensively with good success. Hecker (1981) and Dennis *et al.* (1984) observed in sheep and O'Farrell (1995) observed in rats that intravascular catheters with rounded tips evoke a less thrombogenic response than straight or bevel-cut tips and thus last longer and have less physiological impact (i.e. are less likely to cause catheter clotting, vessel thrombosis and thromboemboli). Radiused tips are not fabricated easily by researchers but are available from laboratory animal catheter manufacturers (Figure 23.1).

23.2.2.2 *Suture retention bulbs and mesh*

Hand-made and finished catheters often have suture retention bulbs or polyester mesh which are added to the catheter to anchor it within the vessel (bulbs) or fascia (mesh). The anchor prevents the catheter from pulling out of the vein and reduces the tension required of a catheter-retaining pursestring suture.

23.2.2.3 *Connection to subcutaneous access ports*

The subcutaneous access port (Figure 23.2) is accessed percutaneously with a hypodermic Huber needle (Figure 23.3). This catheter design obviates the need for any portion of the catheter to be externalised, thereby reducing infection risks, catheter maintenance, and the need for protective jackets and harnesses. Port sizes are available for use in animals ranging from rats to large animals.

Although the port was initially used for intermittent infusions, it is used increasingly in protracted and continuous infusions for two reasons. First, it allows for easy disconnection from an animal that will be maintained in a colony for subsequent studies. Second, Mendenhall (1997) has reported that, in non-human primates, the use of ports can lead to a decreased incidence of infections compared with the use of externalised catheters partly because of the smaller exit site wound created by the port's Huber needle (22 gauge) versus the larger external catheter wound (13–16 gauge).

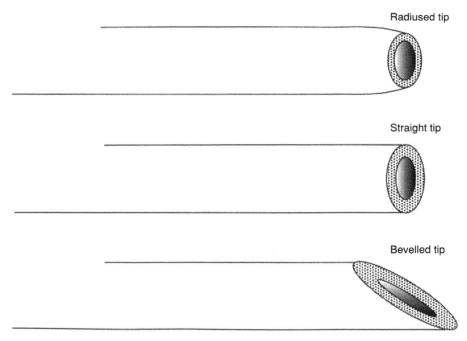

Radiused tip

Straight tip

Bevelled tip

Figure 23.1 Catheter tips.

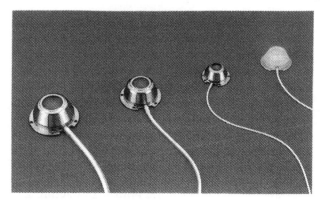

Figure 23.2 Ports (Solomon)

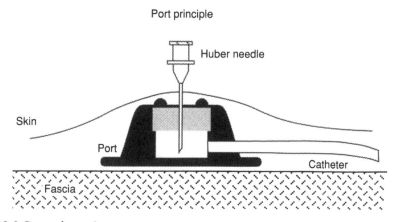

Figure 23.3 Port schematic.

23.2.3 Supply

Hand-made catheters need bulk tubing. The simplest way to acquire bulk rolls (25 ft/ 8 m or more) of laboratory animal sizes of silicone, PVC and PE tubing is from the VWR, Cole Parmer and Fisher Scientific catalogues. The best sources of laboratory animal polyurethane tubing are Access Technologies, Braintree Scientific and Solomon Scientific. Bulk catheter tubing costs about $1/ft.

Sources of standard and customised finished catheters for a variety of applications in small and large animals are Access Technologies, IITC and Solomon Scientific in the US, and Uno Roestvaststaal in Europe. A typical standard or customised catheter is provided packaged sterile for about $10–$40 each. Coated polyurethane catheters are available from Access Technologies (hydrogel), Solomon Scientific (long-term heparin) and Uno Roestvaststaal.

Features to consider for intravenous catheters include:

1 biomaterial haemocompatibility;
2 resistance to infection;

3   distal tip geometry;
4   compatibility with test compounds;
5   ease of insertion;
6   kink resistance;
7   size availability;
8   ease of bonding;
9   tear strength;
10   sterilisation requirements.

Ports are available from sizes for rats (>200 g) to sizes for large animals. Ports are fabricated from plastic or titanium with a variety of catheters available for attachment. Access Technologies and Solomon Scientific both manufacture an extensive line of ports in standard and customised configurations specifically for laboratory animals. Ports range in price from $35 for a plastic rat port to $150 for a large-animal titanium port.

Features to consider for subcutaneous access ports include:

1   appropriate size and shape of port body;
2   largest needle port will accept;
3   number of needle sticks without leaking;
4   dead volume of port body.

23.3 Tethered infusion apparatus

The following components essentially constitute the tethered infusion models:

1   catheter (discussed above);
2   jacket, harness, button or tail cuff;
3   swivel;
4   tether spring and other hardware;
5   pump.

23.3.1 *Jacket, harness, button and tail cuff for mice and rats*

The mouse or rat jacket, harness, button and tail cuff each serve to accomplish the same purpose, which is to protect the catheter externalisation site and to provide, with minimal stress, attachment of the catheter to the swivel (Figure 23.4). These devices, with the exception of the tail cuff, permit externalisation of the intravenous catheter through the scapular region. The tail cuff device permits tail externalisation of a tail vein or femoral vein catheter. The rat and mouse jackets consist of a vest constructed of cloth or nylon with two cut-outs for the front limbs. The jacket wraps around the rib cage and back and has a reinforced area over the catheter exit site to attach to the tether (Figure 23.5).

The harness functions similarly to the jacket (Figure 23.6) but is constructed of a soft, elastomeric dome over the scapular region which is secured to the rat or mouse by two adjustable rubber tubes wrapped around the two front limbs. The button infusion device, made of polyester mesh or polysulphone plastic is sutured to the fascia under the animal's skin allowing catheter externalisation through the button's

Figure 23.4 Mouse tethered infusion apparatus (Instech).

Figure 23.5 Rat and mouse jacket (Lomir).

centre (Figure 23.7). Finally, the tail cuff infusion device is a metal tube surrounding a portion of the tail permitting externalisation of a tail vein or femoral vein catheter (Figure 23.8).

23.3.1.1 Supply

Lomir and Instech are the principal suppliers of these tethering devices, which require little or no customisation. They are also available through Uno Roestvaststaal. A jacket (Lomir Biomedical) or harness (Instech Laboratories) each costs about $20, including a spring tether. The button (Instech) costs about $5 for a disposable button and $30 for a reusable button and spring assembly. Lomir's tail cuff infusion device sells for about $15.

Features to consider for tethered rodent jacket, harness, button and tail cuff include:

1   animal comfort;
2   effect on animal temperature;
3   accommodation of animal growth;
4   ability to clean and reuse;
5   ease of placement and required surgery;
6   ability of animal to damage catheter;
7   protection of catheter from animal.

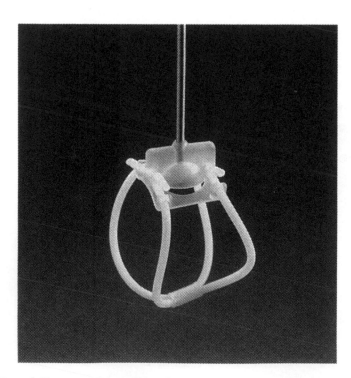

Figure 23.6 Instech/Covance harness (Instech).

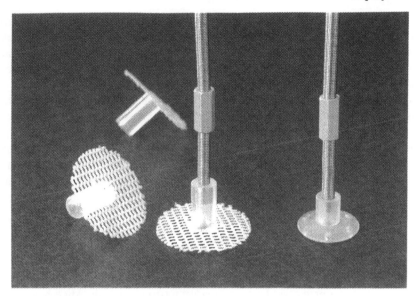

Figure 23.7 Button (Instech).

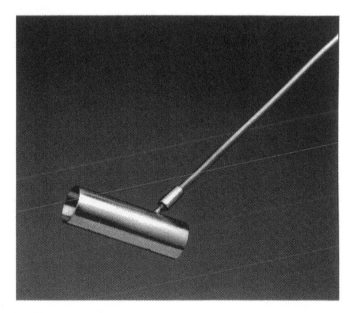

Figure 23.8 Tail cuff (Lomir).

23.3.2 *Large-animal jacket*

There is a much greater variety to be found in large-animal jackets than in rodent jackets. Dogs and non-human primates are the most routinely jacketed large animals (Figure 23.9). The large-animal jackets are typically fabricated from cloth, nylon, polyester mesh and flexible Spandex®-type material. The jackets have zippers or Velcro® straps to permit customised, secure placement of the jacket on the animal.

These jackets are placed over the entire torso/trunk and cover all four limbs, although some models cover only the upper torso/trunk and the upper/front two limbs. As with rat jackets, the catheter is externalised in the scapular region, where the jacket is reinforced to accommodate placement of metal tethering hardware. Some jackets have sleeves for the limbs while others do not.

The large-animal jacket presents a significant design challenge because of the dog's and non-human primate's flexibility and insatiable desire to address externalised devices. As such, the fit about the limbs and neck must be tight enough to prevent intervention by the animal, yet not too tight so as to cause discomfort and chafing. The chief complaint about large-animal jackets is the irritation to the animal's skin about the neck and limbs. Fortunately, the manufacturers are responsive to researchers' requests for customised jackets.

23.3.2.1 Supply

There are several suppliers of large-animal jackets. Each manufacturer provides jackets for all commonly used laboratory animals, including cats, rabbits, minipigs, pigs, dogs and non-human primates. Leading manufacturers include Alice King Chatham Medical Arts, Kent Scientific and Lomir Biomedical. Prices range from $75 to $200 per jacket.

Features to consider for large-animal tethered jackets include:

1   animal comfort;
2   effect on animal temperature;
3   accommodation of animal growth;

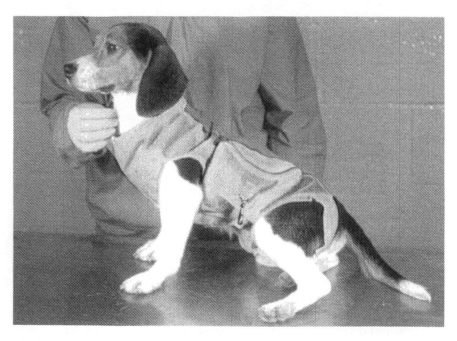

Figure 23.9 Large-animal jacket (Covance Laboratories).

4 ability to clean and reuse;
5 ease of placement and required surgery;
6 ability of animal to damage catheter;
7 availability of customisation.

23.3.3 *Swivels*

Arguably one of the most critical hardware components between animal and pump is the swivel. Without the swivel, the animal would be significantly encumbered in its movement, and the fluid line would become partially or totally occluded. The body of the swivel is usually fabricated from either plastic (Teflon® or polysulphone) or stainless steel, stainless-steel pins, stainless-steel ball bearings and a Teflon® tube. The inert fluid path consists only of stainless steel and Teflon. Specialised swivels with a quartz-lined fluid path are available for use with compounds which are incompatible with stainless steel.

Swivels are categorised as follows:

1 disposable (plastic) versus reusable (stainless steel) body (Figures 23.10 and 23.11);
2 single-channel versus multichannel.

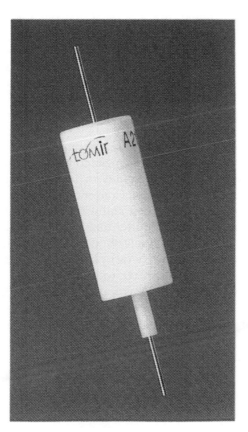

Figure 23.10 Disposable swivel (Lomir).

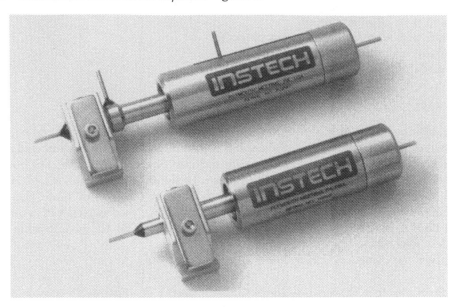

Figure 23.11   Reusable swivel (Instech).

In laboratories with high animal turnover, the disposable swivel is more prevalent than the reusable swivel owing to the high labour costs associated with cleaning and preparing swivels for reuse. Plastic swivels are marketed as disposable but in fact are frequently reused. Stainless-steel swivels represent considerably higher quality, which is apparent in the longer functional life that includes extensive reuse and resterilisation. The end-point in the life of a swivel is when it begins to leak fluid (which also presents a contamination risk), when it becomes clogged or when the force required to turn the catheter becomes too great.

Single-channel swivels are the most widely used, as most intravenous infusion studies are designed to investigate the effects of a single compound. Dual-channel swivels offer the ability to infuse multiple compounds or to infuse while simultaneously sampling fluid or monitoring blood pressure. Conventional swivels with three or more channels are available, though at a high cost and with a propensity to leak. Recently, new devices have been introduced which offer multiple fluid paths for studies of the effect of multiple drugs or to measure multiple physiological parameters while infusing. These new devices are not conventional swivels, but rather are devices to rotate the entire animal apparatus, not just the catheter, to compensate for animal movement.

For any particular swivel design, there is a direct trade-off between torque and leaking. Increased leak protection comes from a stronger seal, but this stronger seal comes at the expense of increased torque and thus more resistance to the animal's movement. Additionally, smaller fluid paths allow for lower torque devices as there is less surface area to be sealed. Furthermore, torque for any swivel increases with the number of channels.

23.3.3.1 *Swivel supply*

Instech and Lomir are the market leaders in the manufacture of single- and dual-channel swivels, in plastic or in stainless steel, for large and small laboratory animals. Two new non-conventional swivel-type products are available to accommodate multiple lines: the Instech Swivelless Swivel, which accommodates dozens of lines (fluid, electrical, optical, etc.), and the Bioanalytical Systems Raturn. Single-channel swivels range in price from $15 to $30 for plastic units to $200 for metal ones. Dual-channel swivels are sold for $175 for plastic swivels and $300 for metal ones. The Swivelless Swivel and Raturn sell for several thousands of dollars.

Features to consider for swivels include:

1 torque;
2 leak rating;
3 dead volume;
4 fluid path compatibility with compounds;
5 ability to clean fluid path;
6 number of uses and sterilisations;
7 number of channels;
8 outlet sizes and configurations for attachment to catheters.

23.3.4 Other hardware linking animal to pump

23.3.4.1 Spring tether

In the rat and mouse, the spring is a flexible coil of stainless steel wire ('spring stock'), usually 30 cm long, attached to the animal's jacket, harness, button or tail cuff, which covers the tubing between the animal and the swivel. The spring functions to protect the tubing from gnawing by the animal. Spring features include inner diameter, flexibility, length and material. The spring for large-animal tethers is a stainless-steel conduit that is much thicker and larger than the rodent spring stock. It is usually 90 cm long depending on the dimensions of the cage.

23.3.4.2 Tubing

Two or three sets of tubing are needed to connect the animal to the pump. In some infusion models, the animal catheter extends from the intravascular tip all the way to the swivel. In other models, the animal catheter's proximal tip terminates immediately outside the animal skin or at a button. Here, tubing is needed to join the animal catheter to the swivel. All models require tubing to connect the swivel and pump. As with catheters, care should be taken in selecting the proper tubing biomaterials.

23.3.4.3 Counterbalance arm

The counter-balance arm is necessary for mouse infusions (Figure 23.12). To enhance unrestricted movement, the counterbalance arm serves to keep slack out of the spring and to remove the weight of the spring from the mouse.

23.3.4.4 Supply

Swivel manufacturers supply these other hardware components, either separately or bundled with the swivels.

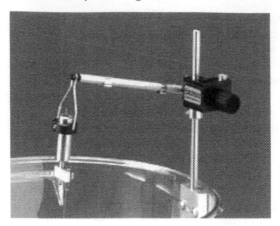

Figure 23.12 Counterbalance arm (Instech).

23.3.5 Pumps for tethered infusion models

The pump discussion can be long and detailed as these devices are increasingly sophisticated electromechanical devices, complete with microprocessor-based programmability and PC hook-ups. There is quite an array of pump choices to make. It is useful to distinguish between low-flow pumps (syringe pumps) and high-volume pumps (peristaltic, piston or diaphragm). Syringe pumps are usually suited for low-flow and/or low-volume infusions (approximately 0.001–1.000 ml/h), whereas high-volume pumps are typically utilised for high-flow and/or high-volume infusions (1–100 ml/h).

23.3.5.1 Syringe pumps

Syringe pumps accept generic plastic and glass syringes, so their cost of use is low. Glass syringes provide increased accuracy owing to the reduced friction between the plunger and the barrel. Hamilton syringes are capable of accurate delivery of very small volumes (μl/h). Syringe pumps can further be classified as laboratory bench-top or hospital pole mount. Laboratory bench-top pumps include the popular Harvard Apparatus pumps, which are marketed specifically for laboratory research. Hospital pole mounts include pumps, such as the Baxter pump, marketed primarily for human use in hospital settings. The selection of syringe pumps is vast. In addition to pumps made, serviced and marketed exclusively for laboratory animal infusion studies, many human-use pump companies target laboratory animal infusion as a secondary market for their products. The pump described below is from the leader in laboratory animal syringe pumps, Harvard Apparatus.

HARVARD APPARATUS 11 SYRINGE PUMP SERIES

The 11 syringe pump series from Harvard Apparatus (Figure 23.13) is a common tool for rodent applications requiring low flow. It is an affordable pump that is quite easy to use. Only two pieces of data need to be entered to begin operation – syringe diameter and flow rate. It is not possible to interface this model to a PC but other similar models have this feature. Two syringes can be attached to the pump at once, though other models allow for arrays of many more syringes for simultaneous infusions.

This series of pumps has the following specifications:

flow rate: 0.001–45 ml/min
syringe sizes: 0.5 μl to 50 ml
PC interface: none (available on other models)
power source: AC
accuracy: ± 1%.

23.3.5.2 High-volume pumps

Peristaltic and piston pumps provide pumping of higher volumes. Because most syringe pumps need to be refilled after the syringe is emptied, an infusion would be limited to 10 or 20 ml between refills, which is not practical for most large-animal infusion studies. High-volume pumps draw compound from a fluid bag that holds up to 1,000–2,000 ml of compound. The tubing set for each high-volume pump is dedicated to that pump and can be costly ($10–$20 each). Thus, the cost of use of a peristaltic pump is greater than that of syringe pumps as syringes are quite inexpensive. High-flow pumps are designed to deliver fluids at about 0.1–100 ml/h, making them suitable for most large-animal studies, but not as attractive for small animal studies. High-volume pumps also are available from a variety of sources, including laboratory animal and human-use pump suppliers. One of the more common laboratory animal high-volume pumps from Instech Laboratories is profiled below.

Figure 23.13   Harvard syringe pump (Harvard Apparatus).

INSTECH MODEL 720

The Instech 720 is a rotary peristaltic pump with a wide range of flow rates (Figure 23.14). It is available either as a stand-alone pump or attached to a rate controller in groups of five. With modifications, the flow rate range can be expanded dramatically. This pump also has a battery back-up.

This pump has the following specifications:

flow rates: 0.5–120 ml/h
reservoir sizes: Luer attachments connect to any size reservoir
tubing set: dedicated, disposable set; silicone rubber
PC interface: none (available on other models)
power source: AC with DC back-up
accuracy: ± 5%.

Features to consider for tethered infusion pumps include:

1 range of flow rates;
2 price of pump sets;
3 syringe or reservoir sizes accepted;
4 accuracy;
5 precision;
6 ease of use;
7 extent of programmability;
8 DC and AC capability;
9 ability to network to PC.

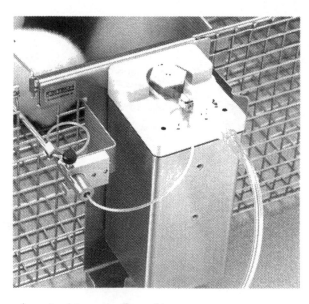

Figure 23.14 Instech peristaltic pump (Instech).

23.4 Tetherless infusion apparatus

The tetherless jacketed infusion model (often called a 'jacketed' or 'ambulatory' model) consists of a catheter or subcutaneous port, a jacket, a pump contained in the jacket's pouch and tubing joining the pump to the catheter or port. All components for this model are housed in or covered by the jacket. In the case of studies in dogs, researchers often add a cervical collar to prevent the animal from disturbing the infusion apparatus.

The most common species for this model are the dog and the non-human primate. Other species, including swine and sheep, are studied in this model, and a new, smaller tetherless pump has enabled tetherless infusions in rabbits and smaller non-human primates ($\leq$ 3 kg). Rats, mice and other small laboratory animals are precluded from the tetherless model because the available pumps are too large to carry.

Many laboratory animal researchers believe that the tethered models are more stressful than untethered models. This belief especially applies to the non-human primate, which is more mobile than the dog. Adams *et al.* (1988) reported that jacketed non-human primates (*Macaca fascicularis*) had elevated heart rates when tethered compared with when they were untethered. At a 1996 infusion and access roundtable attended by 45 laboratory animal veterinarians, technicians and investigators from North America, the overwhelming majority of participants stated the belief that tetherless models were less stressful to dogs and non-human primates than tethered models.

23.4.1 Tetherless jacket

The tetherless animal jacket is the same jacket as used in the tethered infusion model with minor modifications. These include the addition of one or two pouches to contain the pump and a bag containing the test infusate. In most dog jackets, two pouches are added to the jacket, one pouch for the pump and compound infusate and the other pouch for a bag of fluid to counter the weight of the pump. For non-human primates the jackets usually contain only one pouch to hold the pump and the infusate compound. Ideally, the pouches are sized to match the dimensions of the pump and infusate bags so that the devices are snugly secured. Additional modifications are made to the jacket to ensure proper tubing connections between the pump and the catheter or port.

23.4.2 Catheter or port

The catheter in the tetherless model usually terminates into a female Luer lock, which allows for easy attachment to the pump's tubing set. When ports are utilised, a right-angle Huber point needle with an 8–30 cm tubing extension is attached to the pump set.

23.4.3 Tetherless pump

Like the syringe pumps, there is a significant selection of tetherless ('jacketed' or 'ambulatory') pumps to choose from, all of which were developed for the human-use medical market. Two of these pumps are targeted to the laboratory animal infusion market – CADD from Sims-Deltec and Pegasus from Solomon Scientific. There are three mechanisms by which tetherless pumps function: (1) peristaltic, (2) piston and (3) syringe. The tetherless syringe pumps are small and light, but only use 5–10 cc syringes, thereby significantly limiting their use in laboratory animal infusions.

Tetherless pumps are quite sophisticated, and offer varying levels of programmability. Some pumps offer a PC connection, which permits programming of the pumps and allows for the downloading of 'diary' files from the pumps. These diary files contain a complete history of the pump's use during an infusion study and can be used to support GLP record keeping. Additionally, CADD and Pegasus pumps have telemetric hook-up capabilities to programme and monitor the pump without disturbing the animal. In a large room with numerous instrumented animals, pump telemetry simplifies the task of checking pump status, responding to pump alarms and changing infusion programmes.

Tetherless pumps are less accurate than syringe pumps, as evidenced by pump trumpet curves (Figure 23.15). This difference is because tetherless pumps are operated by battery-powered 'stepper' motors, which deliver fluid in relatively large pulses. Syringe pump infusion rates are usually accurate to ± 1%. Micropiston tetherless pumps are accurate to ± 2%, and tetherless peristaltic pumps are accurate to ± 6–8%. However, the pump accuracy feature goes beyond discussion of '± x%.' In fact, over shorter time intervals lasting several minutes, a tetherless pump's accuracy can actually deviate by more than 15%. The trumpet curve is a useful tool to determine pump accuracy over varying time intervals.

Electromechanical pumps infuse fluids in small pulses. Syringe pumps dispense doses in tiny increments: smaller than 0.01 µl. Micropiston tetherless pumps infuse in 0.4-µl pulses. Peristaltic tetherless pumps infuse in 100-µl pulses. The issue of pulse size is critical in applications with a low flow rate and/or short compound half-life. For example, if a peristaltic tetherless pump that infuses in 100-µl pulses were set to deliver at a rate of 0.2 ml/h, the pump would pulse only twice (two boluses) in 1 h. As such, the resulting flow would be intermittent and not continuous. With this flow, the concentration of the short half-life compound in the animal will cycle with the pulses.

The two tetherless pumps marketed and supported specifically for laboratory animal infusion studies are profiled below.

23.4.3.1 CADD

The CADD peristaltic pump is used for infusion of chemotherapy, parenteral nutrition, analgesia, and other compounds for therapeutic applications in humans (Figure 23.16). It is also used extensively in dogs and larger non-human primates for infusion studies. Its strengths are ease of programming, reliability and low power consumption. Its weaknesses are its size, accuracy and large pulse size. Many models are available, including a low-flow syringe pump.

CADD Legacy specifications are as follows:

size: $4.1 \times 9.5 \times 11.2$ cm
weight: 284 g (without batteries)
flow rates: 0.1–125 ml/h
pulse size: 100 µl
reservoir: 50–1000 ml cassettes
batteries: two AA

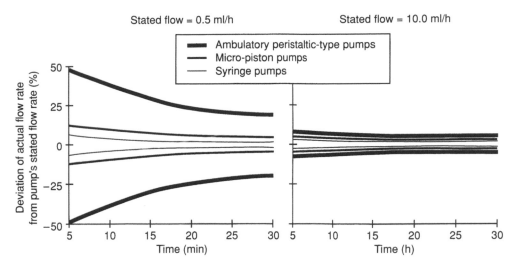

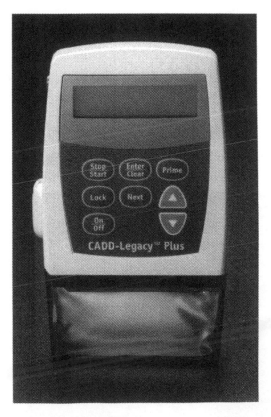

Figure 23.15 Pump trumpet curve.

Figure 23.16 CADD pump (Sims-Deltec).

infusion profiles: continuous or intermittent; immediate or delayed start time and date
telemetry: yes
PC interface: yes
accuracy: ± 6%
alarms: low and empty battery, low and empty fluid bag, downstream and upstream occlusion, air detection.

23.4.3.2 *Pegasus*

The LogoMed Pegasus is a micropiston pump used for the same human therapies as CADD (Figure 23.17). It is used in most large animals, including rabbits and smaller non-human primates, because of its small size and light weight. Other features of the pump are good accuracy, extensive programmability and small micropulses. Disadvantages are high battery consumption and low maximum flow rate. A second model is available with flow rates up to 100 ml/h.

The specifications of this pump are as follows:

size: 3.1 × 6.2 x 8.4 cm
weight: 180 g (with batteries)
flow rates: 0.01–15 ml/h
pulse size: 0.4 μl
reservoir: 25–300 ml bags
batteries: two AA
infusion profiles: continuous, intermittent, circadian, vast customisation, delayed start time and date, specified stop time and date
telemetry: yes
PC interface: yes
accuracy: ± 2%
alarms: low and empty battery, low and empty fluid bag, downstream occlusion.

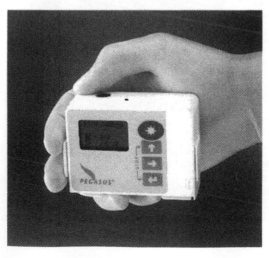

Figure 23.17 Pegasus pump (Solomon).

Features to consider for tetherless pumps include:

1   size and weight;
2   range of flow rates;
3   price of pump sets;
4   accuracy as shown on trumpet curve;
5   pulse size;
6   extent of infusion programmability;
7   battery life at various flow rates;
8   ability to network to PC;
9   telemetric capability;
10 alarms.

23.5 Implanted pumps

Regrettably, this section is the shortest of the chapter. The implanted pump could be a highly advantageous continuous infusion device for laboratory animals as it is entirely subcutaneous – no jacket, swivel, tether or external pump is necessary. Animal interference with the apparatus is essentially eliminated. The infection risk drops considerably with the absence of a chronic wound opening, and one could infer from Adams (1988) that the implanted pump would be the least stressful of all the infusion models. In certain models, the implanted pumps might permit group housing of animals on study, which could be a considerable improvement in animal well-being.

To offset these advantages, there are multiple disadvantages, not the least of which is the scarcity of implanted pumps suitable for laboratory animal use. Currently only the osmotic ALZA minipump discussed in a previous chapter is widely available and proven for small and large laboratory animal use. However, the minipump is only useful for infusing very small volumes of fluid, and thus is used primarily in rat and mouse studies. There are some human-use pumps available, but they are far too costly for widespread utilisation and are only appropriate for use in very large animals. Other obvious disadvantages with implanted pumps include difficulty in accessing the pump for monitoring, troubleshooting, repair and replacement.

23.6 Conclusion

Barring dramatic advances in technology (e.g. infusion devices or *in vitro* drug screening methods), new requirements by the regulatory agencies that approve new drugs or a sea change in the rules governing laboratory animal care and use, it is likely that the current, mature models for infusion will continue to prevail. The 30-year-old tethered infusion model and the newer tetherless jacketed model will continue to be the primary platforms for intravenous infusion in safety studies. We will see several incremental improvements in laboratory animal infusion technology in the few years ahead. Already, the small companies serving this niche market are developing telemetric jacketed pumps, increasing the automation and networking of pumps, providing improved biomaterials for implanted catheters and developing new devices and systems for infusion in mice.

References

Adams, M.R., Kaplan, J.R., Manuck, S.B., Uberseder, B. and Larkin, K.T. (1988) Persistent sympathetic nervous system arousal associated with tethering in cynomolgus macaques. *Laboratory Animal Science*, 38: 279–281.

Dennis, M.B., Cole, J.J., Jensen, W.M. and Scribner, B.H. (1984) Long-term blood access by catheters implanted into arteriovenous fistulas of sheep. *Laboratory Animal Science*, 34: 388–391.

Hecker, J.F. (1981) Thrombogenicity of tips of umbilical catheters. *Pediatrics*, 57, 467–471.

Mendenhall, H.V. (1997) A long term study on the use of vascular access ports in multiple species. *Abstracts of 1997 Laboratory Animal Long-Term Access Roundtable*.

O'Farrell, L. (1995) Optimal central venous catheter design for long-term blood sampling – a thesis in laboratory animal medicine.

Appendix: manufacturers of laboratory animal infusion apparatus

Access Technologies

7350 N. Ridgeway
Skokie
IL 60076, USA
Tel: 847 674 7131
Fax: 847 674 7066
www.norfolkaccess.com
Products: bulk tubing, catheter, jacket, pump, subcutaneous port

Alice King Chatham Medical Arts

11915–17 Inglewood Avenue
Hawthorne
CA 90250, USA
Tel: 310 970 1063
Fax: 310 970 0121
Products: jacket

Alza Corporation

950 Page Mill Road
Palo Alto
CA, USA
Tel: 650 962 2251
Fax: 650 962 2488
www.alza.com
Products: implanted osmotic pump

Bioanalytical Systems

2701 Kent Avenue
West Lafayette
IN, USA
Tel: 765 463 4527
Fax: 765 497 1102
www.bioanalytical.com
Products: multichannel swivel system, pump

Braintree Scientific

PO Box 850929
Braintree
MA, USA
Tel: 781 843 2202
Fax: 781 982 3160
www.braintreesci.com
Products: bulk tubing, catheter

Harvard Apparatus

84 October Hill Road
Holliston
MA 01746, USA
Tel: 508 893 8999
Fax: 508 420 5732
www.harvardapparatus.com
Products: bulk tubing, button, catheter, harness, jacket, pump, subcutaneous port, swivel, tether

Instech Laboratories

5209 Militia Hill Road
Plymouth Meeting
PA 19462,USA
Tel: 610 941 0132
Fax: 610 941 0134
www.instechlabs.com
Products: button, harness, multichannel swivel system, pump, swivel, tether

IITC

23924 Victory Boulevard
Woodland Hills
CA 91367, USA
Tel: 818 710 1556
Fax: 818 892 5185
www.iitcinc.com
Products: catheter

Kent Scientific

325 Norfolk Road
Litchfield
CT 06759, USA
Tel: 860 567 5496
Fax: 860 567 4201
www.kentscientific.com
Products: jacket

Lomir Biomedical

95 Huot
Notre-Dame Ile Perrot,
Quebec, Canada J7V 7M4
Tel: 514 425 3604
Fax: 514 425 3605
www.lomir.com
Products: jacket, pump, swivel, tail cuff, tether

Razel Scientific

100 Research Drive
Stamford
CT 06906, USA
Tel: 203 324 9914
Fax: 203 324 5568
razelsci@aol.com
Products: pump

Solomon Scientific

8038 Wurzbach, Suite 360
San Antonio
TX 78229, USA
Tel: 210 692 9908
Fax: 210 692 9907
www.SolSci.com
Products: bulk tubing, catheter, pump, subcutaneous port

Sims-Deltec

1265 Grey Fox Road
St. Paul,
MN, USA
Tel: 612 633 2556
Fax: 612 638 0364
www.deltec.com
Products: catheter, pump, subcutaneous port

Strategic Applications Inc

1220 Hunters Lane
Libertyville
Illinois 60048
USA
Tel: 847 680 9385
Fax: 847 680 9837
scdenault@worldnet.att.net
Products: pumps, jackets, swivels, catheters, subcutaneous ports

Uno Roestvaststaal

Postbus 15
6900 AA Zevenaar
The Netherlands
Tel: 31 316 524451
Fax: 31 316 523785
unorvs@worldonline.nl
Products: bulk tubing, catheters, jackets, pumps, swivels, tail cuffs, tethers, subcutaneous ports

Index